Petrology and Geochemistry of Continental Rifts

NATO ADVANCED STUDY INSTITUTES SERIES

Proceedings of the Advanced Study Institute Programme, which aims at the dissemination of advanced knowledge and the formation of contacts among scientists from different countries

The series is published by an international board of publishers in conjunction with NATO Scientific Affairs Division

A Life Sciences B Physics	Plenum Publishing Corporation London and New York
C Mathematical and Physical Sciences	D. Reidel Publishing Company Dordrecht and Boston
D Behavioral and Social Sciences	Sijthoff International Publishing Company Leiden
E Applied Sciences	Noordhoff International Publishing Leiden

Series C – Mathematical and Physical Sciences

Volume 36 – Petrology and Geochemistry of Continental Rifts

Petrology and Geochemistry of Continental Rifts

Volume One of the Proceedings of the NATO Advanced Study Institute Paleorift Systems with Emphasis on the Permian Oslo Rift, held in Oslo, Norway, July 27–August 5, 1977

edited by

E.-R. NEUMANN
Mineralogical-Geological Museum, University of Oslo, Oslo, Norway

and

I. B. RAMBERG
Geological Institute, University of Oslo, Oslo, Norway

D. Reidel Publishing Company

Dordrecht : Holland / Boston : U.S.A.

Published in cooperation with NATO Scientific Affairs Division

Library of Congress Cataloging in Publication Data

NATO Advanced Study Institute Paleorift Systems, with Emphasis on the Permian Oslo Rift, Oslo, Norway, 1977.

(NATO advanced study institutes series: Series C, Mathematical and physical sciences; v. 36)

Bibliography: p. Includes index.

1. Rifts (Geology)—Congresses. 2. Petrology—Congresses. 3. Geochemistry—Congresses. I. Neumann, Else-Ragnhild, 1948- II. Ramberg, Ivar B. III. Title. IV. Series.

QE606.N37 1977 551.1'4 78-29

ISBN-13: 978-94-009-9805-6 e-ISBN-13: 978-94-009-9803-2

DOI: 10.1007/978-94-009-9803-2

Published by D. Reidel Publishing Company
P.O. Box 17, Dordrecht, Holland

Sold and distributed in the U.S.A., Canada, and Mexico
by D. Reidel Publishing Company, Inc.
Lincoln Building, 160 Old Derby Street, Hingham, Mass. 02043, U.S.A.

CONTENTS

ACTIVE CONTINENTAL RIFTS

PALEORIFTS

CONTENTS OF

VOL. II TECTONICS AND GEOPHYSICS OF CONTINENTAL RIFTS

INTRODUCTION

PREFACE

The NATO Advanced Study Institute "Paleorift Systems with Emphasis on the Permian Oslo Rift" was held at Sundvollen near Oslo, Norway, 26. July - 5. August, 1977. The meeting included 6 field trips to various parts of the Oslo Region. 70 official participants and 16 observers from 14 countries attended the meeting.

The majority of the invited lectures and short research papers and progress reports presented at the meeting are published in two volumes of which this is volume No. I. Lists of content for both volumes are presently included. The guide to the field trips is being published in the Norwegian Geological Survey Series (1978).

Oslo, 10. November 1977.

Else-Ragnhild Neumann Ivar B. Ramberg

Organizing Committee members:

O. Eldholm G. Grønlie J. Naterstad I.B. Ramberg (chairman)	Geological Institute, University of Oslo
J.A. Dons B.T. Larsen (secretary) E.-R. Neumann (secretary)	Mineralogical-Geological Museum, University of Oslo
K.S. Heier (chairman) S. Huseby B. Sundvoll	Norwegian Geological Survey
M.A. Sellevoll	Seismological Observatory, University of Bergen
K. Storetvedt	Geophysics Institute, University of Bergen
P.M. Ihlen Chr. Oftedahl F.M. Vokes	Geological Institute, University of Trondheim
E.S. Husebye	NORSAR, 2007 Kjeller

This volume is Scientific Report No. 39 of the Geodynamics Project.

The Geodynamics Project is an international programme of research on the dynamics and dynamic history of the Earth with emphasis on deep-seated foundations of geological phenomena. This includes investigations related to movements and deformations, past and present, of the lithosphere, and all relevant properties of the Earth's interior and especially any evidence for motions at depth. The programme is an interdisciplinary one, coordinated by the Inter-Union Commission on Geodynamics (I.C.G.) established by the I.C.S.U. at the request of I.U.C.G. and I.U.G.S., with rules providing for the active participation of all interested I.C.S.U. Unions and Committees.

PALEORIFT SYSTEMS WITH EMPHASIS ON THE PERMIAN OSLO RIFT OPENING ADDRESS

K.S. Heier

University of Oslo, or
Geological Survey of Norway,
7001 Trondheim, Norway

It is with considerable honor, pleasure and pride that I welcome you to this Study Institute. As students and lecturers from 14 countries we are together here with one common interest; the study of world rift systems. After the 1974 NATO sponsored meeting "Iceland and midocean ridges", we were unofficially approached by members of working group 4 (WG 4) of the Inter-Union Commission on Geodynamics (ICG) to consider undertaking a similar conference on Paleorift systems.

World rift systems are today considered by earth scientists as providing significant clues to the development of the earth's crust and to relations between continental and oceanic crust, as well as between crust and mantle. World rift systems are also of interest as possible source regions of mineralization in the crust. This makes an opportune moment to review our present knowledge of the Permian Oslo Rift and discuss it in the context of world rift systems in general.

In 1810 the German geologist Leopold von Buch first described parts of what we know today as the Oslo region. The first map and regional description, Das Christiania Übergangsterritorium, was published by B.M. Keilhau in 1838. During the period 1855-1880 Th. Kjerulf produced several maps and gave correct interpretations of several tectonic features of the region. He also gave the correct interpretation of the relationships between the volcanics and the underlying Cambro-Silurian rocks. However, the Oslo region became well known from the works of W.C. Brøgger during more than 50 years. I refer here to his series "Die Eruptivgesteine des Kristiania- (Oslo-) gebietes" in seven volumes. The first one in 1894, and the last one in 1934. In 1943 a new series "Studies on

E.-R. Neumann and I.B. Ramberg (eds.), Petrology and Geochemistry of Continental Rifts, xiii–xv.

the igneous rock complex of the Oslo region" was initiated by Olaf Holtedahl. Contributions to this series were all published in "Skrifter utgitt av Det Norske Videnskaps-Akademi i Oslo" and today consist of 24 numbers.

The most recently published thorough discussion, is by Ivar Ramberg "Gravity interpretaion of the Oslo Graben and associated rocks" which was published as a monograph in NGU Bull. 325 during 1976. Also the final report on the Norwegian geotraverse project in 1977 which was the Norwegian contribution to the IGP contains previously unpublished data on the Oslo rift.

The age of the Oslo rocks was long uncertain. The commonly held belief was that they were of Devonian age and relatively closely associated in time with the Cambro-Silurian rocks. Early age determinations by Fajans on thorite from the syenite pegmatites in the Langesundfjord indicated a much younger age of 225 m.y. On a student excursion led by Professor Olaf Holtedahl in 1931 fossils were found in a sediment sequence closely associated with the lava series dating the sediments as lower Permian. More recent age determinations on the plutonic rocks indicates that major emplacement occurred around 270 m.y. With improved analytical techniques we decided not only to determine the relative ages of the different rocks, but also approach the problem of how long the Oslo rift was active. This we consider a fundamental problem and our preliminary results will be presented at this meeting. In the initial stages of this study we received considerable support and assistance from the Isotope laboratory at the Geology Department, University of Oxford.

Of course the Oslo province and its structures has long been known, and it is considered one of the "classical" provinces in the world. It is also conveniently located in the most densely populated parts of Norway, and one would be justified in assuming that this is an area where truly all geological observations worth while have already been made. This is far from the case. We are still frustrated by our lack of knowledge. However, the last thirty years have improved conditions for work in the region. We now have excellent aerial photographs, topographic maps on the scale of 1:5 000, and last but not least, a good system of roads, constructed for the purpose of forestry. Thus we feel that now, and the near future is when we will make a major advance in our understanding of the area. NATO, through their Advanced Study Institute Programme, is the major single contributor to this meeting, and without whose assistance the meeting could not be arranged. The Universities of Oslo, Bergen and Trondheim, and the Geological Survey of Norway also contributed generously both directly and indirectly. Financial assistance was also received from:

The Norwegian Research Council for Science and Humanities (NAVF)
Amoco Norway Oil
Den norske stats oljeselskap A/S
Elkem-Spigerverket A/S
Esso Exploration and Production Norway Inc.
Norsk Agip A/S
Norsk Geofysisk Forening
Norsk Hydro A/S
Phillips Petroleum Company Norway
Saga Petroleum A/S & Co.
Sydvaranger A/S

This assistance is most heartly acknowledged.

The organizing committee of 16 persons has relied heavily on the efforts of Professor I.B. Ramberg, Mr. B.T. Larsen and Mrs. E.-R. Neumann. For clerical assistance we are indebted to Mrs. Inger Holm who will also act as secretary during the meeting. Though it is difficult to single out individuals I am certain the organizing committee agrees in that the credit for the success of this meeting largely should be given to these four individuals.

Finally it is my hope that this meeting will contribute to the understanding of Paleorifts in general, and to an increased knowledge of the Oslo rift in particular. I hope the proceedings of the meeting will turn into a classical reference work on the subject, and that the guides to our excursions to be published as a special NGU-Bulletin will benefit a large number of students to come.

MEMORIAL OF BRUCE HEEZEN

Dr. R.W. Girdler

School of Physics
The University of Newcastle upon Tyne NE1 7RU, U.K.

It is a pleasure to open the first General Session of this NATO Meeting on Palaeorifts. The topic for this morning is General Aspects of Rifting but before opening the Meeting I have been asked to say a few words about Professor Bruce Heezen.

As many of you will know, Bruce Heezen died suddenly on the 21st June on board the U.S. nuclear submersible NR-1 whilst preparing to explore the Reykjanes Ridge, south of Iceland. His death at such an early age is a great loss, not only to those of us with special interests in rifts, but to the whole world of marine geology and geophysics. We have a very full programme, but nevertheless, it is right and appropriate to pay tribute to his work relating to the subject of this Meeting.

In 1956, he published with Maurice Ewing a pioneering paper which gave us "the world rift system". This paper combining a study of earthquake epicentre locations and topography, mapped out the extensional rift zones of the world thus locating the world mosaic of boundaries which we now call "creative plate margins" and "transforms".

Bruce Heezen was one of the first people to recognize that the age of the sea floor increases with distance from the active rifts. He argued this by looking at the youthfulness of the topography in the rift zones and comparing it with the different topography of the neighbouring ridge mountains, taking into account the increasing thicknesses of sediments as one proceeds towards the ocean basins. This was quite remarkable and predated by several years the interpretation of magnetic anomalies in terms of sea floor spreading.

E.-R. Neumann and I.B. Ramberg (eds.), Petrology and Geochemistry of Continental Rifts, xvii–xviii.

I had the good fortune to meet him first in 1959 and to have many lively and exhilarating discussions with him. He was an enthusiast for explaining the rift zones by the expanding Earth whereas I preferred mantle convection. As you know, the arguments concerning the expanding Earth are still going on (two books have recently appeared) and I suppose there is a chance that he could still be right, at least in part!

In the 1960's he was Chairman of the International Upper Mantle Committee Panel on the World Rift System, the fore-runner of Geodynamics Working Group 4 which is also meeting here in Sundvollen.

There is not time to recall his many other contributions to oceanography and submarine geology but I feel I must mention and pay tribute to the most beautiful maps of the sea floor which he produced with Marie Tharp. They have been shown many, many times at meetings such as this and adorn the walls of science departments all over the world. I think we can truthfully say that Bruce Heezen revolutionised studies of the world's oceans and their rifts.

Ladies and gentlemen, as is customary, I would like to ask you to stand and join with me in a few moments silence in memory of Professor Heezen.

GENERAL ASPECTS OF RIFTING

PALEORIFT SYSTEMS - INTRODUCTION

I.B. Ramberg[(1)] and E.-R. Neumann[(2)]

(1) Geological Institute, University of Oslo, Oslo, Norway
(2) Mineralogical-Geological Museum, University of Oslo, Oslo, Norway.

Within the continent of Europe there are a number of well documented intracontinental grabens and rifted continental margins which can be grouped into several rifting systems (Fig. 1). In addition there are numerous less well known, more tentatively postulated rifts. It would appear that the central European craton, flanked by the Alpine fold belt in the south, the Ural Mountains in the east and the Caledonides in the west has been a geological battleground, scarred by rifts which subdivide it into a complex pattern of cratonic fragments. These rifts do not represent a single period of extensional tectonics, but a series of events occurring over a wide range of geological time, from Precambrian to Cenozoic.

As the various parts of the European continent became cratonized, it seems that they also became sites of rift formation [1]. Among the products of the various taphrogenic stages are many embryonic and aborted rifts, which in many cases are now buried beneath sediments, epicontinental seas, or both.

An early generation of possible paleorifts was formed in Early to Mid-Proterozoic time, among them are the Pechenga and Imandra-Varzuga fault-troughs within the oldest parts of the Baltic shield in the southern Kola peninsula, USSR [1]. Subsequent important rifting epochs, include (1) the Late Precambrian (Riphean), (2) the Mid to Late Paleozoic and (3) the Mesozoic to Cenozoic. - The subdivisions of rifts into a number of "generations" or epochs may be more one of convenience than of tectonic significance. It is yet to be seen whether the apparent preferred episodes of rifting identified within the European continent, can be confirmed and matched by similar periodicity from other regions, or whether continental rifting on a global scale is a purely

E.-R. Neumann and I.B. Ramberg (eds.), Petrology and Geochemistry of Continental Rifts, xix–xxvii.

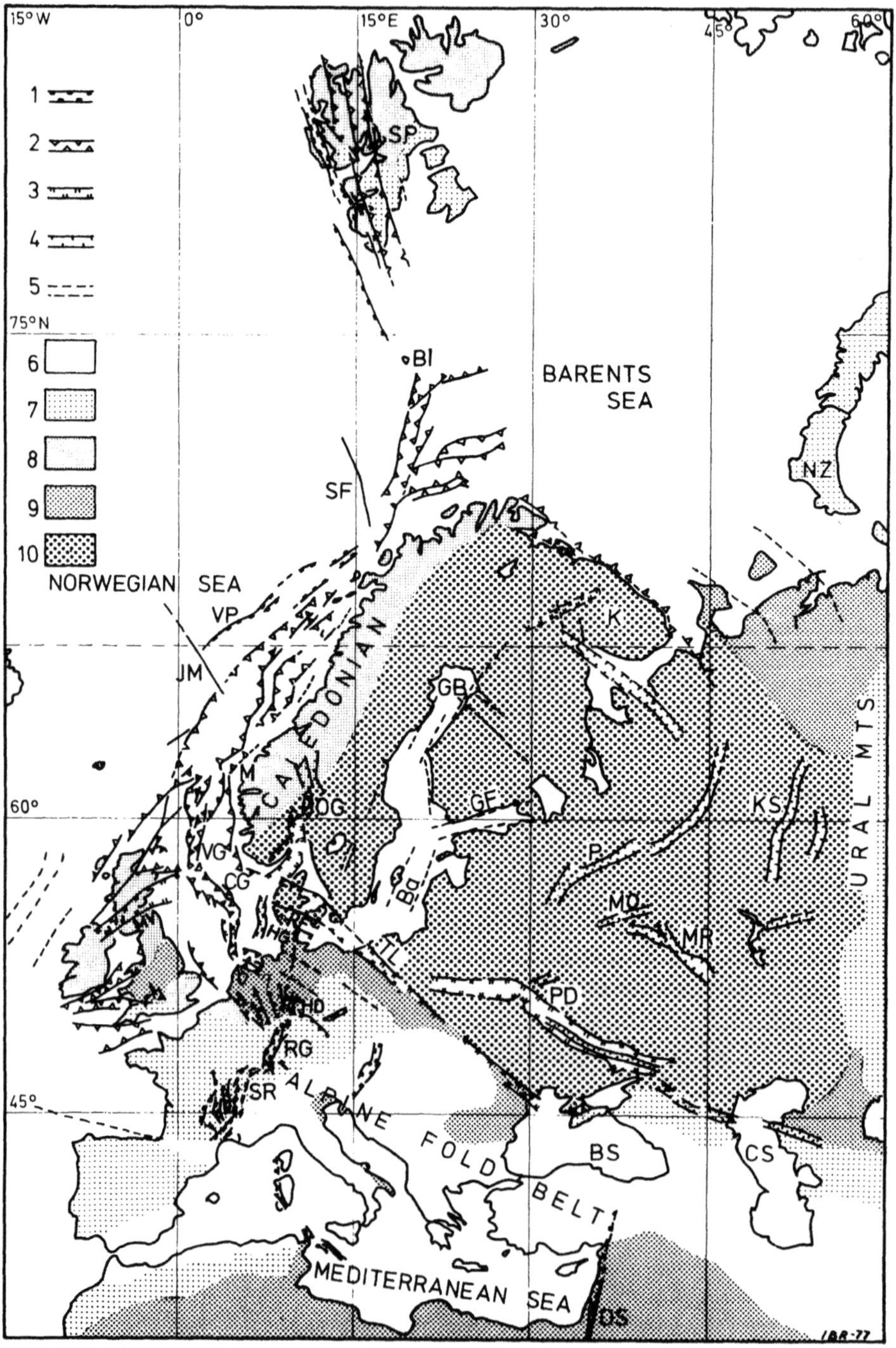
15°W
0°
15°E
30°
45°
60°
75°N
60°
45°
1
2
3
4
5
6
7
8
9
10
NORWEGIAN SEA
BARENTS SEA
SP
BI
SF
VP
JM
CALEDONIAN
NZ
K
GB
M
OG
GE
KS
URAL MTS
VG
CG
Bd
P
MD
MR
HG
TL
PD
HD
RG
SR
ALPINE FOLD BELT
BS
CS
MEDITERRANEAN SEA
DS
IBR-77

random or even a continuous phenomenon.

The best known examples of the Late Precambrian generation are the Riphean aulacogenes of the Russian platform. The fault-bounded "Sparagmite" sedimentary basin occurring to the north of the much younger Oslo Graben may belong to the same epoch [2]; see Fig. 1. These early features represent the most ancient structures similar to the Cenozoic epiplatform fissure-like rift zones.

The second epoch, which was initiated at the beginning of the Hercynian orogeny, includes major rift systems such as the largely Devonian Pripiat-Donets and Kazan-Sergiev aulacogenes. These features are themselves superimposed on older (Riphean) depressions [1,3]. The central zone of the Kola peninsula [4] and the Midland Valley of Scotland [5], both associated with alkaline-basic magmatism, are considered to be rift-related features of the Mid to Late Paleozoic phase. Towards the end of this phase, another series of (Permo-Carboniferous) grabens formed, mainly to the west of the older ones, including the strongly magmatic Oslo Graben [6], the Horn Graben and possibly also the southernmost part of the Viking Graben in the North Sea area [7]. Subsidence of the Danish-Polish depression [8] along the southwestern boundary of the East European craton, the socalled Tornquist line, was also initiated at this time. The Midland Valley, the East Greenland and the Spitsbergen grabens are all examples of epiorogenic continental rifts [3] overlying and parallelling young fold belts, in this case the "newly" formed North Atlantic Caledonides.

The third epoch of rifting began early in Mesozoic time. The sharply increased rifting activity at this time was accompanied by a significant expansion of the areas affected. Thus, major rifting and subsidence resumed in the North Sea region [7,9] and

---

Fig. 1. Continental rifts and rifted continental margins in Europe. Subdivision into cratonic regions or generations after Khain [1]. 1-Cenozoic rifts, 2-Mesozoic rifts, 3-Paleozoic rifts, 4-Precambrian rifts, 5-Inferred rifts and fractures, 6-Neo-Europe (Alpine fold belt), 7-Meso-Europe (area of Epi-Hercynian cratonization), 8-Paleo-Europe (area of Epi-Caledonian cratonization), 9-Proto-Europe (area of Epi-Baikalian cratonization), 10-Eo-Europe (Ancient craton). Ba-Baltic Sea, BI-Bear Island, BS-Black Sea, CS-Caspian Sea, DS-Dead Sea, GB-Gulf of Bothnia, GF-Gulf of Finland, HD-Hessen Depression, HG-Horn Graben, JM-Jan Mayen Fract. Zone, K-Kola Peninsula, KS-Kazan-Sergiev aulacogen, M-Möre, MO-Moscow aul., MR-Mid-Russian aul., MV-Midland Valley, NZ-Novaya Zemlja, OG-Oslo Graben, PD-Pripiat Donets aul., RF-Ringköbing-Fyn High, RG-Rhine Graben, SF-Senja Fract. Zone, Sp-Spitsbergen, SR-Saone Rhône Graben, TL-Tornquist Line, VG-Viking Graben, VP-Vöring Plateau.

along its southern periphery in the Netherlands, West Germany [10] and England [11]. Rifting also occured along the North Atlantic Seabord, and many earlier structures (for instance the Oslo Graben, the Midland Valley, the Great Glen fault, and the Tornquist line) were reactivated. The Mesozoic represented a quiet period in the central part of the craton, that is to say in the area of the Russian platform, but numerous grabens were formed to the east of the Urals, in the Trans-Uralian region and in the Black Sea area [1].

The central North Sea rift system, which merges towards the north with the rifted continental margin of Mid-Norway and East Greenland [12], was mainly formed during the Triassic. Subsidence continued, however, in a series of distinct phases that have been correlated with similar rifting phases in the northern North Atlantic [7,13,14]. One major rifting phase at the Jurassic-Cretaceous boundary (the Late Kimmerian) affected the entire North Sea system. Mesozoic rifting may have extended northwards to the Barents Sea, where several fault-bounded basins formed at this time [14-16].

Subsequent Cretaceous rifting to the west and southwest of the British Isles constitutes part of the continental margin now flanking the North Atlantic ocean. A last but distinct taphrogenetic phase which took place in the North Sea area in early Paleocene [7], coincided with, or slightly predated, the onset of sea floor spreading between Greenland and Europe [12,17], and in the Eurasia Basin in the Arctic [18].

Continental rifting again intensified abruptly in early Cenozoic time with the formation of the Rhine graben, the Saone-Rhône graben and possibly also the Algeria-Provence grabens [10]. This system which formed mostly within the region of 'Neo-Europe' (Fig. 1), extends northwards to the North Sea and southwards to the Mediterranean. An approximately synchronous but less well defined system is found in the eastern Mediterranean area [1].

In addition to the well documented rifts, many of the major morphological features of Europe may be interpreted as traces of incipient, early rifting. Examples include the central part of the Baltic Sea, the Gulf of Bothnia, the Gulf of Finland and several inland sea depressions.

From this summary it is possible to draw a number of conclusions:

(a) The area reviewed (Fig. 1) contains both active rifts and rift relicts of various ages and stages of development, thus offering a natural laboratory for the study of paleorifts.

(b) Most of the individual grabens, whether they are active or aborted (paleo-) rifts, appear to be just portions of more extensive rift systems.

One such system is the impressive Möre-Mediterranean (or West European) rift system of more than 2200 km length. Near Möre (offshore West Norway at about 62°N), it merges with the rifted continental margin. The transitional Möre area thus represents a triple junction consisting of two arms that developed into oceanic spreading axes and a third failed arm that was initiated somewhat earlier than the other two.

(c) The European craton and surrounding region appear to provide evidence of spasmodic lateral migration of rift systems from areas of earlier to those of later cratonization [1]. An example of such migration is the rift sequence which starts with the Riphean Moscow rifts in the central part of the Russian platform, and moves westward to the Late Paleozoic Oslo rift, then to the West Norwegian trough and main North Sea rift system, and finally to the continental margin and subsequent Norwegian Sea spreading axis. Another sequence, migrating from N to S, also starts with the Central Moscow rifts, but moves via the Black Sea depressions to the marginal Cenozoic rifts of the Mediterranean [1]. - It should, however, be emphasized that there are a number of observations which appear to be in conflict with the hypothesis of spasmodic rift migration, and the evidence merit further evaluation.

(d) On the basis of (c) the Tornquist line (or Fennoscandian border zone) should be regarded not only as a boundary between a highly "fragmented" Europe to the SW and a more cohesive craton to the NE [9], but also as a time marker (or marker of crustal age). The conclusion being that NE of this line the lithosphere was old and hence strong enough to withstand post-Permian rifting.

(e) A number of individual cases demonstrate unequivocally that rift zones are subjected to repeated tectonic activity. Major taphrogenesis connected with younger, more marginally located rift belts may cause rejuvenation of older rifts. "New" rifts may completely overprint or propagate further along the axial direction of older ones.

(f) Rifting phases may correspond to periods of generally high tectonic activity, which in other areas may give rise to compressive deformation, or volcanic activity.

(g) All areas of reduced crustal (and lithospheric) thickness within Europe and the surrounding areas coincide with regions of past or present rifting activity [1,6,7].

These hypotheses, based on careful studies of one small portion of the earth's surface need careful scrutiny and evaluation for general applicability. If generally valid, they would be important factors in the work to unravel the fundamental mechanics of rifting.

Geochemical studies of the magmatic products of rifting may also contribute to our understanding of rift genesis. A direct relationship between mafic rock composition and tectonic setting has been observed in several rifts. For some cases a shift from alkali basaltic to tholeiitic magmatism occurs with time [20]; in other cases geographic position is the major factor with extrusion of tholeiitic basalt occurring near the assumed "center of underlying mantle plume" or near the axis of a rift, while alkali basalts are extruded on the flanks [21-27].

Many attempts have been made to set up models of magma genesis in rifts to account for observed distributions. Such models are subjects to a number of constraints imposed by 1) our knowledge of the composition of the mantle, 2) studies of partial melting under various conditions and 3) studies of partitioning of both major and trace elements.

The models favoured at the present time can for the most part be placed under one of the following three general types:

(a) different degrees of partial melting at progressively shallower depths of one type of mantle material [20,28,29].

(b) different degrees of partial melting at different depths of an inhomogeneous mantle [30].

(c) mixing of primordial hot mantle plume (PHMP)(relatively richer in large ionic lithophile elements) and low velocity layer (DLVL) material (depleted in large ionic lithophile elements) [24,31-33].

These models explain systematic shifts of magma composition in a rift area from one chemical extreme to another. Detailed studies of magmatic rocks from rifts provide, however, increasing evidence for random alternations between various mafic rock types within a single rift province (e.g. 26, 34-36). This cannot be readily explain in terms of the models postulated so far, so there is a need for more detailed studies of magmatism in and along rifts in order to obtain the data needed to revise the models.

On the basis of a compilation of data on crustal rocks, Engel et al. [37] have reported a systematic overall change in the K_2O/Na_2O ratios of both magmatic and sedimentary rocks from Archean to Mesozoic time. Engel et al. have related these petrochemical differences to changes in crustal thickness and tectonic style. No similar petrochemical change has been reported for continental rift rocks. Even the oldest continental rifts recognized

(1100-2200 Ma [38-40]) seem primarily to have had alkaline magmatism. It is, however, difficult to recognize Precambrian rifts, and one of the criteria used for identification is the presence of alkaline magmatic rocks. There may therefore exist an unrecognized petrochemical difference between very old and young rift magmatic rocks.

The objective of the conference was to promote comparative studies of paleorifts and their active counterparts since it is data of this type which are now needed to test the validity of the deductions and hypotheses which can be advanced today on the basis of our present state of knowledge. Comparative studies may reveal whether we can relate differences in age, structure and composition of the host rock, lithospheric thickness, and/or time and type of rifting.

To achieve a better tie-in to plate tectonics and hotspot hypotheses and paradigms one should ask whether or not, for example, "rrr" or "Y" type rift systems form over areas of high mantle temperature and upwelling below plates. One must also question whether all rift systems, including aborted ones, have at some time interconnected with familiar "plate boundaries". In other words, do intracontinental rifts form in response to tension within plates (i.e. tension not large enough to create true plate boundaries?), or do they represent extension of "spreading" boundaries into continental lithosphere? If the latter is true, how is the strain accommodated in the area between the propagating or leading end of the rift and the rotation pole of the incipient plates? These questions relate, for instance, to the question of whether we can apply plate tectonics to the Oslo rift and how the strain is accommodated north of its northern end. Why does a rift, like the Oslo rift, "abort" at all? Do the changes in continental rift activity (post-Triassic) correspond to changes in plate motion? The rapid developments in marine geophysics and geology during the last few decades, have made such correlations not only a possibility but one of the most compelling tasks in modern rift studies.

These questions, and others, should be seen in conjunction with general synthesising hypotheses. For instance, as evidenced from Fig. 1, it is hypothesized that at any given time, a rifting episode is most likely to affect younger continental lithosphere. Insofar as continental lithosphere accretes concentrically, this means that one would expect any area of rifting to show (a) a radial outward decrease in age of rifting, and (b) a tendency for rifts to be concentrically distributed about cratons.

ACKNOWLEDGEMENT

Sincere thanks are extended to many colleagues and to participants of the NATO paleorift symposium for stimulating discussions, to Peter R. Vogt and Brenda B. Jensen who contributed significantly to this paper by critical reading of the manuscript, and to NATO, the Norwegian universities and research councils, and other contributers who together made the meeting and the present Proceedings possible.

REFERENCES

1. V.E. Khain, The new international tectonic map of Europe and some problems of structure and tectonic history of the continent, in: Europe from Crust to Core, ed. by D.V. Ager and M. Brooks, John Wiley & Sons, London, 1977.
2. I.B. Ramberg and B.T. Larsen, Tectonomagmatic evolution, in: The Oslo Paleorift, ed. by J.A. Dons and B.T. Larsen, Norges Geol. Unders. (in press).
3. E.E. Milanovsky, Continental rift zones, their arrangement and development, in: East African Rifts, ed. by. R.W. Girdler, Elsevier, Amsterdam - London - New York, 1972.
4. A.A. Khukarenko, Alkaline magmatic activity in the eastern part of the Baltic Shield [in Russian], Zap. Vses. Min. Obsts, 96, 547, 1967.
5. E.H. Francis, this volume.
6. I.B. Ramberg, Norges Geol. Unders., 325, 1, 1976.
7. P.A. Ziegler, GeoJournal, 1, 7, 1977.
8. W. Pozaryski, Zeitchr. angew. Geol., 8, 427, 1962.
9. A. Whiteman, D. Naylor, R. Pegrum and G. Rees, Tectonophysics, 26, 9, 1975.
10. J.H. Illies and K. Fuchs (eds.), Approaches to Taphrogenesis Schweizerbart'sche, Stuttgart, 460, 1974.
11. A. Wittaker, Geol. Mag., 112, 137, 1975.
12. M. Talwani and O. Eldholm, Geol. Soc. Am., Bull., 88, 969, 1977.
13. P.A. Ziegler, North Sea basin history in the tectonic framework of North-Western Europe, in: Petroleum and the Continental Shelf of NW Europe, I, Geology, ed. by A.W. Woodland, Applied Science Publ., London, 131, 1975.
14. H. Rønnevik and T. Navrestad, GeoJournal, 1, 33, 1977.
15. O. Øvrebø and E. Tallerås, GeoJournal, 1, 47, 1977.
16. O. Eldholm and M. Talwani, Geol. Soc. Am. Bull., 88, 1015, 1977.
17. Y. Kristoffersen, Sea floor spreading and the early opening of the North Atlantic. E.P.S.L. (in press).
18. A.M. Karasik, Probl. Geol. Polyarnikh Oblastec Zeml, NIDRA (In Russian), Leningrad, 24, 1974.
19. W.H. Ziegler, Outline on the geological history of the North Sea, in: Petroleum and the Continental Shelf of NW Europe, I,

Geology, ed. by A.W. Woodland, Applied Science Publ., London, 165, 1975.
20. I.G. Gass, Phil. Trans. Roy. Soc. London, A, 267, 369, 1970.
21. R.W. Girdler, (Ed.), East African Rifts, Elsevier Publ. Comp., Amsterdam - London - New York, 1972.
22. B.H. Baker, P.A. Mohr and L.A.J. Williams, Geol. Soc. America Spec. Paper, 136, 1972.
23. A. Pilger and A. Rösler, Afar Degression of Ethiopia, Vol. I and II, E. Schweiserbartsche, Stuttgart, 1975.
24. J.-G. Schilling, Nature Phys. Sci., 242, 2, 1973.
25. F. Barberi, R. Santacroce and J. Varet, Bull. Volc., 38, 755, 1974.
26. B.H. Baker, P.A. Mohr and L.A.J. Williams, Geol. Soc. Am. Spec. Pap., 136, 1972.
27. P.W. Lipman, Geol. Soc. Am. Bull., 80, 1343, 1969.
28. I.G. Gass, Phil. Trans. Roy. Soc. Lond. A, 271, 131, 1972.
29. D.H. Green, Phil. Trans. Roy. Soc. Lond.A, 268 , 707, 1971.
30. P.W. Gast, Geochim. Cosmochim. Acta, 32, 1057, 1968.
31. J.-G. Schilling, Nature, 242, 565, 1973.
32. J.-G. Schilling, Nature, 246, 141, 1973.
33. S.-S. Sun, M. Tatsunobu and J.-G. Schilling, Science, 190, 143, 1975.
34. P.W. Weigand, Skr. Norske Vidsk., Akad. I Mat.-Naturv. kl., Ny Serie, 34, 1975.
35. B. Zanettin and E. Justin-Visentin, Tectonical and volcanological evolution of the western Afar margin (Ethiopia), in: Afar Depression of Ethiopia, ed. by A. Pilger and A. Rösler, Schweizerbart'sche Verlagsb., 1974.
36. A.M. Kudo, K.-I. Aoki and D.G. Brookins, Earth Planet. Sci. Lett., 13, 200, 1971.
37. A.E. Engel, S.P. Itson, C.G. Engel, D.M. Stickney and E.J. Cray Jr., Geol. Soc. Am. Bull., 85, 843, 1974.
38. W.F. Fahrig, E.H. Gaucher and A. Larochelle, Can. J. Earth Sci., 2, 278, 1965.
39. P. Hoffmann, Phil. Trans. Roy. Soc. Lond. A, 273, 547, 1973.
40. K. Burke and J.F. Dewey, J. Geol., 81, 406, 1973.

CONTINENTAL RIFTING AND MANTLE DEGASSING

D. K. Bailey

Department of Geology, University of Reading, Reading, England, RG6 2AB.

ABSTRACT. Various lines of evidence have previously revealed the necessity for a vapour phase as an active ingredient of continental rift valley magmatism. Further constraints are now placed on the magmatology by the consistently low H_2O contents of natural peralkaline felsic glasses. The H_2O data confirm that these "highly evolved" liquids cannot be the products of closed system fractional melting or crystallization. They show, too, that the vapour coexisting with rift valley magmas must be deficient in H_2O. The vapour must be strongly reducing, with high concentrations of carbon and halogen gases, and probably nitrogen. Rift valley magmas result from the fluxing action of this vapour as it escapes from the deep mantle and passes through the lithosphere segment below the rift. Initially, the vapour produces metasomatic carbonates, and halogen bearing silicates in the upper mantle and lower crust: subsequently it buffers the activities of Na, Fe, Si, F and Cl, and trace elements in the ensuing magmas. The physical expression of the activity is, first, metasomatic and thermal uplift; followed by devolatilization, melting, eruption and subsidence as the flux cycle climbs through the lithosphere.

1. INTRODUCTION

Escape of juvenile gases through continental rift zones is perhaps so typical, and so obvious, that it is too easily taken for granted. Yet the nature, quantities and functions of these gases must be known before rifting can be understood, because the volatiles and oft-associated magmas provide the only <u>direct</u> evidence of the composition and state of matter below the rifts. Evidence about the volatiles can be derived and appraised in different ways, in

E.-R. Neumann and I.B. Ramberg (eds.), Petrology and Geochemistry of Continental Rifts, 1-13, terms of: (a) geotectonics and rift magmatology (see [1] for review); (b) effects on magma emplacement [2] and magma generation [3] ; (c) effects on chemical variations in peralkaline liquids ([4] for major elements; [5] for minor elements and halogens); (d) experimental melting and crystallization of natural obsidians [6] ; (e) metasomatism of ultramafic nodules [7]. From the ultramafic nodules we could deduce that in deeper parts of the mantle, in regions of low geothermal gradient, the escaping gases were rich in CO_2, and carrying alkalis, H_2O, Al, Fe, Ca, Mn, Ti, Rb, Sr, Ba, Zr, Nb, Y, and La. At the stage of felsic magma generation we could deduce that the gas was carrying Na, Fe, Si, F, Cl, Rb, Zr, Nb, Y, La, Ce, Zn. This paper presents a further step, a consideration of the role of H_2O in peralkaline felsic magmas: and, from this, what may be deduced about other attributes of the gas phase in rift magmatism, and the constraints imposed on magma genesis. The results have important implications also for all kinds of felsic magmatism in all tectonic regimes and lithosphere environments.

2. H_2O CONTENTS OF MAGMAS

Before considering the felsic magmas it is necessary to present the evidence of intrinsic H_2O contents of basaltic melts, because these are traditionally assigned a parental role in the genesis of trachytes, phonolites and rhyolites. In recent years, glassy basalts from different localities, and modes of eruption, have been providing H_2O values that can be seen, by various tests, to be valid measures of the intrinsic H_2O dissolved in the magma prior to eruption. Anderson [8] has reviewed the evidence, and typical values are quoted here in Table 1, where it may be seen that the kinds of basalts common in continental rifts (transitional, or mildly alkaline, to alkaline) may be expected to have intrinsic H_2O contents in the range 0.5 to 1.0 weight percent. Many of the peralkaline felsic volcanics that will be referred to later come from the Quaternary-Recent volcanic fields around the topographic culmination of the Kenya rift (between volcanoes Menengai and Suswa). The range of H_2O contents of contemporaneous transitional basalts from this region are also shown in Table 1. The most glassy basalt gives the highest H_2O content, 0.75, almost exactly what might be expected for the intrinsic H_2O in this basalt type. The lower H_2O values in the other basalts possibly reflect losses by degassing on eruption, or during crystallization.

Table 2 lists the H_2O, F and Cl contents of this glassy basalt, and all peralkaline felsic glasses available at the time of writing. As noted by Macdonald and Bailey [9] the latter are consistently low in H_2O, and we now find that they are mostly lower than the intrinsic H_2O to be expected in any basalts with which they may be associated. This is true regardless of the level of silica

Table 1

A. Intrinsic H_2O in basalts (Anderson, 1975) (Weight percent) [8]

Tholeiite	Transitional	Alkaline
0.2 — 0.5	0.66 — 0.74	about 1.0

B. H_2O in Quaternary-Recent basalts from Badlands, near Eburru volcano, central Kenya rift.* (Weight percent)

	Range	Mean	No. of samples
Lower silica group (~48% SiO_2)	0.10 — 0.35	0.23	5
Higher silica group (~53% SiO_2)	0.13 — 0.75**	0.51	3
All specimens	0.10 — 0.75	0.34	8

*Analysts: S. A. Malik and D. A. Bungard, University of Reading.

**The glassiest specimen (Spec. No. 374) has the highest H_2O.

Table 2

H_2O, F, and Cl in glassy lavas (weight percent)

		H_2O	F	Cl
A. Basalt 374 (Table 1B). Contemporaneous with D.*		0.75	0.10	0.03
B. Peralkaline oversaturated, compilation of available analyses to 1972 (Macdonald and Bailey, 1973)	MIN.	0.00	0.06	0.05
	MAX.	1.18**	1.30	0.82
	MEAN	0.24	0.28	0.28
	SAMPLES	107	67	88
C. Peralkaline oversaturated, analysed in Reading since 1972*	MIN.	0.00	0.03	0.02
	MAX.	0.60	0.95	0.80
	MEAN	0.13	0.40	0.28
	SAMPLES	117	132	132
D. Peralkaline oversaturated from Eburru volcano, central Kenya rift*	MIN.	0.00	0.13	0.07
	MAX.	0.25	0.86	0.47
	MEAN	0.12	0.50	0.32
	SAMPLES	24	29	29

*Analysts: S. A. Malik and D. A. Bungard, University of Reading.

**The second highest value is 0.67 so that this sample must be highly exceptional.

Table 3

Contemporaneous basalt (A) and pantellerite (B) from Eburru region, Kenya. All values in p.p.m.

	Zr*	H_2O	F	Cl
A	130	7500	970	280
B	3058	1000	7800	4290
Concentration Factor $^B/_A$	23.5	0.13	8.0	15.3

*Analysts A.W.H. Bowhill, G. Smith and N. Tarrant, University of Reading.

saturation, or geographic or tectonic environment. Apparently calc alkaline obsidians, too, commonly have low H_2O contents [R. Macdonald, *pers. comm.* 1976]. Evidently, this is a fundamental characteristic of felsic volcanic liquids, which must be accounted for in their petrogenesis.

3. IMPLICATIONS FOR MAGMA GENESIS

Assuming that the accepted intrinsic H_2O contents of basalts are real, it follows that felsic magmas with low H_2O contents could not be related to basalts by any closed system (isochemical) evolutionary process. By either fractional melting or fractional crystallization, the H_2O concentration should be greater in the so-called "highly evolved" felsic magmas. Indeed, these melts do contain high concentrations of incompatible elements such as Zr, and other volatiles such as F and Cl, which only serve to emphasise the enigma of low H_2O. A sequence of magmas derived by increasing degrees of melting of the same source requires that H_2O should correlate with the incompatible and volatile elements. Closed system crystallization could provide the observed relationship *only* if a hydrous phase were fractionating *throughout*, and no hydrous minerals have been observed in the rocks (nor indeed, have they ever been advocated by proponents of fractional crystallization for these rocks). If these magmas are in evolutionary relationship, therefore, the magma system was open.

In earlier discussions, when I have raised the problem of low H_2O in felsic glasses, it has sometimes elicited the suggestion that the obsidians may not be fully representative samples, implying that the felsic magmas originally contained H_2O commensurate with derivation by fractional crystallization from a basaltic parent, but that this H_2O has been lost from the melt prior to, during, or after eruption. This suggestion at least allows that the magma system was open to losses! Fortunately it is susceptible to quantitative examination. Trace element variations in a series of 15 obsidians from Eburru volcano, Kenya rift, have been described elsewhere [5], and the contents of H_2O, Zr, F and Cl from the most enriched pantellerite may be compared with the contemporaneous glassy basalt cited in Tables 1 and 2. The results appear in Table 3, from which it may be seen that if these two magmas are parts of an evolutionary sequence, various degrees of concentration have been produced in the Zr, F and Cl. If we assume that nothing has been added during fractionation then the maximum concentration factor must set a minimum limit for the extent of crystallization, i.e. in order to achieve a x 24 concentration the residual liquid cannot have been more than $^1/_{24}$ of the initial liquid. The lower concentration factors of the other constituents can only be ascribed to losses (either to the crystal fraction, or to an escaping vapour). As there are no hydrous phenocrysts in either the basalt or the pantellerite it must be assumed that H_2O was subject to the same degree of concentration as Zr, but has been lost as vapour. The required degree of concentration, starting from a minimum of 0.75 wt. percent H_2O in the basalt would have produced 18 wt. percent H_2O in the pantellerite! Using Luth's curve for the solubility of H_2O in a synthetic granite melt [10] we can infer that a pantellerite containing 18 percent H_2O would be saturated at a pressure of about 11 kb, approximately 40 km deep, and oversaturated with water at any shallower depth. Throughout this depth range (the whole thickness of continental crust) this water-saturated pantellerite would have a solidus with a negative dT/dP, and be incapable of rising towards the surface without crystallisation and exsolution of vapour. The pantellerite obsidian in question is devoid of crystals and non-vesicular (in common with many others from all parts of the world), and it is, moreover, oversaturated with the other volatiles. In a series of melting and crystallization experiments [11] the natural sample was found to be oversaturated in volatiles at least to 0.3 kb. In its natural state it had, therefore, been effectively quenched (i.e. closed to volatile loss) from a pressure of 0.3 kb, or greater. At 0.3 kb a synthetic granite melt could dissolve 2.5% H_2O [10] from which we must infer that this obsidian, when quenched, was strongly undersaturated in H_2O (see [6] for more details). So far as I am aware, no mechanism has yet been described that would allow a massive, differential escape of H_2O from a crystallizing magma system, yielding a superliquidus, dehydrated residual melt.

Other factors also indicate that the low H_2O levels in the obsidians in Table 2 cannot be due to non-systematic losses from hydrous melts. Consistently low H_2O is found despite the following variability of the samples:

1. Bulk compositions from phonolite to comendite, i.e. wide variations in degree of differentiation.
2. Samples from all geographic settings.
3. Wide variations in vesicularity.
4. Wide variations in phenocryst types and contents.
5. Wide variations in eruptive mode. Samples from flow margins (top and bottom); flow interiors; angular blocks in agglomerate; angular fragments in pumice deposits; fiamme from welded ash flows; splatter; and agglutinated spatter.

All these factors can be satisfied only if there has been systematic constraint on the H_2O contents of the melts being generated in a felsic magma system. In the absence of any mechanism for systematic H_2O loss that would give consistently low levels (below saturation) during fractional crystallization, we must look to some alternative control on the melt composition.

4. CHARACTERISTICS OF THE SOURCE ROCKS

One way by which the low H_2O content of the different magmas might be controlled by the environment is if they were not related through any line of liquid evolution, but produced by the melting of different source rocks. In the magmas considered here each source would need to be effectively dehydrated. The various melts produced in a peralkaline complex would thus be related only through a common melting cycle which has operated on different sources. But the patterns of peralkaline magmatism are so strong, and repetitive in space and time, that haphazard melting of heterogeneous materials in different lithosphere sections cannot be entertained. What is needed is an environmental control over melt compositions that has general and widespread application.

If each batch of magma in a peralkaline complex has been produced by isochemical melting, then the chemical coherence of peralkaline magmas generated in all parts of the world would require that the lithosphere below each complex has been chemically preconditioned prior to melting. The case for mantle metasomatism below stable cratons has been made before [12,7] and seems to be finding favour through other lines of evidence [13] . Metasomatism would seem to be the only way of producing convergence of lithosphere compositions so that when melting ensues the melt chemistries are strongly coherent, and repetitive through geologic time [14] . All other suggested magmatic processes below cratons must be irreversible, leading to progressive exhaustion of hyperfusibles and volatiles in the lithosphere mantle. It was previously shown [7] that carbonates, biotites and amphiboles could be expected to form in the cratonic lithosphere, as volatiles, migrating towards the Earth's surface from greater depths, successively encountered the PT stability regions of these minerals. A mantle and deep crust selectively enriched in this manner could also provide an explanation for the observed variations in carbonatitic, potassic, and sodi-potassic magmatism. In the metasomatic calculations based on biotites and amphiboles it was necessary to use the data for the hydrous varieties of these minerals because these were the only ones available. Halogen-bearing varieties would be more consistent with the observed chemistry of continental magmatism: the densities would be similar to the hydrous minerals but the thermal stabilities should be higher. A vertically zoned metasomatised lithosphere containing carbonates, halogen-bearing biotites and amphiboles could thus provide different magmas, variously enriched in CO_2, F and Cl (and effectively deficient in H_2O) as a subsequent melting cycle worked its way upwards through the zones. It is conceivable that the melting of each source could be isolated from the others, thus producing a batch of magma by essentially isochemical melting. Such an arrangement seems hardly likely to have general application, however, and it seems more realistic to develop a system of petrogenesis that is not wholly reliant on such conditions. The metasomatic preconditioning of the lithosphere may have taken place through the action of a separate vapour (rather than by diffusion) in which case the scheme of magma genesis should take account of the following:

(a) the mineralisation of the lithosphere will have been determined by the chemistry of the vapour:
(b) the continued passage of vapour through structural channels in the lithosphere could itself lead to melting [3] . Magma production is thus the culmination of the degassing process:
(c) the chemistry of the vapour will condition the chemistry of the melt because it will control the temperature of melting, and the proportions of the solid phases entering the liquid: and
(d) the vapour will maintain a buffering action on the melt during its subsequent history, and passage to the surface.

5. CHARACTERISTICS OF THE VAPOUR

Earlier studies have given some indications of the chemical characteristics of the vapour coexisting with peralkaline over-saturated melts in the Kenya rift zone [4,5]. It was possible to deduce that if these melts were related, either by crystallisation or by melting, some additions from a coexisting vapour would be necessary to maintain the chemical variations in the obsidians. For example, if the magmas were related by progressive crystallisation then Na must be removed in the vapour and Rb added: the converse would apply if the magmas were related by progressive melting. Whatever the process a vapour is required to buffer the chemistry of the magmas, and so far it has been possible to establish buffering for Na, Fe, Si, F, Cl, Rb, Zr, Nb, Y, La, Ce and Zn. Subsequent studies [15] have indicated the need for similar vapour activity in the development of the lavas of the peralkaline trachyte volcano Longonot. It is now clear from the H_2O contents that the coexisting vapour must have the potential to maintain the source rocks, and the subsequent melts, at low concentrations of H_2O: this requires that the P_{H_2O} of the vapour must be low because H_2O is very soluble in silicate melts and the P_{H_2O} of the vapour will correlate with the H_2O concentration in the melt. It follows that H_2O must be heavily diluted in the vapour by gases which have low solubilities in silicate melts, thus leaving little record of their previous activity in the final volcanic glass —— the most obvious candidate among the common volcanic gases is CO_2.

High concentrations of CO_2 are most evident in the carbonatite activity which is a hallmark of rift magmatism, but it is worth pointing out that CO_2 is much in evidence in other ways. Carbonates are characteristic of nephelinite magmatism, and from the only active nephelinite volcano, Nyiragongo, the continuous gas emission consists largely of CO_2 (see [16] for discussion). What is perhaps less well known is that there is continuous emission of CO_2 along the East African rift even in areas of silicic magmatism. Juvenile CO_2 is considered to be the main heating agent in many places where the geothermal activity has been studied, and in at least two localities in Kenya (Esageri and Uplands) CO_2 of high purity is obtained commercially from shallow wells at pressures up to 60 lbs. per square inch. Specially relevant is the fact that the best described cases of juvenile CO_2 activity are in the same central part of the Kenya rift that has provided the peralkaline obsidians described in this paper [17,18,19]. Little evidence has been found by these authors for juvenile H_2O, in fact McCall [18, p.85], expressly states that "there is no evidence of juvenile water".

The limited amount of experimental evidence on the partitioning of CO_2 and H_2O between silicate liquid and vapour [20] indicates that in order to maintain low concentrations of H_2O in the liquid, the vapour would have to be largely CO_2 (almost anhydrous). Prior to melting, such a vapour (with very low P_{H_2O}) would severely reduce the thermal stability of hydrates in the source rock, to the extent that pure hydrates would be unstable in the mantle in regions of normal geothermal gradient. The higher F and Cl in natural glasses suggests that halogen bearing silicates would be favoured. In this way a CO_2-rich vapour would also provide the appropriate preconditions for the formation of anhydrous magma. Evidence of high CO_2 and low H_2O is directly available in the modern alkali carbonatite lavas of Oldonyo Lengai at the southern end of the Gregory rift [21] which have remarkably low H_2O, and hydrate on exposure to the atmosphere. It is probably significant, too, that the characteristic zones of metasomatised country rocks around carbonatite and alkaline intrusions show little evidence of hydration, although many have introduced carbonates. Analyses of phlogopites from intrusive carbonatite complexes show these to be rich in F [pers. comm. T. Deans, 1977] which, in view of the relatively low temperatures of formation, would again suggest that low H_2O is possibly a characteristic of intrusive carbonatite.

There are thus several lines of evidence indicating that the characteristic vapour phase in rift magmatism is rich in CO_2 and deficient in H_2O, but it must be borne in mind that CO_2 and H_2O are almost certainly the low temperature, low pressure, near-surface expression of the actual vapour in the magma system: at greater depths and at higher temperatures it is to be expected that the C and H will be present in the form of other species. Evidence from high temperature volcanic gas collections [22] , fluid inclusions, and calculated gas compositions in alkaline rocks [23] suggest that CO, CH_4 and H_2 may assume greater importance at higher temperatures and pressures. The low oxidation states of peralkaline obsidians [9] would also be consistent with less oxidised states of C and H in the vapour. It should also be borne in mind that the relatively high concentrations of fluorine and chlorine in the lavas suggest that these gases will also be helping to diminish P_{H_2O} in the vapour by their presence as elemental gases, and through chemical combinations with available hydrogen.

Finally, one gas which is often neglected in discussions of volcanic activity must here be considered, namely, nitrogen. In volcanic gas analyses there is always a risk of some ambiguity in accounting for nitrogen because of atmospheric contamination, but as Anderson [8] points out, some nitrogen values cannot be disposed of in this way: furthermore ammonium salts are frequently recorded in fumarole deposits. The presence of nitrogen in the deep mantle is unquestionable because it occurs as inclusions in diamonds [24] and the relevance of this occurrence to the present case is clear from the genetic links between kimberlites and carbonatites. Nitrogen is also recorded, along with CO_2, as a juvenile gas in the central volcanic region of the Kenya rift and is especially noteworthy in the gases emanating from the axial region of the peralkaline trachyte volcano Menengai [18] . A later Reading University collection of gases from Menengai also contains samples with significant amounts of nitrogen that cannot be explained by atmospheric contamination. This would suggest that the vapour coexisting with the silicic magmas of the Kenya rift was rich in carbon gases, halogens, and nitrogen. Hydrogen may be an important constituent but not in the form of H_2O.

6. SOURCE OF THE VAPOUR

Mysen has deduced from experimental evidence that an ultramafic melt with dissolved CO_2 and some H_2O will, during cooling and decompression, emit a gas phase that is strongly enriched in CO_2 [20] . Nephelinitic magma enriched in CO_2 is a characteristic eruptive in the rift zone, and slowly rising and cooling masses of such material at depth could be providing part of the CO_2 emission at the surface today. The release of hot CO_2 from a crystallising melt in the mantle might conceivably cause melting at higher levels to produce silicic liquids. But to attribute the silicic magmatism to the release of volatiles from an ultramafic magma at greater depth simply begs the question of the origin of the vapour. The ultimate source of the vapour must be in the deep mantle, and the metasomatic formation of carbonates and their eventual melting to produce CO_2-rich magmas, are simply stages in the progress of volatiles from the deeper parts of the Earth to the surface.

7. CONSEQUENCES OF CONTINUED VOLATILE FLUX THROUGH THE RIFT ZONE

A flux of gases through the lithosphere at first reduces the density by metasomatic and thermal expansion [7] . Persistent and extensive fluxing of gases through a preferred channel system, such as a rift zone, if sustained for a long period, must ultimately lead to thermal decomposition of volatile-bearing minerals, and melting [3] . Extensive magmatism has been taking place along the East African rift zone from Miocene to Recent, during which time volatile-bearing minerals in the rift segment of the lithosphere will have been gradually eliminated as the geothermal gradient has steepened and the magmatism has worked its way through to higher levels [1] . By this process it would be expected that the originally expanded lithosphere has been collapsed into denser, residual solids, together with the expulsion of less dense magmas and volatiles. This transformation could be the main cause of the subsidence of the rift prism in the middle of a regional uplift, and would leave a residual positive gravity anomaly superimposed on the regional negative anomaly. The denser rocks necessary to provide the axial positive anomaly along continental rifts are thus simply the dense, collapsed residua from a sustained cycle of heating, devolatilisation, and partial melting. The popular explanation of the positive anomaly by intrusion of a basic "batholith" along the rift axis, poses severe problems, not least of which is providing room in the continental lithosphere, and especially in the continental crust. The gravity model favoured by Searle [25; Model 4] postulates a basic intrusion above the 19 km level in the continental crust, with an approximate vertical cross-sectional area of 320 km^2, i.e. 320 km^3 per unit length of rift. The calculated 5 km extension across the Kenya Rift [26] even if it were entirely explained by plate separation, could provide a maximum free space above the 19 km level of only 95 km^3 per unit length of rift. Most of the required basic material must, therefore, have replaced pre-existing continental crust and there appears to be nowhere for this pre-existing crust to go. This difficulty applies also to the Oslo graben, and, after a detailed appraisal of possible solutions, Ramberg is led to ask if crustal anatexis might "ease the room problem" [27] . The alternative suggested here, metasomatism and melting of the continental crust has the attraction of a consistent mass, volume, chemical, and thermal budget, as well as fitting the surface observations.

8. SYNOPSIS OF EVENTS IN CONTINENTAL RIFTING AND MANTLE DEGASSING

An outline of the process is as follows:

1. Horizontal movement of the anisotropic continental plate re-activates earlier zones of weakness, leading to flexuring,

fissuring, and the channelling of gas release from the deeper mantle. The zones of weakness are typically the most recent orogenic belts, girdling the old cratonic nuclei, and the stress release along these zones provides the initial decompression, which localizes the escape of volatiles.

2. The gas phase is rich in carbon gases and characterised by its content of halogens: it is strongly reducing, and little of the hydrogen is in the form of water: nitrogen may be important as an inert species modulating the activities of the other gases.

3. Transit of the volatiles along low to normal geothermal gradients leads to mantle metasomatism, and especially the formation of carbonates and halogen-bearing alkali silicates. The metasomatised upper part of the mantle constitutes a "holding zone" for alkalis and volatiles prior to melting.

4. During the metasomatising period there will be continental uplift by thermal and metasomatic expansion of the lithosphere.

5. Continued heating by the gas flux ultimately leads to thermal decomposition of some of the volatile-bearing minerals, and melting.

6. Melt compositions are limited in amount and buffered by the gas phase (still escaping from the deep mantle, but now complemented by volatiles from thermal decomposition).

7. Gradual ascent of melting regime to higher levels in lithosphere.

8. Extensive devolatilization and melt extraction from the deeper parts of rift prism leads to increase in density, causing subsidence and the axial positive gravity anomaly.

9. Cycle may fail, abort, or die out at any stage without lithosphere separation, but it can also pass into complete lithosphere separation.

10. If there is lithosphere separation, then convective wedging of the mantle and oceanic type magmatism follow. The volatiles and alkalis characteristic of continental rift magmatism are swamped by extensive melting.

Acknowledgements. A.W.H. Bowhill kindly allowed me to use the information on the Badlands basalts which he is studying. Special thanks are due to W.F. Price for his help and stimulating discussion during the preparation of the manuscript.

REFERENCES

1. D.K. Bailey, Continental rifting and alkaline magmatism, in: The Alkaline Rocks, ed. by H. Sørensen, John Wiley and Sons, New York, 148, 1974.
2. D.K. Bailey, Carbonatite volcanoes and shallow intrusions in Zambia, in: Carbonatites, ed. by O.F. Tuttle and J. Gittins, J. Wiley and Sons, New York, London and Sydney, 127, 1966.
3. D.K. Bailey, Geol. J. Special Issue No. 2, 177, 1970.
4. R. Macdonald, D.K. Bailey and D.S. Sutherland, J. Petrol. 11, 507, 1970.
5. D.K. Bailey and R. Macdonald, Min. Mag. 40, 405, 1975.
6. D.K. Bailey, J.P. Cooper and J.L. Knight, Anhydrous melting and crystallization of peralkaline obsidians, in: Oversaturated Peralkaline Volcanic Rocks, Special Vol. Bull. Volc. 38, 653, 1974.
7. F.E. Lloyd and D.K. Bailey, Phys. Chem. Earth 9, 389, 1975.
8. A.T. Anderson, Reviews Geophys. Space Phys. 13, 37, 1975.
9. R. Macdonald and D.K. Bailey, U.S. Geol. Surv. Prof. Paper, 440N-1, 1973.
10. W.C. Luth, Granitic Rocks, in: The evolution of the crystalline rocks, ed. by D.K. Bailey and R. Macdonald, Academic Press, London, 335, 1976.
11. J.C. Cooper, Experimental melting and crystallization of oversaturated peralkaline glassy rocks, unpubl. Ph.D. thesis: University of Reading, 1975.
12. D.K. Bailey, J. Earth Sciences (Leeds) 8, 225, 1972.
13. C. Brooks, D.E. James and S.R. Hart, Science 193, 1086, 1976.
14. D.K. Bailey, J. geol. Soc. Lond., 133, 103, 1977.
15. S.C. Scott, The Volcanic Geology and Petrology of Mount Longonot, Central Kenya, unpubl. Ph.D. thesis: University of Reading, 1977.
16. D.K. Bailey, Nephelinite and ijolite, in: The Alkaline Rocks, ed. by H. Sørensen, John Wiley and Sons, New York, 53, 1974.
17. A.O. Thompson and R.G. Dodson, Rep. 55, Geol. Surv. Kenya, 1963.
18. G.J.H. McCall, Rep. 78, Geol. Surv. Kenya, 1967.
19. J. Walsh, Rep. 83, Geol. Surv. Kenya, 1969.
20. B.O. Mysen, Carn. Inst. Yr. Bk., 74, 460, 1975.
21. J.B. Dawson, Oldonyo Lengai --- an active volcano with sodium carbonatite lava flows, in: Carbonatites, ed. by O.F. Tuttle and J. Gittins, John Wiley and Sons, New York, 1966.
22. A.T. Huntingdon, Phil. Trans. R. Soc. Lond., A. 274, 119, 1973.
23. R.H. Mitchell, Phys. Chem. Earth 9, 903, 1975.
24. W. Kaiser and W.L. Bond, Phys. Rev., 115, 857, 1959.
25. R.C. Searle, Geophys. J.R. astr. Soc. 21, 13, 1970.
26. B.H. Baker and J. Wohlenberg, Nature, 229, 538, 1971.
27. I.B. Ramberg, Norges geologiske undersøkelse, 325, 1976.

A NOTE ON THE BEHAVIOR OF INCOMPATIBLE TRACE ELEMENTS IN ALKALINE MAGMAS*

B. H. Baker

Center for Volcanology, Department of Geology, University of Oregon, Eugene, Oregon, USA

1. INTRODUCTION

Several recent studies have shown notable regularities in the behavior of incompatible trace elements in alkaline volcanic suites [1-5], and there has been some discussion of the significance of these features, especially of linear correlation (or lack of it) between such trace elements [5]. This contribution is not an exhaustive treatment of trace element behavior, the principles of which have been applied to alkaline rocks by others [4,6,7]. Rather it is proposed to discuss some petrogenetic uses to which such data may be put.

The incompatible trace elements discussed here (hereafter called the ITE) are Zr and Hf, Nb and Ta, Th, Rb, La and Sm. These elements do not enter readily into common igneous minerals and therefore tend to accumulate in the liquid component of silicate systems. Zr^{4+} and Hf^{4+} (r = 0.80Å), Nb^{5+} and Ta^{5+} (r = 0.72 Å), and Th^{4+} and U^{4+} (r = 1.08Å) are crystallochemically nearly identical element pairs whose ratios are similar in chondrites, the mantle and nearly all alkaline rocks. They are incompatible primarily because of their high charge. The larger ions (Rb^{+}, r = 1.57 Å; La^{3+}, r = 1.13 Å; Sm^{3+}, r = 1.04 Å) are less incompatible but also more liable to be mobile in post-magmatic processes [8], and hence are less reliable for petrogenetic interpretation. Some accessory or late crystallizing minerals (zircon, sphene, eudyalite, alkali amphibole) contain significant quantities of these elements, but the small quantities and late crystallization of such minerals in most (oversaturated) alkaline

* This work has been supported in part by National Science Foundation Grants GA 32333 and EAR75-20664, and by NERC.

E.-R. Neumann and I.B. Ramberg (eds.), Petrology and Geochemistry of Continental Rifts, 15-25.

rocks do not influence the distribution of ITE substantially.

2. MODELS OF INCOMPATIBLE ELEMENT BEHAVIOR

The distribution of trace elements in ideal closed system fractionation and partial melting processes may be described by

$$C_\ell = C_o.F^{(\bar{D}-1)}$$

for equilibrium fractional crystallization, by

$$C_\ell = C_o/\bar{D}+F(1-\bar{D})$$

for partial melting, where $\bar{D} = D^a.X^a+D^b.X^b....D^nX^n/1-F$, C_ℓ and C_o represent trace element abundances in the derivative liquid and parent material respectively, F is the mass fraction of the system represented by the liquid, $\bar{D}$ is the bulk distribution coefficient, and $D^a....D^n$, and $X^a....X^n$ represent individual phase/liquid distribution coefficients and the weight fractions of these phases in the total system.

The enrichment factors of ITE for a range of constant values of $\bar{D}$ are plotted against F for closed system (Rayleigh law) fractionation and for Nernst law partial melting in Fig. 1. Except for the cases $\bar{D}_{ITE}$= 0 or 1 fractionation results in larger enrichments of ITE in magmas than does partial melting, the differences becoming very large for small values of F (i.e., for differentiated rocks). F will be approximately proportional to bulk composition of a magma formed either by partial melting or fractionation processes. For salic rock compositions approaching cotectic residua compositions the change of bulk composition is small. In this composition range (about F = 0.02 to 0.1) progressive partial melting will result in small increase in ITE contents of magmas, whereas fractionation would result in much larger increases. Many differentiated alkaline suites show an astonishingly wide range of ITE abundances among salic rocks, for example comenditic trachytes: 500-700 ppm Zr, pantelleritic trachytes (700-1500 ppm Zr), pantellerites (1500-3000 ppm Zr), and proportionally high abundances are also seen for the other ITE. The great range of ITE abundances over a limited range of bulk composition strongly suggest the operation of fractionation rather than partial melting processes. Furthermore, the average abundance of Zr, for example, in sialic crust, is about 250 ppm; consequently many peralkaline rhyolites cannot be formed by partial melting unless $\bar{D}_{Zr}$ for that composition and temperature is 0.1 or less. Evidence is given below that $\bar{D}_{Zr}$ is greater than this.

The foregoing discussion is oversimplified, however, because it neglects the theoretical and experimental evidence to show that distribution coefficients rise substantially with falling temperature [9,10], and are also affected by little understood factors

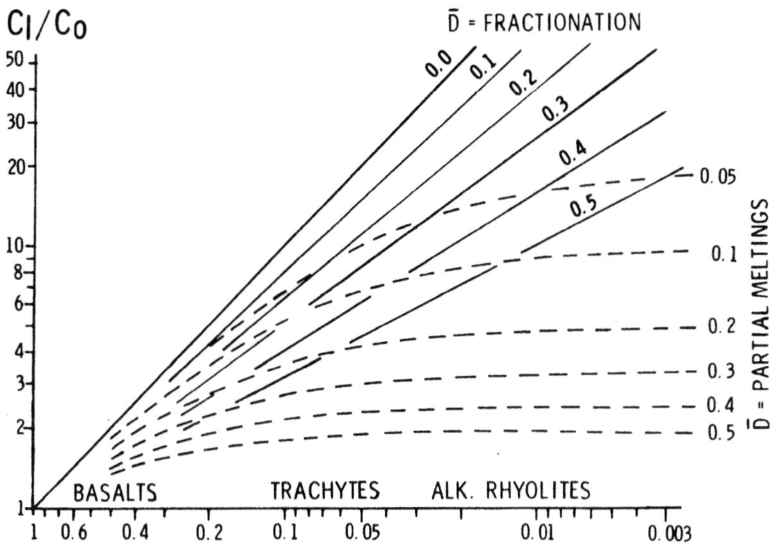

Fig. 1. Concentration factors (C_ℓ/C_o) of incompatible trace elements plotted against F, the weight fraction of liquid in a closed crystal/liquid system. Solid lines show enrichments for surface equilibrium (Rayleigh law) and dashed lines show enrichments for bulk equilibrium (Nernst law).

related to liquid composition and structure [11]. The temperature effect conforms empirically to an expression of the type

$$\bar{D} = \sqrt{A+B/T_k},$$

and for many dispersed elements can result in order of magnitude increases of $\bar{D}$ over the range of magmatic liquidus temperatures. The consequence of this is that the ideal C_ℓ/C_o vs. ℓnF relationship is not exactly as shown in Fig. 1 for D_{ITE} increase from values of about 0.05 for basaltic liquids to about 0.4 for salic liquids (see below).

A further consequence of the rise of $\bar{D}_{ITE}$ is that the use of the inverse of ITE enrichment ratios

$F = C_o/C_\ell$, which is valid if $\bar{D} = 0$, as

a differentiation index [2,6] is valid only for basaltic compositions, and will produce distortions if applied to more salic rocks. Nevertheless the parameter F, whether estimated as above, or computed from major element mass balance modelling [3,12,13] is a superior differentiation index which reveals the inter-relationship of comagmatic rocks in a more meaningful manner than most others. In particular it helps to demonstrate that the Daly gaps which characterize many alkaline suites are largely artefacts of the behavior of silica during fractionation. Fig. 2 shows an"alkalies-" silica diagram for a suite of basalts and trachytes [14], and a silica-'F' plot for the same rocks. It demonstrates that the Daly gap is largely due to a sudden rise of SiO_2 abundance in the intermediate composition range, and that the Daly gap may be insignificant if a seemingly bimodal suite is studied using a more fundamental differentiation parameter.

3. INCOMPATIBLE TRACE ELEMENT RATIOS

The ITE abundances in many volcanic suites commonly show a strong linear correlation [1,2]. The ratio (R_ℓ) of a pair of ITE in a derivative liquid, and that in a parent liquid are related by $R_\ell = Ri.F^{(\bar{D}i-\bar{D}j)}$ for fractional crystallization and by

$$R\ell = Ri \; \frac{\bar{D}i+F(1-\bar{D}i)}{Dj+F(1-Dj)}$$

for partial melting and for simplified constant values of $\bar{D}$. Consequently linear correlation of ITE abundances will be maintained so long as $\bar{D}i = \bar{D}j$. Such correlation does not require that Di = Dj = 0 [1], nor is it diagnostic of fractionation as opposed to partial melting processes [3]. Nor do departures from linear correlation (changes of $R\ell$) necessarily require that open system processes such as additions or subtractions of ITE by vapors have occurred [5]. This is because bulk partition coefficients must vary with changes in the proportions of solid phases in a solid-liquid system, or with a change in the number of solid phases. Only for geochemically identical ITE (Zr and Hf; Nb and Ta) will there be no change in their ratio. For other ITE the degree of inflextion of ITE covariance curves will be proportional to the difference of their $\bar{D}$'s. For example an increase in the proportion of fractionated clinopyroxene or titanomagnetite will tend to increase $\bar{D}_{Zr}$ through limited substitution for Ti^{4+}, but will not affect the $\bar{D}$'s of other ITE. Curvature or inflexion of covariance plots of ITE is to be expected over broad compositional ranges, and can usually be related to changes in refraction phase assemblages. Such features do not require that open system differentiation has occurred [5], and the fact that inflexions of ITE covariance curves often occur at the same composition for many of the elements suggests that phase changes are involved (see Fig. 3 in [5]).

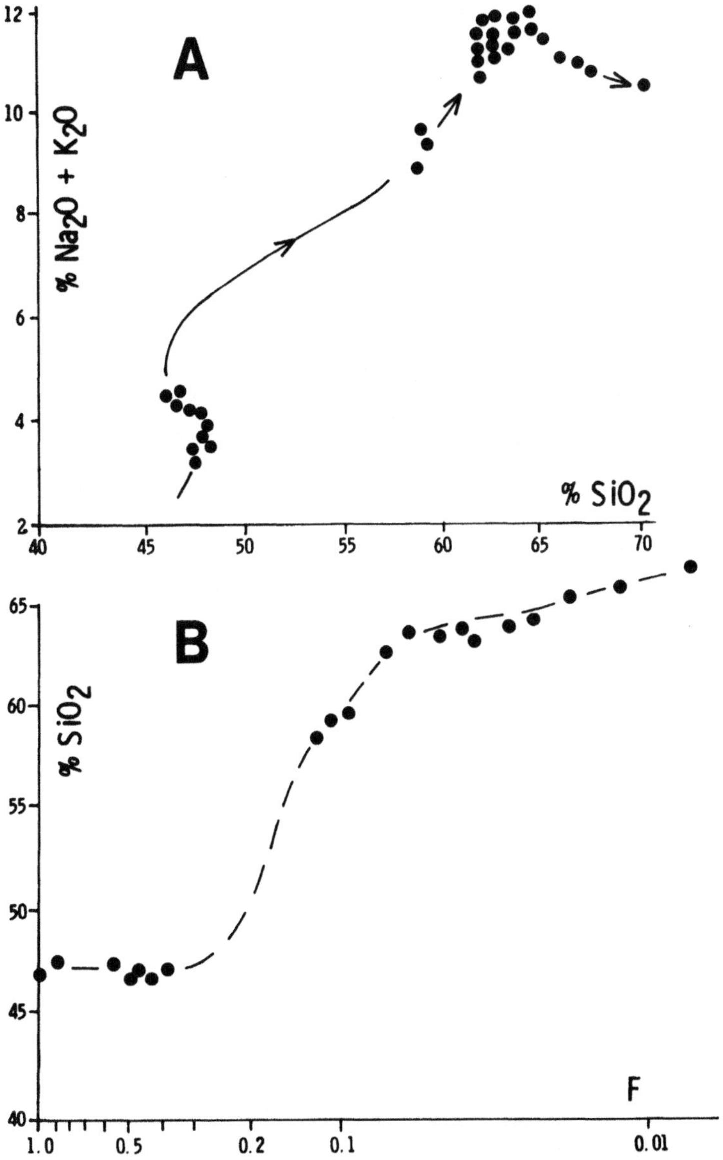

Fig. 2A. Alkalies-silica plot for a basalt-trachyte-pantellerite suite of the Kenya rift valley. Fig. 2B shows the sharp rise in silica abundances with increasing fractionation (1-F) which is partly responsible for Daly gaps in such suites.

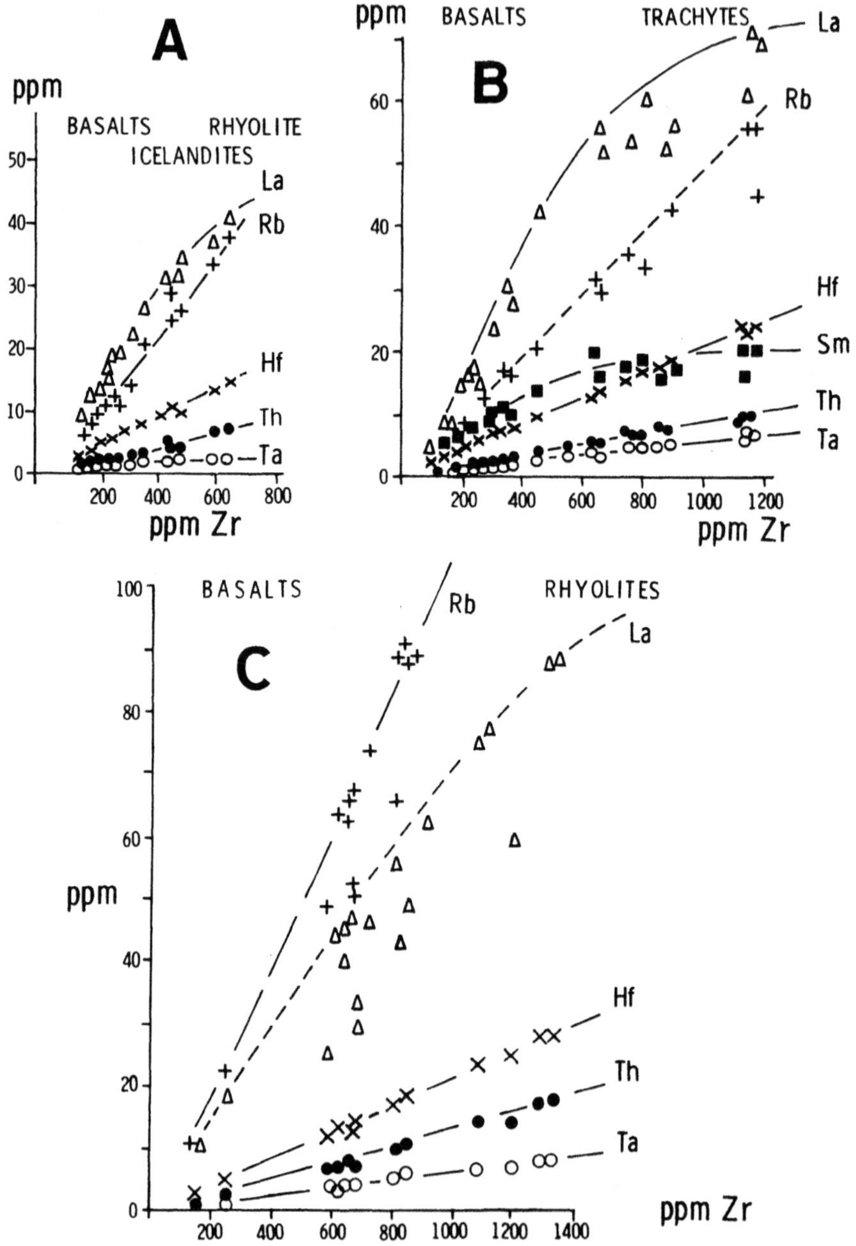

Fig. 3. ITE covariance of: A-Pinzon tholeiitic suite; B-Santiago alkaline suite, Galapagos [15]; C-transitionally alkaline suite from Isla San Benedicto, Mexico [16].

Covariance of ITE with Zr abundances in ocean island volcanic suites are shown in Fig. 3. These are a basalt-trachyte suite from the Galapagos group [15], and a basalt-trachyte-rhyolite suite from San Benedicto, one of the Socorro group, west of Mexico [16]. Figure 4 shows similar covariance curves for two series of continental rift lavas from the Kenya rift valley--a basalt-trachyte-pantellerite and a trachyte-pantellerite suite, both represented by voluminous flat-lying flood lavas [14]. All of these suites exhibit a substantial degree of correlation of ITE abundances, with Zr, Hf, Nb, Ta and Th showing the best linear correlation. La, Sm and Rb show more erratic behavior, especially among more silicic rocks, with a tendency for the enrichment of these elements to fall off with differentiation relative to Zr. Since none of these suites crystallize more than trace amounts of apatite that could fractionate La and Sm, their declining enrichment of light REE must be due to a relative large increase in $\bar{D}$ owing to falling temperature and more salic liquid compositions, while the erratic low abundances in many rocks are certainly due to post-magmatic losses that are known to affect the REE [8]. In no case is the behavior of the more stable ITE erratic; there are no increases in slopes of the correlation curves resulting from open system additions of these elements, and behavior of the ITE in oceanic and continental suites is qualitatively the same.

4. OPEN SYSTEM AND VAPOR PHASE PROCESSES

Because of the scarcity of experimental and observational data on open system and vapor phase processes in magmatic differentiation they have served as *deus ex machina* processes to be invoked when other explanations are lacking, especially for genesis of the peralkaline rocks. Processes like wall rock reaction or contamination cannot, for thermodynamic reasons, influence the geochemical evolution of large volumes of magma. It remains to consider the possible effects of supercritical aqueous liquids or volatile phases such as are the probable agents of fenitization. Regrettably little evidence about trace element behavior during such processes is available, but some studies strongly suggest that large variations of ITE ratios, even of geochemically coherent pairs like Zr/Hf, Nb/Ta, do occur in rocks affected by such processes [17,18,19]. It is reasonable to suppose that the distribution of trace elements between silicate solids or liquids and aqueous or gas phases are drastically different from those prevailing between solid and liquid silicate components. Conceivably kinetic effects render ionic mass more important than radius or charge in such processes. Preliminary results of geochemical study of fenitization of granodiorite bears out these conclusions, for the ratios of ITE in these fenites varies by an order of magnitude [20]. By contrast cogenetic suites of alkaline rocks show remarkably constant coherent ITE ratios [1,14,21], and the variations that occur are

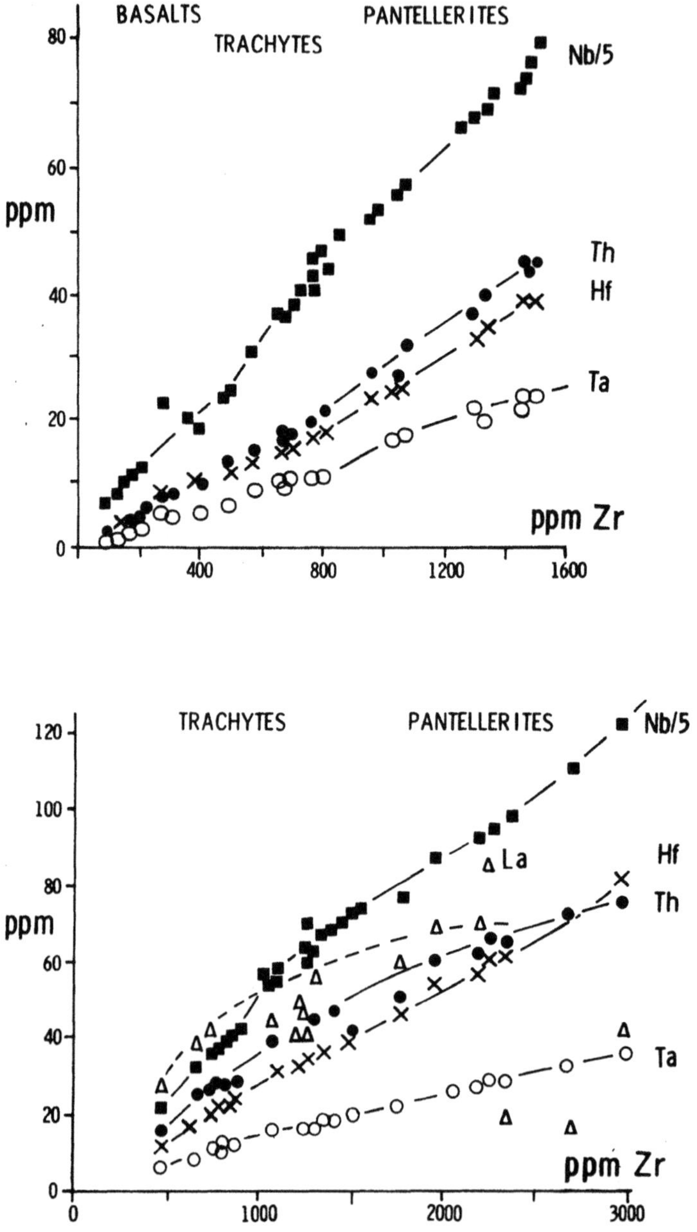

Fig. 4. Covariances of ITE in: A-basalt-trachyte-pantellerite suite (Plateau trachytes)[14], and B-trachyte-pantellerite suite (Limuru trachytes) from the south Kenya suite.

explicable in terms of fractionation processes. Nevertheless, there is a need for detailed study of well-documented cases of aqueous/volatile metasomatic processes that have occurred at magmatic temperatures.

5. PETROGENETIC MODELLING

In modelling fractionation processes that might have formed a suite of rocks the unknown parameters are F, the weight fractions of the minerals making up the fractionating mineral assemblage (X^a, X^b etc.; see equations on p.). If the fractionated phases are like the phenocrysts contained in the rocks, and can be analyzed, then a series of mass balance fractionation calculations can model the evolution of the line of liquid descent, yielding parameters F and X [12, 14]. The parameter F is computed using all the major element data of the rocks including mineral compositions, and is a differentiation index to be preferred over others that use more limited data, or are excessively model-dependent. Given a set of F values for rocks that supposedly represent a line of liquid descent, bulk distribution coefficients for trace elements may be calculated, and using X values broken down into estimated mineral D's, especially for elements that predominantly enter only one phase of the fractionating mineral assemblage. A more complete procedure includes the trace element analysis of phenocryst mineral separates and of groundmass [13], but this is often either difficult or liable to contamination errors.

An example of the result of such computations is given in Fig. 5 for a suite of trachytes and alkali rhyolites from the South Kenya rift valley [14]. Plotted against F the enrichment factors of the ITE define nearly straight lines that conform closely to expected behavior during fractionation. The curves fit the Rayleigh fractionation law for values of $\bar{D}$ between 0.3 and 0.5, and are unlike the curves to be expected from segregation of partial melts (see Fig. 1). Figure 5 also shows the remarkable extent of fractionation within the trachyte-pantellerite composition range, which must be assisted by the high solubility of water and halogens in peralkaline melts, resulting in reduction of viscosity and lowering of solidus temperatures. The major element fractionation calculations from which the F values were obtained show that the predominant fractionating phase was alkali feldspar, and the extreme depletion of the pantellerites in Sr and Ba tends to support this view.

6. CONCLUSIONS

Incompatible trace element data provide a valuable means of testing the following questions relating to volcanic petrogenesis: 1) whether the rocks selected for study are likely to be comagmatic.

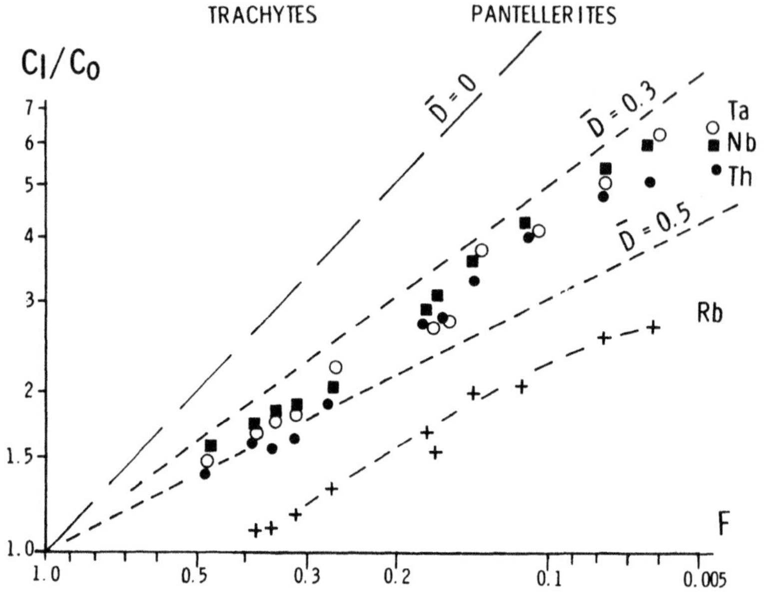

Fig. 5. Concentration factors of incompatible trace elements plotted against F for the Limuru trachytes, south Kenya rift. C_ℓ/C_o for Ta, Nb, Th (and Zr and Hf, not shown) conform to values of $\bar{D} = 0.4$. The Rb line is off-set for clarity. Dashed lines show Rayleigh law fractionation trends at constant $\bar{D}$.

This will be suggested by relative constancy of ITE ratios such as Zr/Nb, Hf/Ta, Th/Nb, which tend to vary from one suite to another, but which vary little with low-pressure differentiation [1,14,22]; 2) whether the rocks have been contaminated by wall-rock reaction, by volatile transfer processes. Regularity of ITE abundance variation and of ITE ratios suggest that open system processes have not been significant. Losses of Na, Rb and light REE suggest the effect of post-magmatic alteration [8]; 3) to distinguish between closed system fractional melting or crystallization processes. Provided a broad compositional range is represented different patterns of ITE enrichment may distinguish these processes; 4) to provide a check on the validity of major element mass balance calculations based upon some quantifiable petrogenetic process, given that realistic estimates of distribution coefficients can be made; to compare the behavior of ITE in volcanic suites that might have formed by radically different processes. A comparison of ITE behavior in two continental and two oceanic alkaline suites suggests that their behavior is qualitatively identical in these two environments.

REFERENCES

1. S.D. Weaver, J.S.C. Sceal and I.L. Gibson, Contr. Min. Petrol., 36, 181, 1972.
2. F. Barberi, R. Santacroce and J. Varet, Bull. Volc., 38, 755, 1974.
3. F. Barberi and others, J. Petrol., 16, 22, 1975.
4. F. Ferrara and M. Treuil, Bull. Volc., 38, 548, 1974.
5. D.K. Bailey and R. Macdonald, Min. Mag., 40, 405, 1975.
6. C.J. Allegre, M. Treuil, J.F. Minster and F. Albarede, Contr. Min. Petrol., 60, 57, 1977.
7. J.F. Minster, J.B. Minster, M. Treuil and C.J. Allegre, Contr. Min. Petrol., 61, 49, 1977.
8. B.H. Baker and L.F. Henage, J. Volc. Geotherm. Res., 2, 17, 1977.
9. S. Banno and Y. Matsui, Chem. Geol., 11, 1, 1973.
10. M.J. Drake and D.F. Weill, Geochim. Cosmochim. Acta., 39, 689, 1975.
11. D.J. Lindstrom, Experimental study of the partitioning of the transition metals between clinopyroxene and coexisting silicate liquids. Ph.D. dissertation, University of Oregon, 1976.
12. W.B. Bryan, L.W. Finger and F. Chayes, Science, 163, 926, 1969.
13. A. Ewart, W.B. Bryan and J.B. Gill, J. Petrol., 14, 429, 1973.
14. B.H. Baker, G.G. Goles, W.L. Leeman and M.M. Lindstrom, Geochemistry and petrogenesis of a basalt-benmoreite-trachyte suite from the southern part of the Gregory rift, Kenya, Contr. Min. Petrol., (in press).
15. M.M. Lindstrom, Geochemical studies of volcanic rocks from Pinzon and Santiago Islands, Galapagos Archipelago. Ph.D. dissertation, University of Oregon, 1976.
16. S.M. Endrodi, Petrology and geochemistry of Isla San Benediete, Mexico, M.S. thesis, University of Oregon, 1975.
17. I. Nekrasov, Geochem. Int., 7, 378, 1970.
18. E.M. Yeskova and A.F. Yefimov, Geochem. Int., 7, 121, 1970.
19. N.I. Tikhomirova, Geochem. Int., 8, 81, 1971.
20. M.J. LeBas, B.H. Baker and G.G. Goles, Geochemistry of fenitization around the Sakarume ijolite (in preparation).
21. V.I. Beloussov and others, East African Rift System, vol. III, Nauka, Moscow, 1974.

GEOLOGICAL AND GEOPHYSICAL PARAMETERS OF MID-PLATE VOLCANISM

I.G. Gass, D.S. Chapman*, H.N. Pollack** and R.S. Thorpe

Dept. of Earth Sciences, The Open University,
Milton Keynes MK7 6AA, UK.

ABSTRACT. Volcanism occurs mainly at oceanic ridges and subduction zones, but also takes place within continental plates. Mid-plate Cenozoic volcanism is widespread in Africa, where it is alkaline in composition and characteristically associated with uplift and faulting. There is a clear time correlation between the pause at ca. 45 Ma in the African apparent polar wander path and the outbreak of volcanism at ca. 35 Ma, suggesting that the slowing or coming to rest of the plate may be an important antecedent to the outbreak of volcanism. Moreover, the Cenozoic volcanism is almost completely restricted spatially to non-cratonic areas. Cratons in Africa are characterized geologically by tectonic stability for at least the past 1200 Ma, and geophysically by low elevations, heat flow less than 45 mW m^{-2}, lithospheric thicknesses greater than 200 km higher than average seismic velocities. In contrast, non-cratonic regions have undergone tectono-thermal events within the past 1200 Ma, stand higher, have greater heat flow and lower lithospheric thicknesses and seismic velocities.

The dynamic model which we envision is one of a variable thickness lithosphere passing over thermally anomalous regions of the underlying asthenosphere. The interplay in various proportions of plate thickness and velocity, and vigor of the asthenospheric thermal disturbance yield lithospheric thinning by thermal erosion,

---

* Dept of Geology & Geophysics, University of Utah, Salt Lake City, Utah 84112, U.S.A.

** Dept of Geology & Mineralogy, University of Michigan, Ann Arbor, Michigan 48105, U.S.A.

E.-R. Neumann and I.B. Ramberg (eds.), Petrology and Geochemistry of Continental Rifts, 27–28.

uplift by thermal expansion, and volcanism following penetrative magmatism. We examine quantitative conductive models that determine the thermal disturbances in the lithosphere explicitly in terms of perturbation strength, plate thickness and plate velocity. Thick and/or rapidly moving plates are relatively imperturbable and durable, while thin and/or slowly moving plates are more susceptible to thermal perturbations which further thin the lithosphere and lead ultimately to volcanism. For example, for a plate over 200 km thick, a relatively modest movement (>2 cm yr^{-1}) will suppress the upward propagation of most sub-lithospheric thermal anomalies thereby precluding mid-plate volcanism. Where such thermal anomalies are confined to the base of the lithosphere, uplift alone, without surface volcanism, would result. Substantial thinning of the lithosphere can be accomplished by deep, strong thermal perturbations or by upward migrating, penetrative perturbations of lesser strength. The latter is consistent with configurations for the East African-Ethiopian lithothermal systems derived from geophysical and geochemical data. Mesozoic volcanism of Gondwanaland was more widespread and voluminous, and erupted through both cratons and non-cratons, thereby suggesting that the causative thermal anomalies were much more vigorous than those in the Cenozoic.

The full text of this paper appears in Phil. Trans. Roy. Soc. (London), Series A, 1977.

TECTONIC AND MAGMATIC EVOLUTION OF THE SOUTHERN PART OF THE KENYA RIFT VALLEY*

B.H. Baker[1], R. Crossley[2] and G.G. Goles[1]

[1]Center for Volcanology, Department of Geology, University of Oregon, Eugene, Oregon, USA
[2]Chancellor College, P.O. Box 280, Zomba, Malawi

ABSTRACT. The southern Kenya rift sector is a 68 km wide complex graben flanked by uplifted plateau 1500-2000 m in elevation. It exhibited a progressive transition from strongly alkaline to less alkaline volcanism during the last 15 my. An asymmetrical graben was formed 10 my ago, and graben development has continued from 4 my to 0.6 my BP. Faulting and igneous activity have tended to migrate from the margin to the center of the rift. Geophysical data suggest that the rift is underlain by anomalous low density mantle and that the crust under the rift floor contains elongated basic rock bodies. A model of rift formation calls upon early formation of strongly alkaline magmas at depth, and later formation of crustal magma chambers from which voluminous salic differentiates were derived. Evidence is given in support of cycles of alternating flood volcanism and faulting. The rift is the result of mantle diapirism, crustal updoming and faulting, formation of crustal magma reservoirs, and graben formation, followed by repeated episodes of alternating volcanism and faulting. The salic peralkaline rocks represent preferential eruption of liquids formed in cupolas by crystal-liquid fractionation.

1. INTRODUCTION

The Kenya rift valley traverses a broad region of late Cenozoic upwarping [1,3], and is a graben which has evolved since early Pliocene times by periodic downflexing and gravity faulting [4-7]. Its volcanic rocks show a transition from strongly alkaline prerift

*This work has been supported in part by National Science Foundation Grants GA 32333 and EAR75-20664, and by NERC.

E.-R. Neumann and I.B. Ramberg (eds.), Petrology and Geochemistry of Continental Rifts, 29–50.

igneous activity to less alkaline volcanism that was largely confined to the developing rift [6,8]. Enough is known of the evolution of this rift to explore inter-relationships of tectonic and magmatic events, to outline rift development, and to relate this development through the interpretation of geophysical data [9] to lithothermal processes in the mantle.

2. STRUCTURAL OUTLINE

The southern Kenya rift sector extends 120 km from latitude 1°S to the Tanzania border (Fig. 1), to the south of which the rift changes its character [1,3]. The flanking plateaux are at elevations of 1700-2000 m in the west and 1500-1700 m in the east, and consist of Precambrian rocks overlain towards the north by Miocene

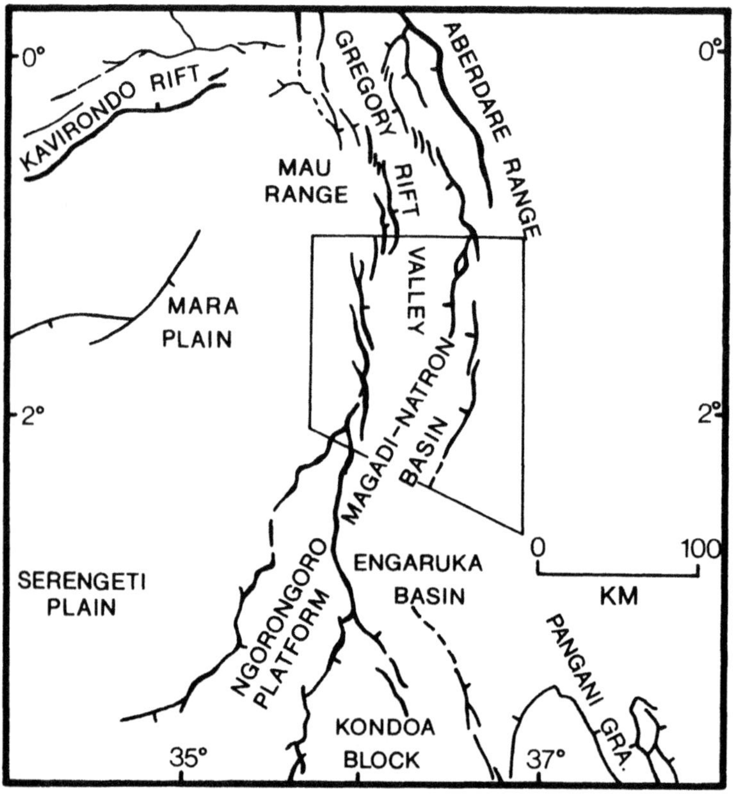

Fig. 1. Structural map of the southern half of the Kenya rift valley, showing the area described.

and Pliocene pre-rift lavas, most of which have flowed into the area from the north. The rift valley is a complex graben 72-73 km wide, with high (1500 m) fault escarpments on its western side, and less high (200-500 m) ones to the east. The rift floor slopes gently southward from elevations of 1500 m in the north to 600 m in the south, and has the character of a smooth plain of flood

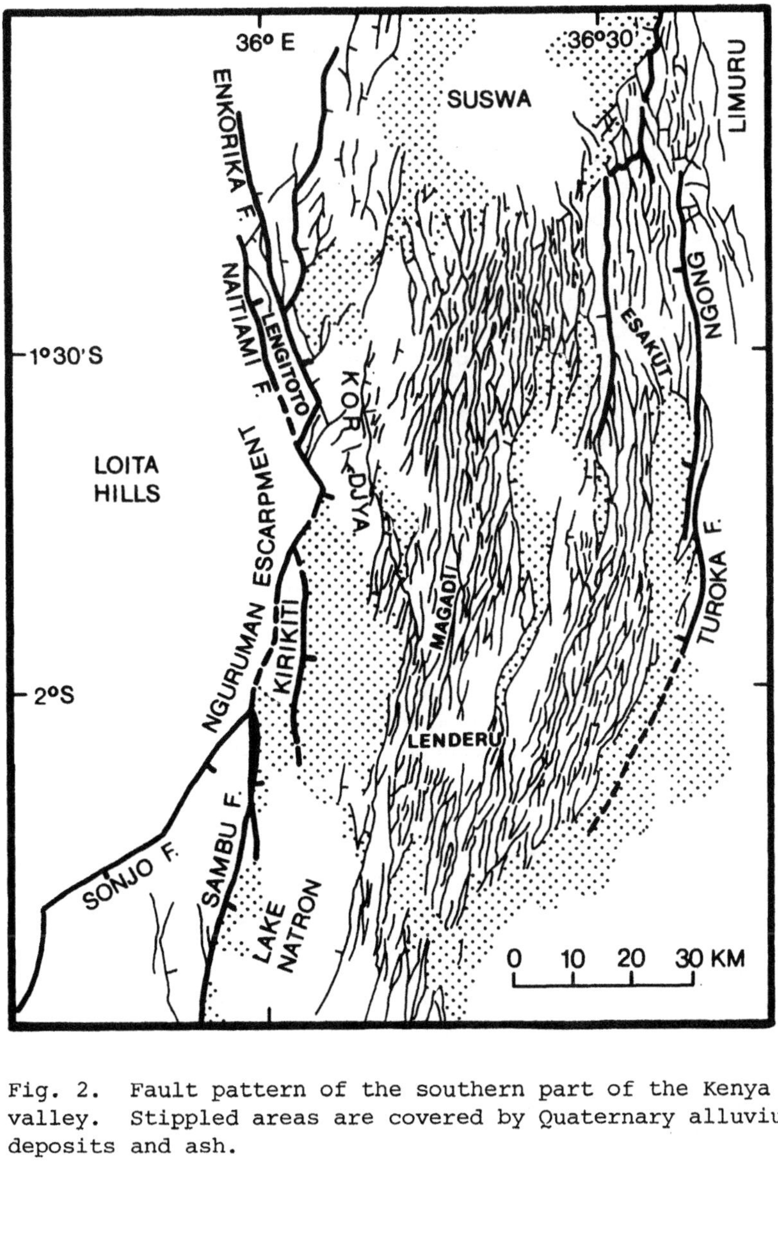

Fig. 2. Fault pattern of the southern part of the Kenya rift valley. Stippled areas are covered by Quaternary alluvium, lake deposits and ash.

lavas which has been broken by many subparallel minor faults that rarely exceed 200 m in displacement.

The major marginal faults form step-fault platforms (Lengitoto and Sambu platforms) and southward sloping 'ramps' 5-17 km wide (Kirikiti and Esakut platforms) that descend to the rift floor (Fig. 2). Some of the marginal fault escarpments are moderately eroded (Sonjo and Turoka faults) but those of the rift floor are perfectly preserved. Between the intermediate platforms and ramps the rift floor is 52-58 km wide, and is cut by many normal faults giving rise to small-scale step-fault, horst-graben, and tilted antithetic fault patterns. Faults of the axial zone tend to be more regular in orientation than are the main marginal faults, have an average spacing of 200 m, and tend to form a discontinuous axial depression along the line Lake Magadi-Suswa. The upper parts of several steep central volcanoes project above the flood-lava infill of the rift floor, and many of the closed depressions are partly filled by alluvium and lake deposits of various ages.

Few faults in the area are actually exposed; two of these are normal faults with dips of about 60° [10,11]. The common occurrence of tilted fault blocks suggests curved fault surfaces with dips of 60° or less, and the total observed (hence minimum) vertical displacement of the faults suggests that minimum horizontal extension of 15 km occurred during 5 my, yielding an average extensional strain rate of 3 mm per year.

The major marginal faults formed early; subsequently the zone of fracturing migrated inward, although rejuvenations of marginal faults occurred. The inward decrease in ages and spacings of faults is best seen at latitude $1^{\circ}30'$S (Fig. 2). In a region undergoing extensional strain the spacing of normal faults is likely to be related to the thickness of the elastic layer that can yield by shear fracture, hence we infer that the depth of the elastic crustal layer under the rift has decreased considerably with time (see section 6, below). This concept is supported by geophysical data [4,12,13].

3. VOLCANIC SEQUENCE AND EVOLUTION

Recent work has established the ages and distributions of volcanic formations (Fig. 3, table 1). Pre-rift volcanic rocks comprise late Miocene melanephelinite lavas and dikes, and clastic rocks erupted from localized central vents in the western part of the area (12-15 my). These were followed by inflow of extensive flood phonolite lavas from the north, the Narok and Kapiti phonolites (12-13 my), and by the lower and middle trachyte (and phonolite) 'divisions' which range down to about 3 my in age on the eastern rift flank [1,14,17].

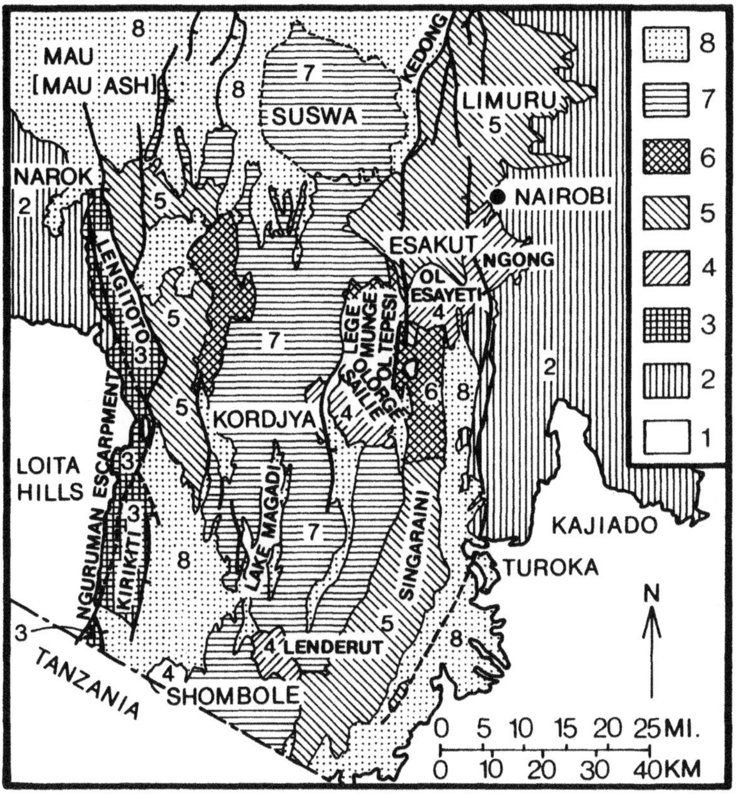

Fig. 3. Simplified geological map of the southern part of the Kenya rift valley. 1-Precambrian gneisses; 2-pre-rift melanephelinites, phonolites and trachytes; 3-basalts, melanephelinites and trachytes of the Lengitoto and Kirikiti platforms; 4-strongly and mildly alkaline rocks of Plio-Pleistocene central volcanoes; 5-Singaraini basalts, Limuru and Mosiro trachytes; 6-Ol Tepesi, Ol Keju Nero and Kordjya basalts; 7-Plateau and Magadi trachytes, Suswa central volcanics; 8-Alluvium, ash and lake deposits.

The Lengitoto and Endosopia trachytes (5-7 my) of the Lengitoto platform in the north-west, and the dominantly nephelinitic lavas of the Sambu platform south-west of the area (5-8 my), both clearly onlap onto the feet of pre-existing escarpments created by movement of the Sonjo and Naitiami faults [3,7]. Probably the whole of this westernmost marginal fault zone was therefore created during the interval 12-8 my, with a minimum displacement of 550 m in the south, decreasing northward, resulting in the formation of an asymmetrical "half" graben. Localized eruptions of central vent olivine melanephelinite lavas (with some phonolites) occurred

TABLE 1 - STRATIGRAPHIC SUCCESSION OF THE SOUTH KENYA RIFT VALLEY

West flank	Kirikiti-Lengitoto platforms	Rift floor west	Rift floor east	Esakut platform	East flank
			Suswa volcanics (0.4-0)		
	Lengorale trachyte (0.6)	*O.D. Nyokie ignimbrites (.65)*	Barajai trachytes (0.4)		
		Magadi trachytes (1.4-0.7)	Plateau trachytes (1.3-0.9)		
		N. Kordjya trachyte	Ol Tepesi basalts trachytes (1.65-1.4)		
			Ol Keju Nero basalts		
		Mosiro trachytes (2.1)	Limuru trachytes (2.0-1.9)	. . . .	Limuru trachytes
Sambu volcanics (2.0)		*Shombole volcanics (2-1.9)*			
		Kordjya basalts	Singaraini basalts (2.3)		
Narosura pyroclastics (2,2.7)			*Olorgesialie volcanics (2.2-2.7)* *Lenderut volcanics (2.6-2.5)*		Middle trachyte Div.
	Kirikiti basalts (3.1-2.5)			*Ol Esayeti volcanics (6.7-3.6)*	
Ol. melanephelinites (5.8-5.0)	Endosapia and Lengitoto trachytes (6.9-5.1)				Lower trachyte Div. *Ngong volcanics (6)*
Narok phonolites					Kapiti phonolite (13.4)
Kishalduga melanephelinites (15.2-12)					
Precambrian gneisses					Precambrian gneisses

on both sides of the Sonjo-Naitimi fault zone (5-6 my), and the strongly alkaline Ol Esayeti and Ngong volcanoes were also formed at this time (6.7-3.6 my).

The Kirikiti basalts (2.5-3.1 my) and the Singaraini basalts are flood transitional basaltic lavas as much as 550 m thick that covered the greater part of the rift floor and are confined by fault escarpments on both sides. Hence the Kirikiti-Enkorika fault was formed during the interval 5-3 my BP, isolating the Kirikiti and Lengitoto platforms, and the Turoka-Ngong fault also developed at about 3 my BP, forming a graben. Displacements were at least 820 m in the west and 500 m in the east. Shortly after formation of this graben the Olorgesailie (2.7-2.2 my), Lenderut (2.5-2.6 my), and Shombole (∿2 my) central volcanoes were built. They consist of nephelinites, alkali basalts, trachytes, and phonolithes, and represent strongly alkaline or mixed strongly and mildly alkaline suites [18,19]. By 2 my BP strongly alkaline volcanism had ceased in this area.

Very extensive thick porphyritic trachyte and pantelleritic flood lavas, the Limuru and Mosiro trachytes (1.9-2.1 my), were erupted on the rift floor to a depth of at least 300 m. These lavas overflowed the eastern marginal escarpment north of Ngong volcano, probably covered the whole rift floor in the Suswa region, and about onto the Enkorika and Ngong escarpments. Subsequent olivine basalt lavas- the Kordjya and Ol Keju Nero basalts (2.2-1.7 my)- locally cover the flood trachytes, and on the eastern side of the rift the basalts rest unconformably on faulted and tilted Limuru trachytes, showing that the Esakut platform was created by movement on the Esayeti and Kedong fault zones during the interval 2-1.7 my BP.

The Ol Tepesi basalts (1.65-1.4 my) contain intercalated very thick porphyritic benmoreite flows that can be correlated with the North Kordjya trachyte [16,17]. These were succeeded by very extensive flood trachyte and pantellerite lavas (Plateau and Magadi trachytes, 1.3-0.9 my) that covered the median zone of the rift floor to depths of at least 190 m, and extend more than 150 km from Suswa into northern Tanzania [10,11,16]. These flood salic lavas cover a faulted surface of earlier volcanic formations, thicken toward the axis of the rift, and are cut by the densest part of the zone of faulting of the rift floor.

The greater part of the 'grid-faulting' of the rift floor was accomplished during the interval 0.9 to 0.4 my BP, because lake deposits (Legemunge beds [20], 0.42 my) were formed in grabens, and the Baragai trachytes (0.4 my) and the Suswa volcanic rocks (0.2 my to recent) are much less displaced by faults. Nevertheless fault displacements as much as 45 m affected rocks 0.4 my old. No extensive volcanism has occurred in the last 0.9 my in this region, the most recent activity being the eruption of localized trachytic agglutinate and ignimbrite cones [2] and the construction of the Suswa trachytic and trachyphonolitic caldera volcano [22,23].

4. ASPECTS OF THE PETROLOGY AND GENESIS OF THE VOLCANIC ROCKS

In the Kenya igneous province two distinct petrological associations exist among the volcanic rocks (Fig. 4), a strongly alkaline nephelinite-Ca-rich phonolite (-carbonatite) suite which forms central volcanoes containing a high proportion of clastic rocks, and a mildly alkaline olivine basalt-trachyte-alkali rhyolite-Ca-poor phonolite suite which occurs both on central volcanoes and as flood lavas [8,14,24-26]. The very voluminous Plateau phonolites of central Kenya appear to belong to the mildly alkaline association [3]. The petrogenesis of metanephelinitic parental melts is still being debated [27], but all the mechanisms

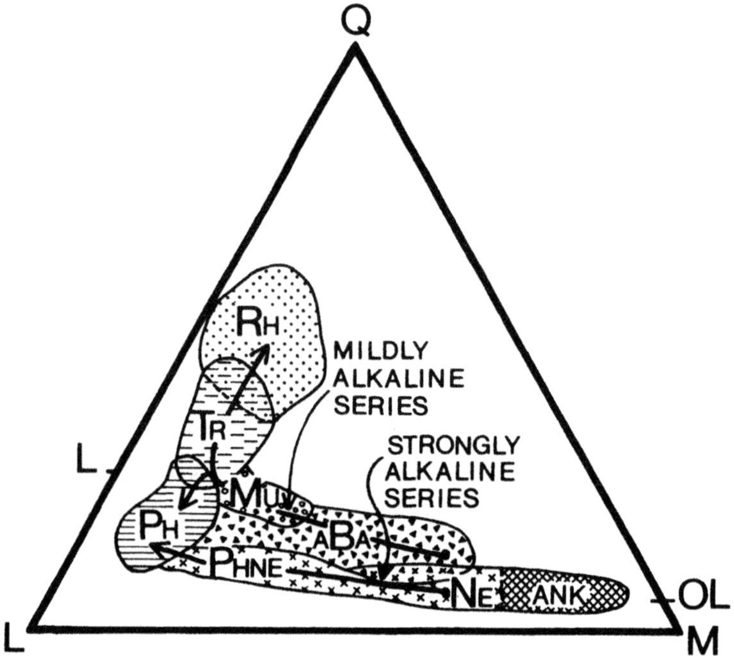

Fig. 4. QLM diagram of volcanic rocks of the southern part of the Kenya rift valley. ANK=ankaratrite; NE=nephelinite; PHNE=phonolitic nephelinite; PH= phonolite; ABA=olivine basalt; MU=mugearite; TR=trachyte; RH=rhyolite.

proposed require formation at pressures of 30 kb or more with little subsequent equilibration at lower pressures [28]. No practical means of deriving alkali basalts from nephelinites has been proposed; hence, the two suites are almost certainly derived independently from the mantle. The great majority of Kenya volcanoes consist entirely of rocks of one or other of these two suites and only six contain representatives of both suites [18]. The transitionally alkaline basalts parental to the mildly alkaline suite are considered to have formed at shallower depths [28,29], and the occurrence of both suites in a few central volcanoes suggests that they were capable of tapping magma from different depths in the mantle. A study of one such volcano, Olorgesailie, suggests that the inferred range of parental mafic magmac was formed by melting of garnet peridotite at pressures ranging from 10-25 kb [19].

These inferences are supported by rare earth element (REE) contents of some Kenyan basaltic rocks. Fig. 5 gives chondrite-normalized REE contents of more-magnesian members of four Quaternary basaltic series of the rift floor, and of Quaternary Chyulu basalts erupted about 150 km east of the rift valley [30,31].

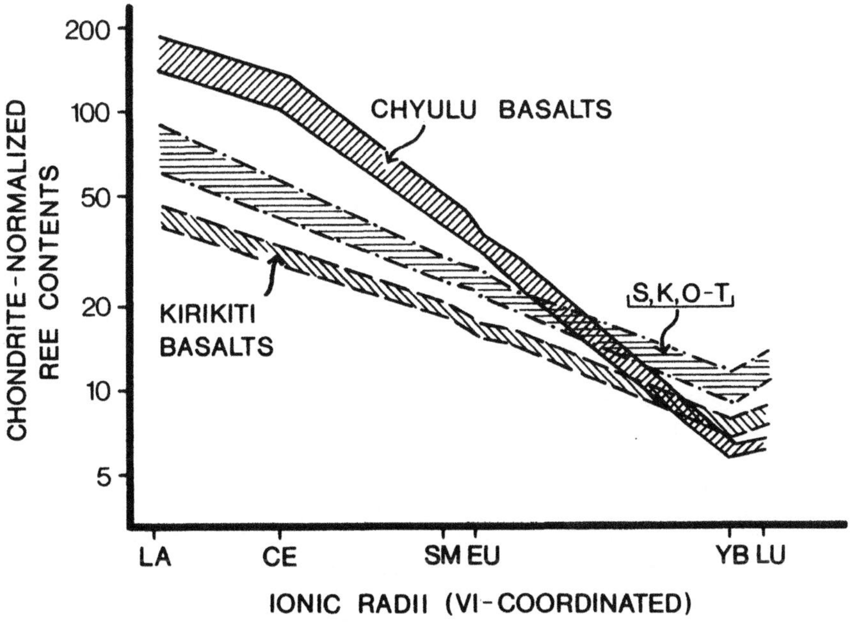

Fig. 5. Chondrite normalized rare earth contents of some basaltic lavas of the southern Kenya rift zone. The Chyulu basalts lie 150 km east of the rift valley. S,K,O-T=Singaraini, Kordjya and Ol Tepesi basalts.

Minor differences in absolute REE contents between the rift-floor series reflect contrasts in loss of liquidus phases from (unobserved) parental magmas. The range of REE contents and REE proportions imply formation of parental magmas by relatively large degrees of partial melting at low pressures. In contrast, Chyulu basalts contain mantle-derived garnet-bearing xenoliths [30; Goles and Bow, in preparation] and evidently erupted from mantle reservoirs. They display REE patterns that are very different from those of the rift floor basalts, with notable light REE enrichments, and probably formed initially by small degrees of partial melting of garnet-bearing upper mantle rocks. Goles [31] has shown that these magmas cannot have undergone much fractionation en route to the surface. Beloussov and others [3] have shown that ratios of light to heavy REE subgroups are significantly greater in the strongly alkaline than in the mildly alkaline series of rocks. These observations suggest that depth of origin is the prime factor determining alkalinity of primary magmas in this region.

In the southern Kenya rift (and throughout the Kenyan alkaline igneous province) volcanism of strongly alkaline character took place mostly before rift formation, or shortly after an initial half graben had been formed. It continued through Quaternary times only in northern Tanzania, south of the rift sector where a true graben was developed (Fig. 1). After graben formation volcanism was dominantly mildly alkaline in character, forming expanses of flood lavas and low shield volcanoes in the rift floor [6,7]. These notable changes in the distribution and kind of igneous activity suggest progressive rise of mantle isotherms under the rift zone, or upward movement of an elongated body of hot mantle in the manner suggested by Gass [32] for the Ethiopian rift.

Considerable debate has centered on the significance of the high proportion of silicic volcanic rocks of the mildly alkaline suite in the Kenya rift [33]. The regional proportion of basaltic to salic rocks is about equal [8], whereas in the southern rift salic rocks are estimated to be twice the volume of basic rocks [16]. The scarcity of intermediate rocks (Daly gaps), and the occurrence of thick, extensive accumulations of trachytic, phonolitic, and alkali rhyolite lavas and ignimbrites have led to proposals that the salic rocks originated by crustal anatexis, and that their peralkalinity is the result of alkali transfer by volatiles [34,35]. This view is based on the assumption that the erupted rocks reflect the proportions of magmas generated, but there are reasons to believe that this is not so [36]. An alternative view, that the large volumes of erupted salic magmas are the consequence of their formation in a rift tectonic environment, is explored further below.

The petrogenetic significance of Daly gaps was debated some years ago [37], but one aspect has been insufficiently discussed.

Daly gaps depend upon silica contents of rocks to the exclusion of other characteristics, and on the assumption that silica content is proportional to differentiation. In the mildly alkaline rock series extensive fractionation of basalt along cotectic lines can occur without increase of silica, but when titanomagnetite fractionation reaches its maximum in intermediate compositions a sudden rise in SiO_2 content occurs. Subsequent extensive fractionation of feldspar can then take place with either a gradual decline in silica content (phonolite trend), or an increase in silica content (rhyolitic trend), depending upon the SiO_2 content of the intermediate magma at the stage at which feldspar fractionation becomes dominant. Use of more meaningful differentiation parameters such as the concentration factors of incompatible elements frequently drastically reduces the width of Daly gaps in basalt-trachyte-alkali rhyolite suites [38].

The peralkalinity of the salic rocks does not require input of alkalies, for once the peralkaline condition is reached as a result of the plagioclase effect [39] the fractionation of feldspar will accentuate it. The fact that some peralkaline salic rocks do not plot exactly along a feldspar control line in the SiO_2-Al_2O_3-Na_2O + K_2O system [40], may well be because the rocks are not truly cogenetic, or some ferromagnesian phases are fractionating with the feldspar, or post-magmatic alteration has caused variable compositional alteration (mainly sodium loss) in the rocks [41,42].

With very few exceptions the entire range of Kenyan alkaline effusive rocks are found on oceanic volcanoes [43]. Detailed studies of phonolites, trachytes, and alkali rhyolites strongly suggest that these rocks were formed by low pressure closed-system fractionation of minerals similar to the phenocrysts that they contain [3,19,26,38,42]. This view is supported by isotopic studies [45,46]. The petrogenetic suites suggested by the volcanic associations and the geochemistry of the rocks are as follows:

1. olivine melanephelinite-nephelinite-phonolite (+carbonatite)
2. basanite-melaphonolite (rare)
3. basalt-mugearite-trachyte ↗ comendite or pantellerite; ↘ trachyphonolite-phonolite (Kenya type)

The degree of enrichment of incompatible trace elements and mass balance calculations of crystal-liquid fractionation show that the percentage of the 'parental' magmas that need to crystallize to produce residual (eutectic) differentiates varies greatly in these suites. Strongly alkaline phonolites (>25% Ne) can form by 60% crystallization of ferro-magnesian phases from nephelinitic parental magma [19], whereas the mildly alkaline trachytes and alkali rhyolites require 97% and 99.3% crystallization respectively

[38]. These contrasts result from the different proportions of feldspar fractionated, and from differences in the compositions and proportions of ferromagnesian phases crystallized along the strongly and mildly alkaline lines of liquid descent, the latter being controlled to a large degree by the oxidation state of iron

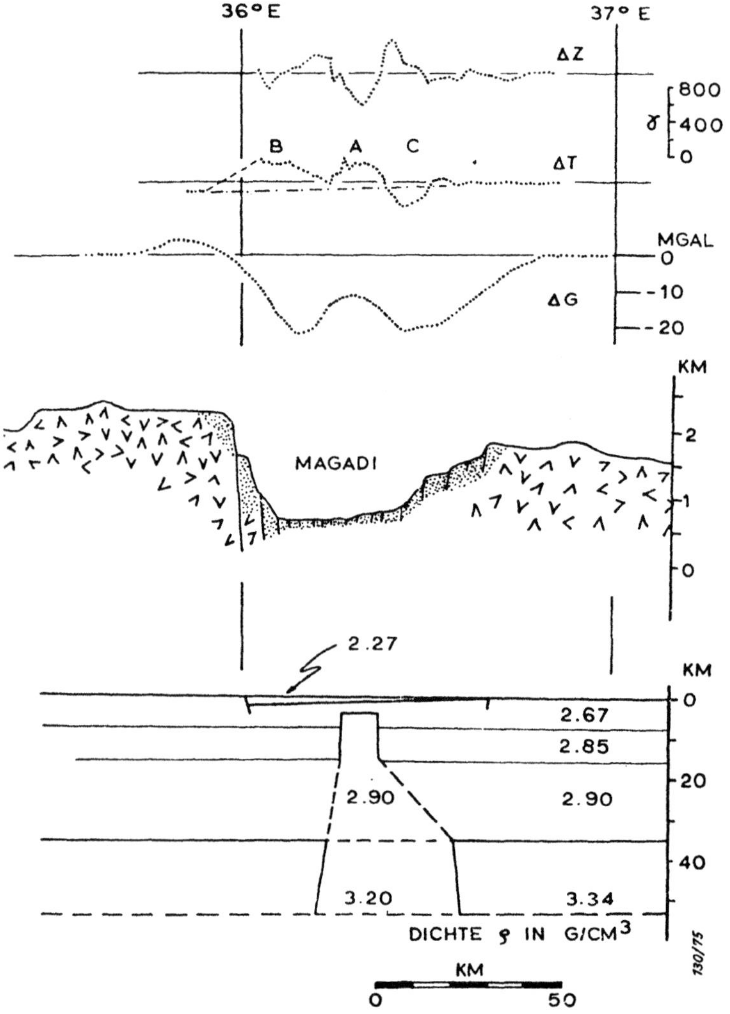

Fig. 6. Magnetic and gravity anomaly profiles across the Magadi region of the south Kenya rift, showing computed lithosphere model (after Wohlenberg [9]).

in the liquid. Probably rates of magma ascent, and the different levels, sizes, and cooling rates of magma bodies, also influence differentiation paths and determine proportions of mafic and salic magmas that are erupted.

The similarity of alkaline volcanic suites and of their salic differentiates in continental and oceanic environments suggests that both are results of similar petrogenetic processes, and that differences in proportions of salic magma erupted result from effects of crustal composition and structure, and from tectonic factors. In our view there is no compelling evidence for open system crustal anatexis as a process for the formation of significant quantities of alkaline salic rocks.

5. LITHOSPHERE STRUCTURE FROM GEOPHYSICAL DATA

Gravity and seismic data for the central and southern part of the Kenya rift show that the crust adjacent to the rift valley is of normal seismic velocity and thickness (36 km) [45], but that the rift zone itself has anomalous lithosphere structure. Gravity profiles across the Magadi and Suswa regions show broad residual lows coinciding in width with the volcanic field, with steep highs of 30-50 GU in the axial zone of the rift [4,9,12,13]. Because of low P_n velocities, S_n attenuation, and station arrival residuals, there is ample evidence that the mantle under the rift floor is of subnormal velocity and is probably hot and partially melted [12,13,47,48]. The broad negative gravity residual has to be ascribed to low density mantle and low density volcanic rocks on the surface, and the axial positive gravity residuals represent a complex of anomalously dense intracrustal intrusive rocks 10-20 km wide and reaching nearly to the surface [12,13], which also apparently gives rise to a magnetic anomaly (see Fig. 6, and [9]). The geophysical data strongly suggest a zone of hot, partially molten mantle under the rift, with emplacement of basic intrusive rocks in the crust, in a manner similar to that postulated for the Oslo rift [49]. The intrusive complex could be due to magmatic stoping [3] but only for a limited range of magma densities lower than those of crustal rocks, yet such that the intrusive rocks that developed from those magmas are now more dense than their wallrocks. An alternative explanation relies on crustal dilation and emplacement of a large volume of basic igneous rocks in the crust under the rift floor, and conversion of the upper mantle to material with properties resembling those of the low velocity zone. It appears that the "missing" basic igneous rocks of the rift are in intracrustal magma chambers and that the greater part of the basic magmas were never erupted but fractionated in situ. It remains to try to explain why such a situation occurs in continental rifts.

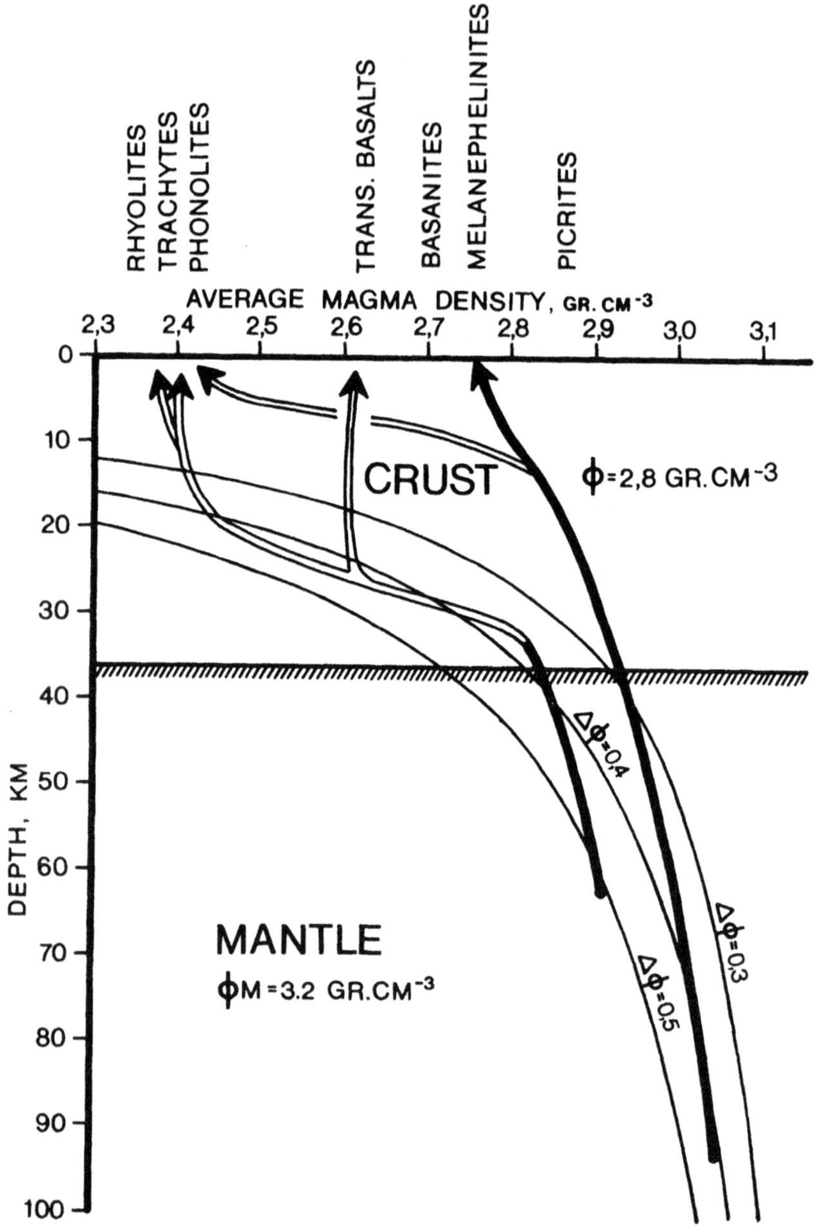

Fig. 7. Magma density-depth of origin curves showing the maximum depth from which magmas of various densities may be driven to the surface by lithostatic load, for three values of the difference between mean crust and mantle densities [Δϕ]. The arrowed lines show inferred paths for strongly and mildly alkaline primary magmas and their derivative magmas.

6. ERUPTIVE MECHANISMS IN A RIFT ENVIRONMENT

The general linear character, abundance of normal faults, and extensional strain characteristic of the Kenya rift zone suggest operation of an anisotropic stress-field with the greatest and least principal stresses oriented vertically and horizontally

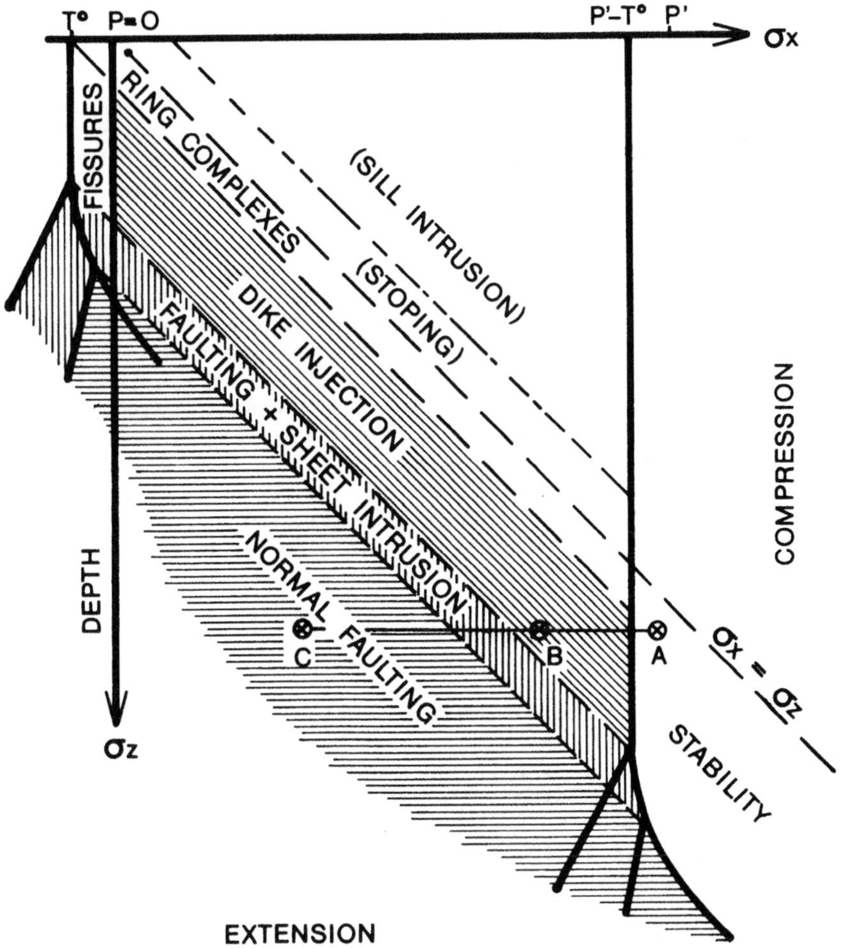

Fig. 8. Plot of horizontal (σ_x) and vertical stresses (σ_z) showing the fields of magma injection and normal faulting produced as a result of a range of magma pressures (P). T^O is the tensile strength of the rocks. High values of P extend the field of dike injection to greater depths. High values of the extensional stress difference ($\sigma_z - \sigma_x$) favors normal faulting rather than magma injection. Modified from Roberts [54].

respectively. Although such a stress field may be formed by up-arching of the lithosphere as a result of the ascent of a large asthenolith in the mantle [50], the amplitude of the uplift is insufficient to account for more than a small fraction of the observed extensional strain [4]; hence, it is necessary to invoke limited lateral spreading of the lithosphere. We infer that the whole period of rift development was characterized by an extensional stress field, and that the difference between vertical and horizontal stresses varied as a result of faulting and magma intrusion or eruption.

From the increasing slopes of the successive erosion surfaces that were cut on both flanks of the Kenya rift [1,2] it appears that the later and major phases of uplift of the Kenya "dome" had a shorter wavelength than the earlier phase. We account for this by a greater depth of the asthenolith in Miocene and early Pliocene times, which correlates with strongly alkaline volcanism and very limited crustal extension. The inferred much shallower depth of the asthenolith in Plio-Pleistocene times correlates with mildly alkaline volcanism, more localized and greater uplift, greater fault displacement, and much greater crustal extension. We envisage melting as a result of adiabatic uprise of mantle material [51], with segregation of primary magma at progressively shallower depths.

The strongly alkaline melanephelinite parental magmas may well be close to the composition of primary partial melts of undepleted garnet peridotite, and require a depth of last equilibration of the order of 80-90 km [52]. Such magmas are rich in volatiles, underwent little fractionation, and probably rose rapidly to the surface creating central volcanoes consisting largely of clastic rocks and lavas of rather uniform composition. The absence of feldspar as a liquidus phase over much of the composition range of this series tends to confirm its deep origin and rapid ascent. By contrast the typical transitional basalts of the rift floor have four phases on the liquidus and must represent a derivative liquid resulting from fractionation at low pressures (depths of 20-40 km). Such basaltic magmas must have had at least a two-stage history- initial segregation at depths of about 60 km, followed by considerable fractionation at shallower depths.

It is generally accepted that ascent of magma in the mantle is governed by buoyancy according to Archimedes' Principle and Stokes' Law [53], or, in the presence of anisotropic stresses and with the possibility of brittle fracture, ascent may be by injection fracturing of the host rocks [54]. In the first case the rate of ascent is governed by the density and viscosity of the magma relative to its immediate host rocks, and in the second it is governed by the density, vertical extent, and pressure of the magma in relation to characteristics of the overlying rocks. The first machanism evidently cannot act to deliver primary basic magmas to the surface through continental crust, so we must examine the implications of the second machanism in order to illuminate further the contrast between distributions of melanephelinite and mildly alkaline magmas. Fig. 7 shows the maximum depth from which magmas of various densities may be driven by 'overburden squeeze' and excess magma pressure. For any reasonable density contrast between the mantle and crust very dense primary magma can reach the surface from the mantle only in favorable structural environments, and all of the mildly alkaline basaltic magmas and their differentiates must have been tapped from intracrustal reservoirs.

The eruptive and structural effects of intracrustal magma reservoirs in different kinds of stress environment have been studied by Roberts [54]. Dike injection is favored by magma pressure greater than the least horizontal stress plus the tensile strength (T^o) of the host rocks, which is taken to be about 200 bars. In a rift environment the principal stresses (σ_z vertical, σ_x horizontal) are unequal, and the stress field is characterized by $\sigma_z > \sigma_x$. This condition favors dike injection, the dikes being formed normal to σ_z (vertical). If the stress difference $\sigma_z - \sigma_x$ becomes sufficiently large, shear failure (normal faulting) may occur. Figure 8 is a modification of a figure by Roberts [54] showing the fields of stability, dike injection, and normal faulting for a range of extensional stress fields and magma pressures, using the Griffith-Murell criterion of failure for fractured rocks in the presence of a fluid. Figure 8 shows that the fields of intrusion (dike injection), stoping, and normal faulting are separate except for minor overlap. In rocks containing magma bodies under high pressure dike injection is favored, whereas high stress differences and low magma pressure result in normal faulting. In a rift environment the vertical stress (σ_z) at the upper part of a putative magma reservoir is essentially invariant, whereas magma overpressure ($P-\sigma_z$) may vary because P changes as a result of the magma fractionation, or of upward or downward extension of the reservoir for example by temporary connection of the reservoir to deeper levels of the mantle. The horizontal stress (σ_x) will fluctuate in response to faulting and intrusion, and is likely to vary over a much greater range. Hence for small stress differences (point A, Figure 8), increasing magma pressure will induce dike injection and eruption if magma density is low enough. With larger stress differences (points B and C in Figure 8), increasing magma pressure will produce a change from stability to dike injection or to normal faulting, depending on the magnitude of $\sigma_z - \sigma_x$. For fixed intermediate magma pressures decrease of horizontal stress results in transition from stability, to dike injection, to normal faulting. There is no condition under which magma can ascend and normal faulting can occur simultaneously. This conclusion provides a theoretical basis for the observations of alternating episodes of eruption and normal faulting such as characterize the southern Kenya rift [16]. It also explains the extreme rarity of volcanic vents situated upon faults, for simultaneous formation of shear fractures and ascent of magma up those fractures is mechanically improbable. Dikes and aligned volcanic vents are commonly parallel to the trend of rifts, for normal faults and injected planar fractures both have their strike normal to the least principal stress (σ_x). The formation of elongated intracrustal magma reservoirs under rift valleys must influence their structural development and eruptive behavior.

7. TENTATIVE MODEL FOR THE SOUTH KENYA RIFT

We postulate that diapiric uprise of mantle under the rift started in the Miocene, and that by late Miocene times a broad shallow uplift with a gentle axial depression was established. Separation of melanephelinite magma from undepleted mantle at depths of 80-90 km resulted in scattered small nephelinite (phonolite) volcanoes.

At 10 my ($\pm$ 2 my) a substantial western marginal escarpment was formed in part by further uplift of the western rift flank, and was succeeded by eruption of localized trachyte lavas and by build-up of more widespread strongly alkaline central volcanoes (5-7 my; Fig. 9B). During this period asthenoliths of picritic magma began to separate at depths of about 60 km, rose into the basal crust where they stabilized and spread laterally owing to their high density (ρ = 2.9-3.0 g/cm^3). By about 4 my BP further rise of the diapiric mantle body caused regional doming and extension of the upper lithosphere, and graben faulting occurred as a result of reduction of horizontal pressure ($\sigma_z > \sigma_x$, Fig. 9C). Horizontal pressure increased after graben formation and fractionated transitional basaltic magma was able to erupt from dikes in the rift floor (2.3-3.0 my). By 2.0 my rapid eruption of trachytic flood lavas from the upper part of the reservoir lowered magma pressure and σ_z-σ_x, and caused further graben faulting, this time creating an inner graben and isolating the step-fault platforms (Fig. 9D). A subsequent tectonomagmatic cycle (1.65-0.9 my) followed essentially this same pattern, probably initiated by accession of fresh primary magma from the deeper regions of partial melting.

With progressive accumulation of dense basic rocks in the crust under the rift zone the transitional basalt reservoirs and the associated cupolas containing differentiated salic magmas rose higher and higher because of the increased lithostatic load and lithosphere extension, and a broad tabular initial magma reservoir was converted to a narrower zone of vertically elongated bodies. This type of development could explain the progressive narrowing of the zone of faulting with time, and the increasing proportion of salic rocks in the three successive mildly alkaline eruptive cycles.

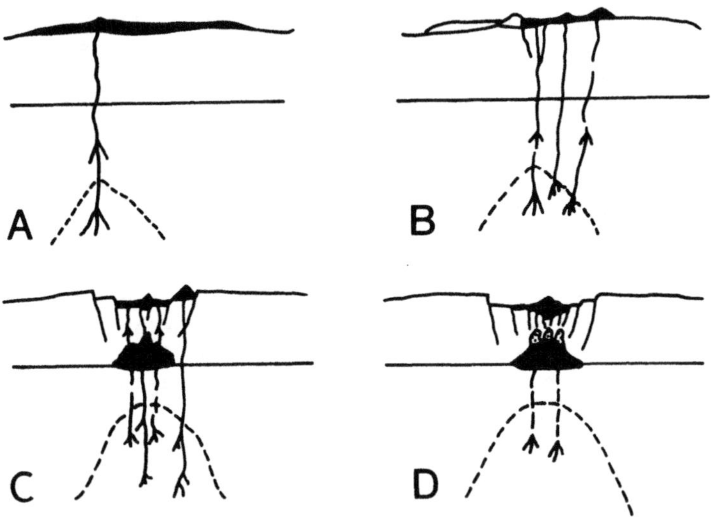

Fig. 9. Diagrams showing suggested stages of ascent of a mantle 'diapir' and of development of the south Kenya rift valley.

The ubiquitous alkaline character of igneous activity follows from the low degrees of parital melting of undepleted mantle subjected to only modest thermal gradients. The alkaline character of the primary melts was not altered by re-equilibration at lower pressures because of efficient segregation and rapid rise of low viscosity alkaline and halogen-rich melts. The strongly alkaline nephelinite magmas fractionated ferromagnesian phases at pressures below the field of feldspar stability to form highly alkaline phonolitic residual liquids, but mostly they rose too fast for this process to occur. Mildly alkaline basaltic magmas of large intracrustal reservoirs crystallized abundant plagioclase, leading by prolonged and extensive fractionation to peralkaline trachytic and alkali rhyolite residual liquids that erupted as lavas with high halogen contents, low viscosities, and unusual extents. After the stabilization of large intracrustal magma reservoirs the crust acted as a density filter [36] and drove to the surface only differentiated magmas with densities less than the average density of the superjacent crust. These features explain the existence of Daly gaps in magma series, and the volumetric predominance of salic eruptives in rift tectonic environments.

8. CONCLUSIONS

The alkaline nature of igneous activity in the Kenya rift (and others) is ascribed to low thermal gradients and limited melting of undepleted mantle. Early nephelinitic igneous activity was located peripherally to the main rift zone. It probably represents rapid rise of highly-carbonated nephelinite magmas which in some cases unmix at shallow levels to yield a small carbonatite fraction. The lavas are deficient in feldspar and underwent little fractionation during ascent.

Late in Pliocene times further slow uprise of a mantle body resulted in formation of mildly alkaline basaltic magma, and the lesser density and depth of formation of these more-voluminous magmas resulted in their accumulation in reservoirs formed by dilation and injection of the crust. This stage coincides with graben formation, and subsequent developments consisted of successive cycles of magma accumulation and graben faulting, low pressure fractionation and eruption of differentiated magmas to the surface, followed by further collapses of the rift floor.

The evidence suggests that salic magma formed by crystal-liquid fractionation, and that their considerable volume is the natural consequence of the continental rift setting in which they occur.

REFERENCES

1. B.H. Baker, P.S. Mohr and L.A.J. Williams, Geol. Soc. Amer. Spec. Paper, 136, 1971.
2. E.P. Saggerson and B.H. Baker, Quart. J. Geol. Soc. London, 121, 51, 1965.
3. V.V. Beloussov and others, The East African Rift System, Nauka, Moscow, 1974.
4. B.H. Baker and J. Wohlenberg, Nature, 229, 538, 1971.
5. L.A.J. Williams, Bull. Volc., 34, 439, 1970.
6. B.C. King and G.R. Chapman, Phil. Trans. Roy. Soc. London, A271, 185, 1972.
7. N.A. Logatchev. V.V. Beloussov and E.E. Milanovsky, Tectonophysics, 15, 71, 1972.
8. L.A.J. Williams, Tectonophysics, 15, 83, 1972.
9. J. Wohlenberg, Geol. J., E-4, 1, 1975.
10. B.H. Baker, Geol. Surv. Kenya Report no. 42, 1958.
11. B.H. Baker, Geol. Surv. Kenya Report no. 61, 1963.
12. R.C. Searle, Geophys. J. Roy. Astron. Soc., 21, 13, 1970.
13. B.W. Darracott, J.D. Fairhead and R.W. Girdler, Tectonophysics, 15, 131, 1972.
14. E.P. Saggerson, Bull. Volc., 34, 38, 1970.
15. J.D. Fairhead, Tectonophysics, 30, 269, 1976.
16. B.H. Baker and J.G. Mitchell, J. Geol. Soc. London, 132, 467, 1976.
17. R. Crossley, The Cainozoic stratigraphy and structure of the western part of the rift valley in southern Kenya, in preparation.

18. L.A.J. Williams, Nature, 224, 61, 1969.
19. L.F. Henage, The geology and compositional evolution of Mt. Ololkisalie, Kenya. Ph.D. dissertation, University of Oregon, 1977.
20. B.H. Baker, L.A.J. Williams, J.A. Miller and F.H. Fitch, Tectonophysics, 11, 191, 1971.
21. B.H. Baker, Bull. Volc., 39, 1, 1975.
22. R.W. Johnson, Phil. Trans. Roy. Soc. London, 265A, 383, 1969.
23. W.P. Nash, I.S.E. Carmichael and R.W. Johnson, J. Petrol., 10, 409, 1969.
24. E.P. Saggerson and L.A.J. Williams, J. Petrol. 5, 40, 1964.
25. L.A.J. Williams, Bull. Volc., 33, 862, 1969.
26. V.I. Gerasimovsky and A.I. Polyakov, Geokhimiya, 4, 457, 1970.
27. D.K. Bailey, Nephelinites and ljolites, in: The Alkaline Rocks, ed. by H. Sørensen, Wiley and Sons, New York, 1974.
28. H.S. Yoder, The Generation of Basaltic Magma, National Academy of Sciences, Washington, 1976.
29. D.H. Green, Phil. Trans. Roy. Soc. London, A268, 707, 1971.
30. E.P. Saggerson, Geol. Rundsch., 57, 890, 1968.
31. G.G. Goles, Lithos, 8, 47, 1975.
32. I.G. Gass, Tectonic and magmatic evolution of the Afro-Arabian dome, in: African Magmatism and Tectonics, ed. by T.N. Clifford and I.G. Gass, Oliver and Boyd, Edinburgh, 1970.
33. R. Macdonald, D.K. Bailey and D.S. Sutherland, J. Petrol., 11, 507, 1970.
34. D.K. Bailey and R. Macdonald, Amer. J. Sci., 267, 242, 1969.
35. D.K. Bailey, Continental rifting and alkaline magmatism, in: The Alkaline Rocks, ed. by H. Sørensen, Wiley and Sons, New York, 1974.
36. R.C.O. Gill, Nature, 247, 25, 1974.
37. F. Chayes, J. Geophys. Res., 68, 1519, 1963.
38. B.H. Baker, G.G. Goles, W. Leeman and M.M. Lindstrom, Contr. Min. Petrol., in press.
39. N.L. Bowen, Amer. J. Sci., 243A, 75, 1945.
40. R. Macdonald, D.K. Bailey and D.S. Sutherland, J. Petrol. 11, 507, 1970.
41. D.C. Noble, Geol. Soc. Am. Bull., 31, 2677, 1970.
42. B.H. Baker and L.F. Henage, J. Volc. and Geothermal Res., 2, 17, 1977.
43. G.D. Borley, Oceanic Islands, in: The Alkaline Rocks, ed. by H. Sørensen, Wiley and Sons, New York, 1974.
44. S.D. Weaver, J.S.C. Sceal and I.L. Gibson, Contr. Min. Petrol., 36, 181, 1972.
45. K. Bell and J.L. Powell, Isotopic composition of strontium in alkaline rocks, in: The Alkaline Rocks, ed. by H. Sørensen, Wiley and Sons, New York, 1974.
46. N.M.S. Rock, Contr. Min. Petrol., 56, 205, 1976.
47. R.E. Long, K. Sundaralongam and P.K.H. Maguire, Tectonophysics, 20, 269, 1973.
48. J.D. Fairhead, Tectonophysics, 30, 269, 1976.

49. I.B. Ramberg, Norges Geol. Under., 325, 1, 1976.
50. H. Koide and S. Bhattacharji, Science, 189, 719, 1975.
51. J. Verhoogen, Trans. Am. Geophys. Union, 35, 85, 1954.
52. R.J. Bultitude and D.H. Green, J. Petrol., 12, 121, 1971.
53. S.A. Fedotov, Bull. Volc., 39, 241, 1975.
54. J.L. Roberts, The intrusion of magma into brittle rocks, in: Mechanism of Igneous Intrusion, ed. by G. Newall and N. Rast, Gallery Press, Liverpool, 1970.

SOME ASPECTS OF THE METALLOGENY OF CONTINENTAL RIFTING EVENTS

Frederick J. Sawkins

Department of Geology and Geophysics
University of Minnesota, Minneapolis, MN 55455, U.S.

ABSTRACT. Continental rifting events involve tectonomagmatic activation of continental areas. Although the intensity of magmatism associated with these events is in general significantly less than that associated with plate convergence, sites of continental rifting provide favorable environments for the formation of a variety of types of metal deposits.

Currently active rift systems such as the East African Rift System and the Rio Grande Rift have fluorite deposits associated with them, but have presumably not developed sufficiently to generate metal deposits, although some may exist in the subsurface. The widespread rifting events that culminated in the breakup of Pangea in post-Paleozoic time presumably led to the formation of many metal deposits (by analogy with older rifting events), but the large majority of these must lie buried beneath the thick miogeoclinal sediment wedges along current Atlantic and Indian Ocean margins. However, minor tectonic activity along the Benue Trough has led to exposure of a 400 km belt of lead, zinc and copper vein deposits of apparent Mississippi Valley affinity.

The important Messina copper deposits in northern South Africa represent a line of breccia pipe and replacement deposits formed in the Limpopo failed rift during Mesozoic continental fragmentation. Recent fluid inclusion and stable isotope studies of these deposits lead to the conclusion that they were formed by circulating heated meteoric waters that obtained copper by downward percolation through overlying Karroo basalts, now largely removed by erosion.

E.-R. Neumann and I.B. Ramberg (eds.), Petrology and Geochemistry of Continental Rifts, 51–54.

The Oslo failed rift, of late Paleozoic age, contains many small metal deposits, in addition to the important Kongsberg silver deposits just beyond its western margin. Many small molybdenum deposits are known and recent exploration activity has revealed the potential for large disseminated molybdenum deposits.

Because continental rifting represents the first stage of the operation of a Wilson Cycle it is only failed rifts or aulacogens that are easily identifiable within the framework of continental geology. Many older rifting events and their products are obscured within the complexities of later continental collision events. This is exemplified by the Devonian-Carboniferous history of parts of Europe. The stratigraphic record indicates a Devonian rifting event that can be traced from the Caspian Sea (Dneiper-Donetz aulacogen) through the Sudetes, the Harz Mountains and the Rhenish Massif to southwest England. Essentially similar rifting in southwest Iberia was initiated by early Carboniferous time. Although the entire area was subjected to strong deformation during the Hercynian continental collision event, the postulated Devonian rifting event is indicated by the presence of mafic-felsic bimodal volcanism, clastic wedges of miogeosynclinal type and adjacent deepwater facies sedimentary rocks. This belt contains a number of polymetallic massive sulfide deposits, including the important Rammelsberg and Meggen ores and the extensive massive sulfide province of southwest Iberia. On this basis a distinct class of massive sulfide deposits related to continental rifting can be recognized.

An important period of continental rifting and tension, perhaps analogous to the breakup of Pangea, can be recognized in the late Proterozoic geology of all the continents. During this series of events, that occurred from 1.2 to 1.0 billion years ago, mafic volcanism, clastic sedimentation and metallogenesis of rift-related type was widespread. To them can be attributed the Keweenawan ore deposits (Norman, this symposium) and copper mineralization in the Coppermine River and Seal Lake areas, Canada, within the Belt Series, western U.S.A. and at Ducktown, Tennessee. In Africa the important copper deposits of the Katangides and Damarides formed in similar tectonic settings during this period, as did the copper deposits of the Adelaide Geosyncline.

The dominant metal in these deposits is copper and the dominant deposit type of stratiform nature. An explanation for the prevalence of such deposits in rift related environments can be found in two observations. Firstly, the basalts generated by the hotspots associated with rifting can be shown to contain two to three times the background levels of copper of average midocean ridge basalts. Furthermore, much of this excess copper appears to reside in discrete sulfide minerals that are amenable to leaching by circulating aqueous fluids. Secondly, thin argillaceous units within the thick sequences of coarse-grained red clastics that characterize continental rift environments represent ideal physico-chemical traps for the extraction of copper from fluids circulating through the volcanic-sedimentary sequences.

In the more remote geologic past the identification of rift-type tectonic environments and their lithologic and metallogenic associations becomes increasingly more hazardous. However, using the guidelines of mafic, bimodal or peralkaline magmatism and the presence of related thick linear clastic sequences, a number of important metal deposits suggest themselves as rift related. These include the large Sullivan lead-zinc deposits of British Columbia, and the lead-zinc-copper deposits of Mt. Isa, Queensland. Both are of Proterozoic age, occur within very thick sedimentary sequences and exhibit a general time-space relationship to mafic magmatism.

The extensive iron formations of the Hammersley Range, western Australia may also have formed within a framework of extension and crustal thinning. Both the thick sedimentary sequences above and below them and the more or less coeval mafic-felsic volcanism support this notion. In southern Africa the huge Bushveld Complex, the nearby copper-rich Palabora carbonatite complex and the Zoutpansberg Trough may all relate to a single period of intracontinental magmatism and tension.

Still further back in the geologic record, the clastic sedimentary sequences of Blind River, Ontario, and the Witwatersrand, South Africa, both contain mafic-felsic bimodal volcanics and reach thicknesses that imply they must have developed on attenuated continental crust. Thus the gold and uranium in these sedimentary sequences may also represent a facet of the metallogeny of continental rifting.

The above suggestions and speculations would seem to indicate that continental rifting, although second to plate convergence in metallogenic importance, is of considerable significance in terms of understanding the formation of many metal deposits. Furthermore a relatively wide spectrum of metal deposits can form in relation to rifting.

From the viewpoint of the exploration geologist, several points of importance regarding metallogenesis and rifting seem to emerge:

1) Continental rift environments are fertile ground for the generation of certain types of metal deposit.

2) The vast majority of the deposits are of stratiform or stratabound type and are thus confined to limited stratigraphic intervals within thick sedimentary and volcanic sequences.

3) Most rift-related deposits require some later tectonic events to unearth them from sites of deep burial.

4) Because of the operation of Wilson Cycles through much of earth history, many metal deposits actually formed as a result of continental rifting become involved in later collision events of variable intensity.

A more detailed discussion of much of the above and an extensive bibliography can be found in:

Sawkins, F.J., 1976, Metal deposits related to intra-continental hotspot and rifting environments: Jour. of Geology, v. 84, p. 653-671.

THE AFAR RIFT JUNCTION*

F. Barberi and J. Varet

Istituto di Mineralogia e Petrografia,
University of Pisa, Italy and Bureau de
Recherches Géologiques et Minières, Orléans, France°

ABSTRACT. The Afar depression (Ethiopia and Gibuti area) is the site of the junction of three major rift zones: two oceanic (Red Sea and Gulf of Aden) and a continental one (East African rift). Approaching Afar some modifications are observed in the structure as well as in the petrology of each rift branch. Gulf of Aden is characterized by WNW spreading segments offset by ENE transform faults; approaching Afar the ridge-transform system leaves the place to "en echelon" disposed narrow rift valleys. The central trough of the Red Sea progressively looses its morphological expression between Jebel at Tair and Hanish islands, the latter being formed by eruptions through fissures transverse to the Red Sea axis. These two spreading rifts are characterized by typically oceanic basalts (light REE depleted, low-K tholeiites). An increase of the alkalinity of basalts is observed in the southernmost part of the Red Sea. The East African rift is the site of continental lithosphere attenuation marked by normal faulting, open fissures and associated volcanism. Volcanologically the main difference with respect to the oceanic rifts is the frequent development of huge central volcanoes and the abundance of salic alkaline and peralkaline products, which occur also in fissural eruptions.

Approaching Afar the most recent activity tends to concentrate along en echelon disposed sets of faults, where basalts

* This work has been supported by CNR (Contract n. 76.00148.05) and CNRS (RCP 341)

° Present address

E.-R. Neumann and I.B. Ramberg (eds.), Petrology and Geochemistry of Continental Rifts, 55–69.

become more abundant and show a progressively less alkalic affinity.

Similar contrasting structural and magmatic situations occur within Afar evolving both in time and space. Short oceanic spreading segments ("axial ranges"), surface expressions of transform faults and fracture zones are recognized and allow the reconstruction of a complex plate geometry within Afar and its evolution with time. Structural as well as petrological and geochemical data are used for a discussion of the plate tectonics and mantle plume models.

1. INTRODUCTION

The Afar triangle is located in the area of junction of three rifts displaying differing geological structures: the Gulf of Aden, typical oceanic rift floored by oceanic crust over a rather wide zone; the Red Sea, characterized by a narrow median valley; the continental East African rift. Some of the characteristics of each of these rifts are found in the Afar depression: it has in fact been shown by various authors - and recently summarized by Barberi and Varet [1] - that whereas northern Afar displays analogies with the Red Sea spreading segments, some geological structures of eastern Afar can be considered as the emerged westerly prolongation of the Gulf of Aden rift; and furthermore the continental Ethiopian rift extends in central Afar up to Lake Abhe. This paper describes briefly how the geology of each of these rifts is progressively modified while approaching Afar. This allows us to follow the evolution of the process of rifting, from continental rift valley to oceanic ridges and to discuss the geodynamic significance of the Afar rift junction.

2. THE WESTERN PART OF GULF OF ADEN

The Gulf of Aden can be described, on the basis of geological and geophysical data, with reference to two major segments (Fig. 1).
-East of longitude 46°30 E, oceanic spreading has been active since the time of magnetic anomaly 5, and the floor of the Gulf is characterized by well defined WNW to NW oriented rift segments displaced by NE transform faults [2]. Transform faults and fracture zones are well expressed, with a well defined transverse trench and axial valley morphology as at 52°E (Alula-Fartak trench). Magnetic anomalies have made it possible to measure a half spreading rate of 0.96 cm.y^{-1} at 45°6 E and 1.11 cm.y^{-1} at 53°3E [2].
-West of longitude 45 E, only magnetic anomalies 1 and 2 are visible and the spreading axis is characterized by a narrow trench of E-W direction and the absence of well defined transform faults.
A zone of transection is found between 45°E and 46°30E.

Bathymetric, magnetic and seismic reflection profiles in the Gulf of Tadjoura indicate that the Gulf of Aden structure continues into the Gulf of Tadjoura. Further to the west, within the Ghoubbet al Karab, bathymetric and seismic reflection surveys show prominent volcanic and tectonic structures demonstrating the continuity of this oceanic structure with the Asal-Ghoubbet rift of eastern Afar. The active nature of this rift is confirmed by an important seismic activity along the axis of the Gulf (Fig. 2). A recent detailed survey of the Gulf of Tadjoura and Tadjoura trench (Ruegg, personal communication) showed the structure of this area to be transitional between a typical spreading zone of the ocean floor and the Afar emerged spreading segments.

A progressive increase of the average depth is observed approaching Afar. Furthermore, the typical rift valley-transform fault morphology is not observed in this segment of the Gulf of Aden ridge, but is replaced by a series of en echelon disposed spreading segments (left lateral arrangement) located on the floor of an E-W trending rift. Locally, as at $43^{o}10$, $43^{o}15$ and $43^{o}40$, discontinuities occur in the trench, which are visible on bathymetric maps, but offsets are always limited - never exceeding the width of the trough - and no clear transform fault is observed. Locally, sea mounts are observed in the Gulf of Tadjoura around the zones of offset. This morphological evolution is accompanied by a petrological evolution. Preliminary data indicate an increase of the content of large size and highly charged cations (such as K, P, light REE) in the basalts erupted along the axis of the Gulf approaching Afar. The geological and petrological characteristics of the Gulf of Aden plate boundary approaching Afar display several similarities with the Reykjanes segment of mid-Atlantic ridge approaching Iceland. There also, a progressive decrease of depth is noticed, as well as the disappearance of the ridge-transform morphology.

3. THE SOUTHERN RED SEA

The Red Sea is characterized, in its median part, by a well defined axial trough, which is a zone of spreading active since 3 to 4 m.y. [3,4,5,6]. Recent detailed bathymetric and magnetic surveys [7,21] have shown that this trench is not a unique spreading segment but a series of short rifts displaced by small, embryonic transform faults. The very limited amount of displacement between rift segments, never exceeding the width of the valley, does not allow definition of the orientation and structure of these transform faults (or fracture zones). Moreover, the high rate of sedimentation and the volcanic activity tend to mask the morphological expression of tectonic structures. In the zone comprised between 14° and 12° Lat. N, the oceanic spreading occurs along two parallel coupled segments limiting the Dana kil (or Arrata)

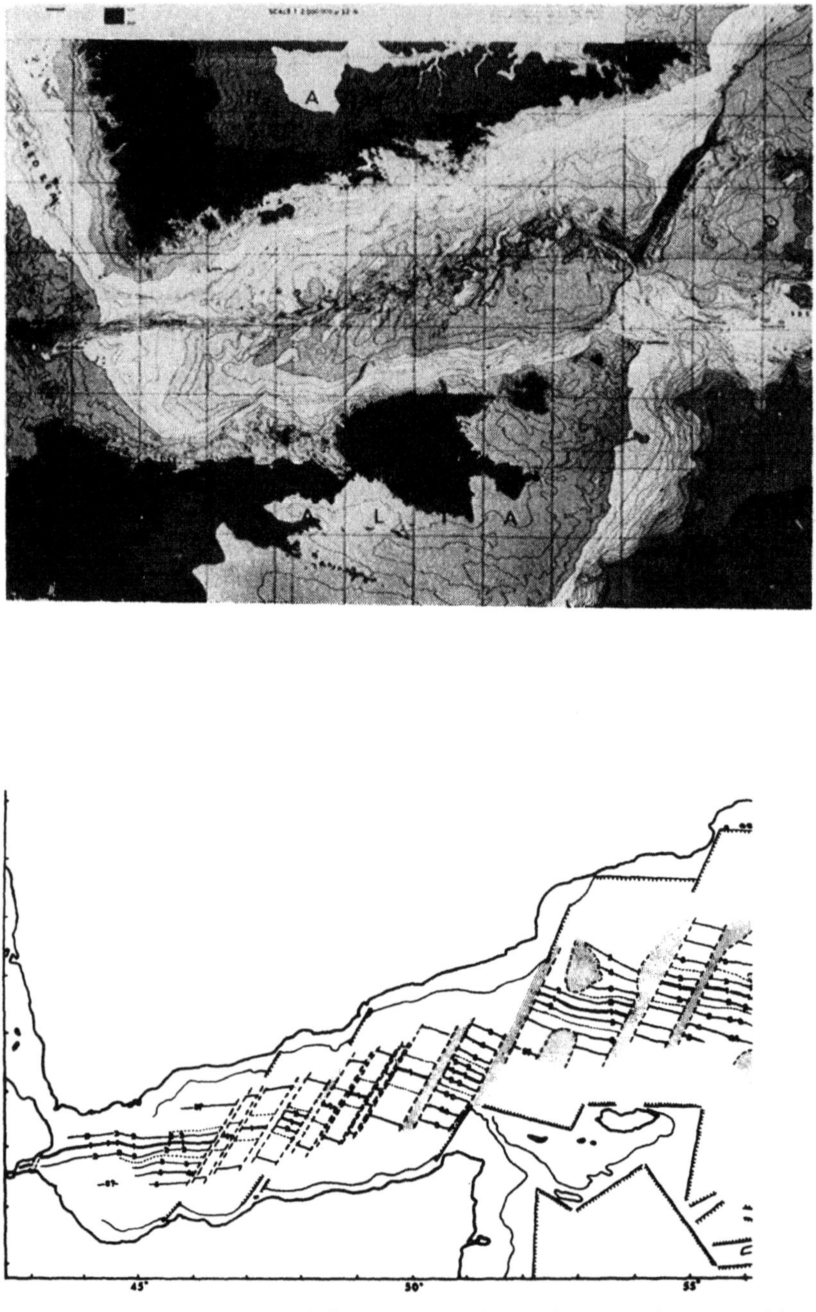

Fig. 1. Bathymetry and interpretation of magnetic anomalies of the Gulf of Aden floor; after [2].

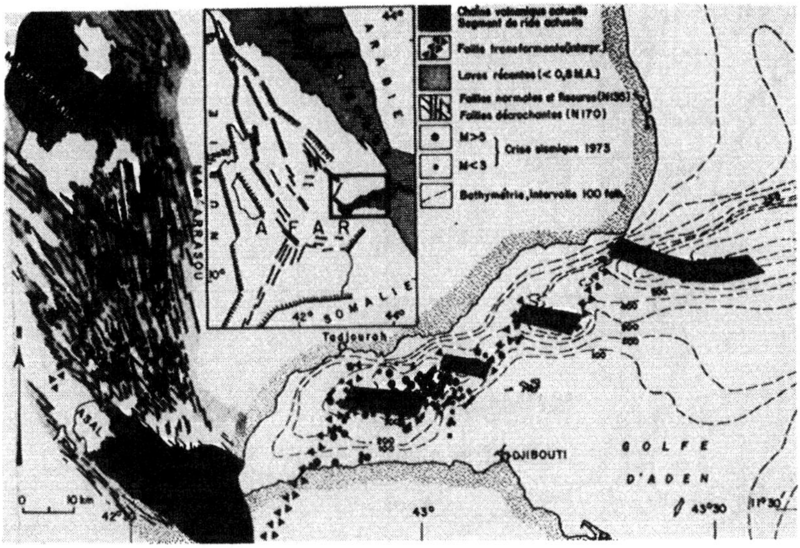

Fig. 2. Schematic representation of the connection between Gulf of Aden, Gulf of Tadjoura and Afar spreading segments (Asal and Manda Inakir); after [15].

microplate [1]. As a whole, a progressive change is noticed in the Red Sea region toward south (Fig. 3): the depth of the floor of the valley progressively decreases; coupled spreading segments occur approaching Afar, where their presence become a characteristic structural feature [1]; volcanic islands are observed, more important and numerous toward south (Jebel at Tair, Zubair, Zukur, Hanish); NW transverse tectonic trend, marked by volcanic activity is progressively more visible and well developed in the Hanish region; basalts become more alkalic and rich in large size highly charged cations toward south [8,9]. These observations emphasize that the structural and magmatic changes in southern Red Sea are analogous to those observed in the Gulf of Aden approaching Afar. The major difference is the occurrence of central volcanoes and the important volcanic activity along transverse structures in southern Red Sea. As already noticed [1], this might be due to the occurrence of "leaky" transform faults related to the rotation of microplates in that part of the rift system where coupled spreading segments occur (see Fig. 5). As shown by Barberi et al. [10] no active rifting presently occurs in the Bab el Mandeb area of the Red Sea, and all the spreading in this part of the rift system occurs within central Afar where geological structures are complex and plate boundaries ill-defined.

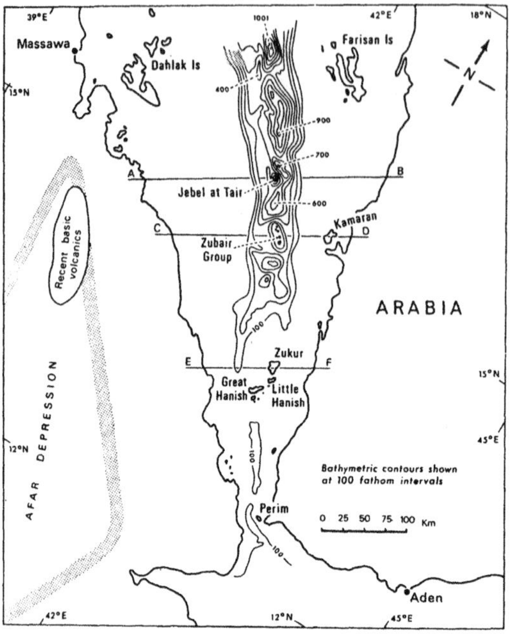

Fig. 3. Map of the southern part of the Red Sea; after [8].

4. THE ETHIOPIAN RIFT VALLEY

As noticed by several authors the Ethiopian rift valley is tectonically characterized by swarms of faults of NNE direction disposed en echelon along the axis of the rift valley. These are normal faults, gaping and emissive fissures which erupted either basalts and silicic products (mainly peralkaline rhyolites and oversaturated trachytes). Central volcanoes with large summit calderas, mainly made of silicic products, dot the rift floor. Two parts can be distinguished (Fig. 4), one south (lake district of Ethiopia) and one north of about 9° Lat. N, where the scarp of the Somalian plateau turns to an E-W direction.

In the lake region [12] the active part of the rift is marked by en echelon swarms of open fissures with associated volcanic activity (both fissural and through central vents). No clear axial graben is observed, however, and silicic volcanic eruptions are widespread. In the northernmost part, and particularly between Ayelu and Lake Abhe, the rift axis is marked by a well defined linear graben of NNE direction. Tectonic and volcanic activity is concentrated along the axis of this graben, where several active fumaroles occur. Volcanism consists both of fissural basalt eruptions and emplacement of intermediate to silicic lava flows, domes or ash-flows from central volcanoes. These volcanoes are

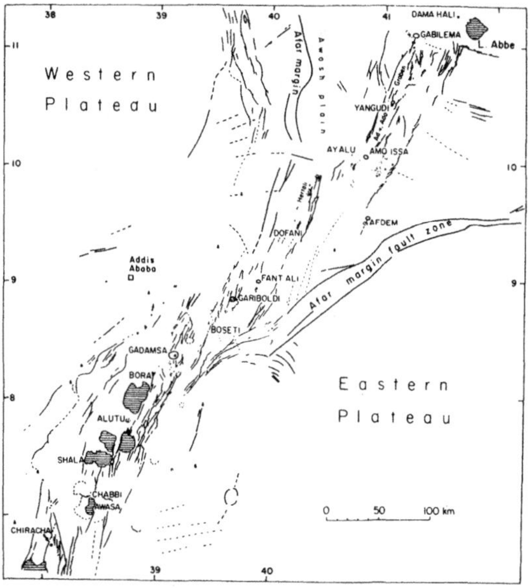

Fig. 4. Schematic representation of the main structural elements and location of Plio-Quaternary central volcanoes of the Ethiopian rift valley; after [11].

regularly spaced (55 Km) [11] and their position seem to be controlled by the occurrence of transverse tectonic lineaments crossing the regional trend. This change from south to north in the geology is accompanied by a variation in the petrology and geochemistry of the lavas, at least in the northernmost portion of the rift. Approaching Afar basic as well as silicic lavas tend in fact to become less alkaline [9]. This was tentatively related [9] to a progressive change in the spreading rate along the Ethiopian rift, in agreement with geophysical data which indicate a progressive increase of the amount of intruded rift material from south to north [14].

In summary the following change is observed along the Ethiopian rift valley approaching Afar: en echelon fissures and faults tend to be replaced by a well marked axial graben where all tectonic and magmatic activity is concentrated, fissural volcanic activity dominates and central volcanoes tend to become less abundant; the alkalinity of lavas decreases. These variations are therefore the reverse of what is observed along the two oceanic rifts of Red Sea and Gulf of Aden approaching Afar.

5. THE MAIN STRUCTURES OF AFAR

Volcanic structures ("axial ranges") analogous to spreading segments have been recognized in Afar; they are built by eruption of basalts with tholeiitic affinity from a linear axial narrow graben and display topographical, seismic, gravimetric and magnetic analogy with mid-oceanic spreading ridges [1]. By analogy with clay experiments it has been shown [15] that some en echelon sets of faults connecting adjacent spreading ridges can be interpreted as the surface expression of transform faults (see Fig. 2). On the now stabilized margins of Afar, in correspondance with offsets of the axial ranges and of the margins, transverse volcanic structures occur. They have many analogies with oceanic fracture zones, like those of the Equatorial Atlantic ocean [16]. They are usually the site of alkalic volcanism, sometimes erupting nodules of ultramafic rocks. Volcanic activity indicates the leaky character of these transverse structures.

The presence of all these structures has been used along with available magnetic and seismic data, to attempt a plate boundary reconstruction of Afar [1]. Results are schematically summarized in Fig. 5. Note the frequent occurrence of coupled spreading segments: the resulting picture is that of a complex mosaic of small microplates located in the area of junction of the three major plates (Arabia, Nubia, Somalia). Intensive deformation with associated volcanic activity affects the microplates even far from their margins. Major and trace elements indicate that basalts having the same composition as those occurring in mid-oceanic ridges (low-K olivine tholeiites with low content of large size lithophile ions) are found in some axial ranges of Afar (Asal graben) and that progressive departure from MOR basalt compositions towards slightly more alkalic varieties are observed in other axial range basalts, that is in those erupted at coupled spreading segments. Spreading segments (axial ranges) within Afar are also characterized by the eruption, at central volcanoes, of complete series of differentiation from basalts to rhyolites (either subalkaline or peralkaline according to the nature of parent basalt). Similar fractionation trends have been observed within the lavas erupted in late Tertiary and early Quarternary time in central Afar and forming the so-called stratoid series, whose formation has been attributed to the earliest episodes of spreading within Afar [1]. Compositional variations are observed within single axial ranges, as well as at a regional scale. They are schematically illustrated in Fig. 6. Sr isotope geochemistry, presently under progress by L. Civetta, indicates that major and trace element variations at the scale of axial ranges (f.i. Erta Ale) are not reflected in the $^{87}Sr/^{86}Sr$ ratio suggesting a common source for these basalts. On the contrary wide regional variations are observed without any apparent relation to trace element geochemistry. No correlation between "alkalinity" and $^{87}Sr/^{86}Sr$ of

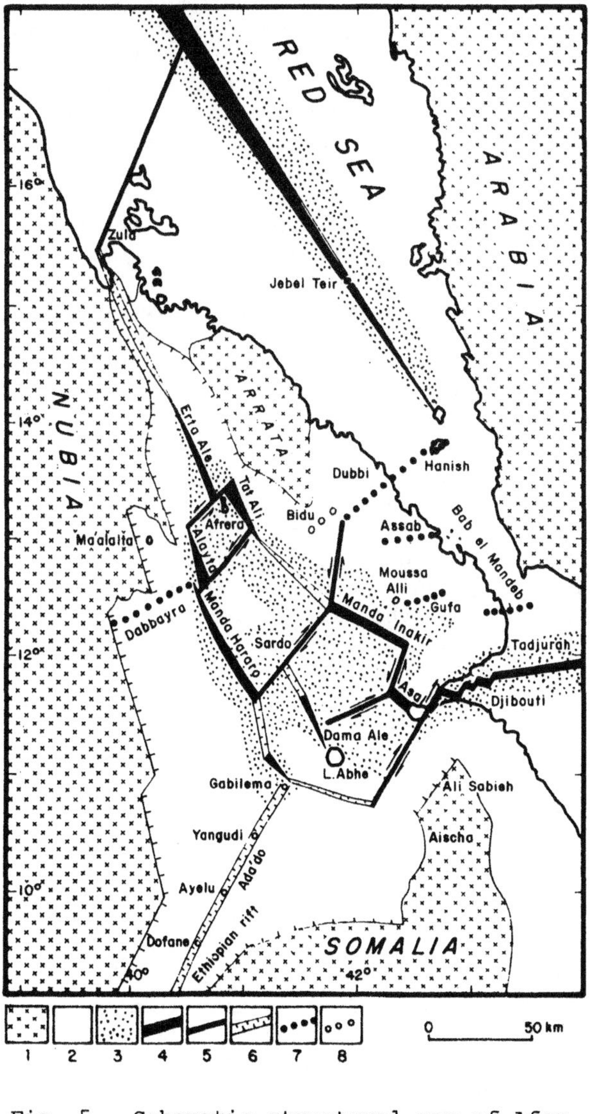

Fig. 5. Schematic structural map of Afar, indicating main geological features relevant to plate boundary determination. 1, outcropping continental basement; 2, continental rift material; 3, oceanic crust formed during past 3 to 4 m.y.; 4, axis of spreading (axial ranges in Afar); 5, relative motion along transform faults, deduced from their surface expression; 6, manifestations of extensional tectonics (large graben without development of axial volcanic ranges); 7, transverse volcanic structure; 8, central volcano.

basalts has so far been recognized. This is a preliminary indication that the proposed genetic models (differing degrees of equilibrium partial melting [9] or mixing of two different sources [17]) cannot explain the compositional spectrum of Afar basalts. Some systematic variations of volcanological and petrological features observed for couples of spreading segments suggest that the spreading rate is not constant along them. The spreading rate apparently decreases approaching the end of the segment near to a transform boundary. This is observed for instance at the southern termination of the Red Sea central trough near the Hanish-Bidu transverse structure, south of Tat'Ali axial range near the presumed NNE transform boundary and in the transition from Manda Hararo to Dama Ale in central Afar, in relation to a possible transform zone (Fig. 5). Spreading variation has a reversed sense passing from one segment to the parallel couplet: spreading increases northwards in the Red Sea and southwards in Erta Ale; northwards in Tat'Ali and southwards in Alayta (see Fig. 5). This variation may be simply explained by the rotation of the "microplate" bounded by the parallel coupled spreading segments. Thus the counterclockwise rotation of the Arrata block accounts for the variations observed along the axial ranges of Northern Afar where volcanology suggests a more important opening southwards (Erta Ale, Alayta, South Boina), and geochemistry broadly indicates a southwards increase of the "tholeiitic" character of erupted basalts. It furthermore accounts for the "leaky" character of the Hanish-Bidu transverse structure.

6. HOT SPOT IN AFAR?

Several features of Afar apparently fit the typical requisites for the occurrence there of a hot spot or the expression of a plume rising from deep in the mantle. These include the large width of the area affected by tectonics and magmatism, the existence of an important swelling (the so-called Afro-Arabian dome [18]), the emerged nature (as in Iceland) of ridge segments, the junction of three rift branches, the occurrence at shallow depth and over a wide zone, of anomalous "hot" mantle, and the rather alkalic nature of volcanic rocks. Some of these arguments have been separately utilized in support of the various models for hot spots [17,18,19,20]. Two of these models specifically refer to Afar:
- the three-stage lithothermal model of Gass [18] which accounts for the upswelling and for the variations of magmatism with time and position across the rift;
- the mantle plume model of Schilling [17] which, by analogy with the Iceland-Reykjanes Ridge, implies directional flows in the asthenosphere with consequent mixing of mantle material from the plume and from the low velocity layer along the ridges.

According to Schilling's model a mantle plume is centered

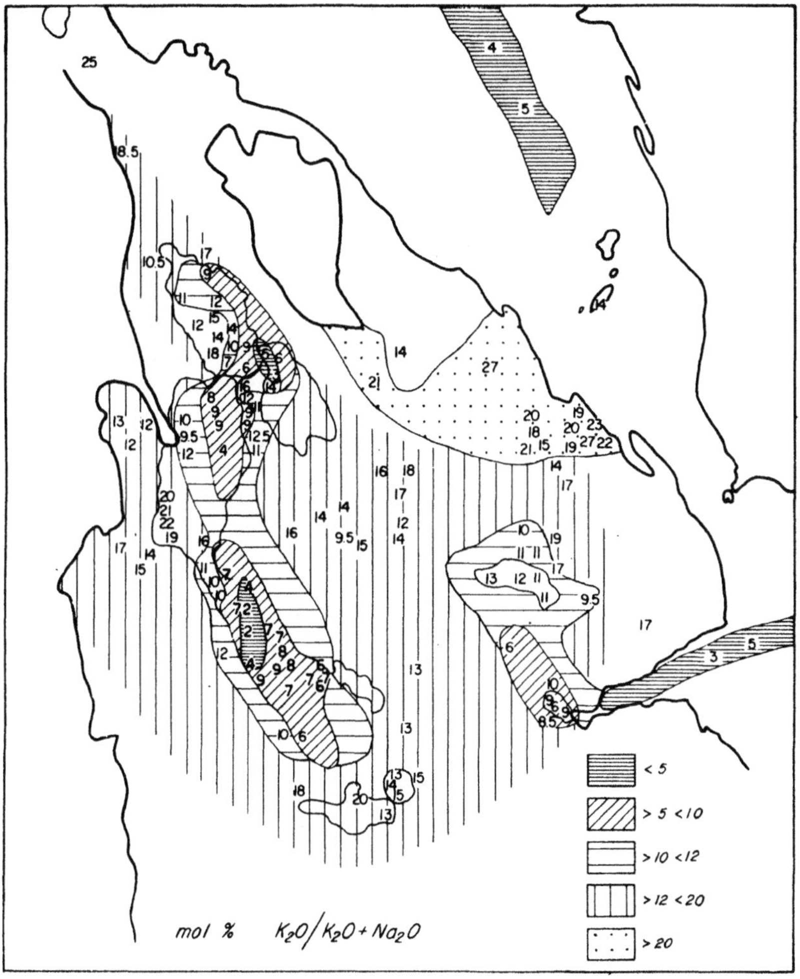

Fig. 6. Regional variation of $K_2O/(K_2O+Na_2O)$ molecular ratio in Afar basalts. Heavy lines are limits of outcropping basement; thin lines are limits of axial ranges. Divisions have been drawn by connecting data points and extrapolating them on a structural basis.

beneath central Afar in the Lake Abhe region near the point where the continental Ethiopian rift joins the oceanic spreading segments of Afar (Fig. 5) and radiates in a star like fashion flowing along three directions which are assumed to represent the three main zones of lithosphere weakness. It implies that magma erupted near the mantle plume are enriched in large size lithophile ions (particularly in light REE) as they are fed by the undepleted material of the hot plume, whereas basalts erupted along the ridge crests (Red Sea - Gulf of Aden) far from the mantle plume, are depleted in the same elements. Magmas erupted in intermediate positions, as Afar basalts, result from mixing of the two previous sources and should be progressively enriched in large lithophile ions as the mantle plume is approached.

Unfortunately it is not possible to reconciliate Afar geology with this three branch flowing pattern, which cuts through the actual tectonic trend along the eastern branch of the star. Let us note that the problem cannot be solved simply by displacing the mantle plume center or by changing the flow geometry, as is easily recognized by looking at the main tectonic trends of Afar (see tectonic maps in [1]). Ages of spreading also argue against a mantle plume in Afar driving lithospheric plates, as indicated by the progressively more recent spreading along the Gulfs of Aden - Tadjoura ridges as Afar is approached. A radial model would not fit any better the Afar structures as shown by the previous discussion on the contrasting structural evolution approaching Afar of the continental rift branch with respect to the two oceanic ones.

On the other hand also the regional variations observed in the nature of erupted basalts are in contrast with Schilling's model. Basalts of central Afar, near the supposed mantle plume, are not particularly enriched in "mantle plume tracing" elements, and the most "depleted" magmas so far recognized in Afar happen to occur near the centre of the star. Similarly, the variation registered going from Afar towards and along the East African rifts, are exactly opposite to those predicted by the plume model [9], and, as previously recalled, available geochemical data are not compatible with a simple mixing model. In conclusion, the proposed plate driving mantle plume model and the related magmagenetic scheme are contradicted in Afar by structural, geochronological and geochemical data.

7. TRANSITION FROM CONTINENTAL TO OCEANIC RIFTING

The previous description indicates that Afar displays structural and magmatic features which are intermediate between those of a typical oceanic ridge, such as Gulf of Aden, and those of a typical continental rift, such as the Ethiopian rift valley. Afar and the three related rifts form a system where various stages of the complex process of evolution from a continental rift valley to oceanic spreading can be observed. Contemporaneously the different stages of the continental fracturing and splitting are in fact observed proceeding along the east African rift towards Afar. Departing from Afar, various processes of oceanic spreading are observed. Present axial ranges have been active the last few hundreds of thousand years in northern Afar, whereas others have been active 3 to 4 m.y. in central Afar. Towards north, in the Red Sea, we observe clearly a stable oceanic spreading for 3 to 4 m.y., whereas in the Gulf of Aden, towards east, the same process has been going on for 10 m.y., with the development of what is commonly considered to be a typical oceanic spreading structure, with segments of rift valley connected by well defined transform faults. If all these stages are connected into a single process, the following model of evolution, from continental break-up to typical ocean floor spreading, can be proposed:

- Stage 1: Continental rift valley (Lake region of the Ethiopian rift). Tensional tectonics form en echelon swarms of normal faults and open fissures with widespread fissural eruptions of basalts and salic products; central volcanoes dominated by salic products are frequent and tend to form where the rift main tectonic trend crosses transverse lines of weakness; magmas are alkalic or transitional with marked enrichment in light REE.

- Stage 2: Transition to an embryonic oceanic spreading (Northern termination of Ethiopian rift in southern Afar). Tectonic and volcanic activity tend to concentrate along a narrow axial graben; central volcanoes are more rare and fissural basaltic volcanic activity tends to dominate; magmas are transitional with moderate enrichment in light REE.

- Stage 3: Development of Proto oceanic spreading structures (Northern and central Afar). The oceanic ridge morphology tends to develop, with a narrow axial rift valley where fissure volcanism and tectonic activity are concentrated. Age of normal faults and basalts increases symmetrically on both sides of the valley. Spreading axes are either arranged en echelon at short distances one from the other, or connected through oblique faulting (embryonic transform zone), basalts with minor iron-rich intermediate differentiates are the only erupted products. Basalts are tholeiitic or transitional with chondritic REE patterns. Small central volcanoes erupting complete sequences of differentiated lavas develop when the spreading rate is very low.

- Stage 4: Oceanic rift (Gulf of Aden). The narrow axial rift valley has been a permanent feature for several m.y. and has produced oceanic crust on both sides over a rather wide zone.

The spreading ridge segments are connected by morphologically well expressed transform faults. Basalts dominate with minor eruptions of intermediate lavas showing a Fe enrichment. Basalts are low-K tholeiites with light REE depleted or flat patterns.

The Afar regions can therefore be considered as a privileged piece of the world rift system where transition from continental to oceanic rifting can be observed together with the transition from a submerged typical oceanic ridge to free-air spreading axes. Several analogies actually exist between the eastern Afar-Gulf of Aden and the Iceland-Reykjanes ridge areas. In both cases morphology, geological structures and basalt compositions seem to be sensitive markers of the transition from a tectonic regime in which energy is dissipated along a narrow zone corresponding to a typical topographically well-expressed spreading ridge, to a regime in which energy is dissipated over a much wider surface and where identification of spreading axes may be less evident. These areas are characterized by a wide lateral extension of a shallow anomalous mantle, by the fact that tensional tectonics are active over large areas up to hundreds of Kms, which largely overpass the width of common diverging plate boundaries, by the coexistence of coupled spreading segments and by a high instability with time of the plate boundaries. Irrespective of whether or not magmatism in these areas is related to a plume rising from deep in the mantle, available data indicate that at least in some areas (Red Sea central trough, northern Afar) departures from the composition of low-K tholeiites with chondritic pattern most likely reflect changes in the mantle partial melting conditions in response to decreases of the amount of energy locally dissipated.

REFERENCES

1. F. Barberi and J. Varet, Volcanism of Afar: smallscale plate tectonics implications, Geol. Soc. Am. Bull., 1977.
2. A.S. Laughton, R.B. Whitmarsh and M.T. Jones, Phil. Trans. Roy. Soc. London, A, 267, 227, 1970.
3. A.S. Laughton, Phil. Trans. Roy. Soc. London, A. 267, 21, 1970.
4. J.D. Phillips, Phil. Trans. Roy. Soc. London, A, 267, 205, 1970.
5. T.D. Allan, Phil. Trans. Roy. Soc. London, A, 267, 153, 1970.
6. R.W. Girdler, The Red Sea: a geophysical background, in: Hot Brines and Recent Heavy metal deposits in the Red Sea, ed. by E.T. Egens and D.A. Ross, Springer-Verlag, 1969.
7. H. Backer, K. Lange and H. Richter, Geol. 7b., D 13, 79, 1975.
8. I.G. Gass, D.E.J. Mallick and K.G. Cox, Jour. Geol. Soc., 129, 275, 1973.
9. M. Treuil and J. Varet, Bull. Soc. Geol. France, 7-15, 506, 1973.

10. F. Barberi, G. Ferrara, R. Santacroce and J. Varet, Structural evolution of the Afar triple junction, in: Afar Depression of Ethiopia, ed. by A. Pilger and R. Røsler, Schweizerbart'sche Verlagsbuchhandlung, 1975.
11. P.A. Mohr and C.A. Wood, Earth Planet. Sci. Lett., 33, 126, 1976.
12. G.M. Di Paola, Bull. Volc., 36, 1972.
13. I.L. Gibson and H. Tazieff, Phil. Trans. Roy. Soc. London, A, 267, 331, 1970.
14. J. Markis, H. Menzel, J. Zimmermann and P. Gouin, Gravity field and crustal structure of north Ethiopia, in: Afar Depression of Ethiopia, ed. by A. Pilger and R. Røsler, Schweizerbart'sche Verlagsbuchhandlung, 1975.
15. P. Tapponnier and J. Varet, C.R. Acad. Sc. Paris, 278, 209, 1974.
16. F. Barberi, E. Bonatti, G. Marinelli and J. Varet, Tectonophysics, 23, 17, 1974.
17. J.C. Schilling, Nature Phys. Sci., 242, 2, 1973.
18. I.G. Gass, Phil. Trans. Roy. Soc. London, A, 267, 369, 1970.
19. W.J. Morgan, Amer. Ass. Petrol. Geol. Bull., 56, 203, 1972.
20. K. Burke and J.F. Dewey, Jour. Geol., 81, 406, 1973.
21. H.A. Roeser, Geol. Jb., D 13, 131, 1975.

PETROLOGY OF PLIO-PLEISTOCENE BASALTIC ROCKS FROM THE CENTRAL RIO GRANDE RIFT (NEW MEXICO, USA) AND THEIR RELATION TO RIFT STRUCTURE*

W. S. Baldridge

Division of Geological and Planetary Sciences**
California Institute of Technology
Pasadena, California 91125 USA

ABSTRACT. Plio-Pleistocene basaltic rocks from the central Rio Grande rift include lavas of both the alkali olivine-and tholeiitic series erupted approximately contemporaneously. The volcanic centers with the greatest diversity of parental compositions and the greatest total range of compositions occur in a complicated transverse structural zone at the offset between the Española and Albuquerque-Belen basins of the rift. The diversity in compositions derives in part from the fact that parental magmas originate at varied depths and in part from fractional crystallization and possibly crustal contamination as magmas move toward the surface. This correlation with structural setting suggests that the transverse shear zone taps partial melts from the mantle diapir beneath the rift at depths greater than along adjacent segments of the rift.

1. INTRODUCTION

The Rio Grande rift consists of a series of north-trending, sediment-filled faulted basins (grabens) arranged en echelon in a north-northeasterly direction extending over 900 km from the upper Arkansas graben near Leadville, Colorado, to the Los Muertos basin in Chihuahua, Mexico. The width of the rift varies from about 15 to 65 km over most of its length, becoming even broader near its southern terminus. The flanks of the rift are defined principally

*This work has been supported in part by the Los Alamos Scientific Laboratory of the University of California and by a grant from the Caltech Ford/Exxon Energy Research Program.
**Contribution No. 2947

E.-R. Neumann and I.B. Ramberg (eds.), Petrology and Geochemistry of Continental Rifts, 71–78.

by normal faults, but in some cases the margins are offset along high-angle reverse faults. Vertical offset of the basins may be at least 5 km [1,2,3]. Inception of rifting is not well determined but probably began at least 18 million years ago and may be as old as 28 m.y.'s in the southern part of the rift [2,4].

The rift has served as the locus of both basaltic and intermediate to silicic volcanism. Yet in comparison to other continental rifts the volume of volcanic material is relatively small ($12x10^3$ km^3) [5]. The main occurrence of intermediate to silicic volcanic rocks is the resurgent caldera forming the present Jemez Mountains [6], located on the western bounding faults of the Española basin (Fig. 1). This caldera constitutes a major study in itself and will not be treated here. Basalts and related lavas were erupted primarily in latest Pliocene and Quaternary time and occur as isolated volcanoes or cinder cone fields at or near the

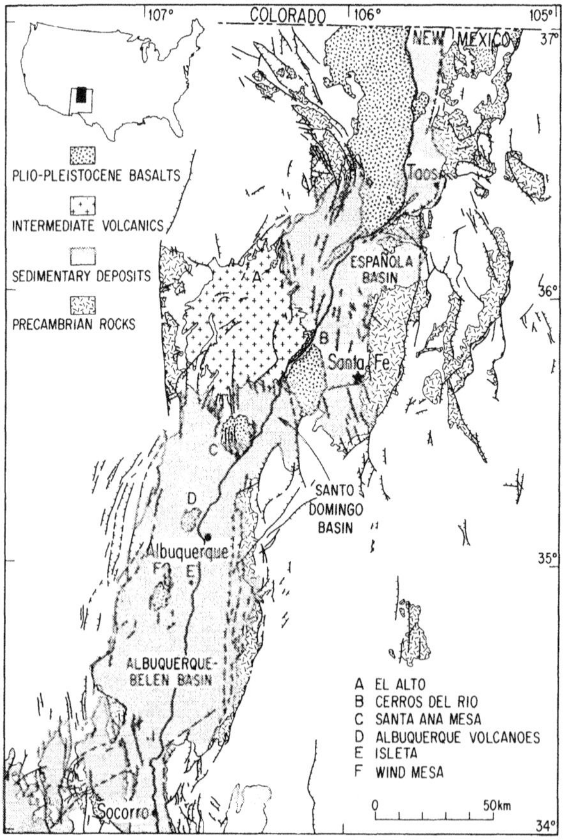

Fig. 1. Generalized structure map. Modified from [14].

present top of the rift-filling sediments. The northern and southern parts of the rift have erupted basaltic magmas of fundamentally different compositions: volcanism in the northern rift has consisted predominantly of aluminous olivine tholeiites and derivatives [7,8], whereas to the south volcanism has consisted primarily of alkali olivine basalts and related rocks [9,10]. Within the central Rio Grande rift both tholeiites and alkali olivine basalts have been erupted approximately contemporaneously [11]. Of these the most important fields in terms of volume of lava and diversity of composition are those occurring within the zone of offset between the Española and Albuquerque-Belen basins. This paper will focus primarily on the volcanic rocks of this offset zone (consisting of the Cerros del Rio and Santa Ana Mesa fields, Fig. 1) and will compare them to the adjacent volcanics immediately to the north and south.

2. RESULTS

2.1 Volcanic rocks of the offset zone

Fig. 2 is an alkalis vs. silica diagram based on 38 new whole-rock major element analyses* of lavas from the Cerros del Rio, Santa Ana Mesa, and El Alto volcanic fields. Analyses from the small El Alto field 40 km to the north are included here for comparison because they are similar to those of the Cerros del Rio and Santa Ana Mesa fields. Lavas from the Cerros del Rio have a very wide range in compositions and include rocks of both the tholeiitic and alkali olivine basalt suites. Analyses from the Santa Ana Mesa and El Alto fields are more limited, with those from Santa Ana Mesa being restricted to olivine tholeiites. No apparent systematic distribution of compositions exists either within the fields or over the time interval of their eruption (approximately 3 to 2 m.y.'s ago). The following features are observed on Fig. 2: (1) Two separate groups of olivine tholeiite are present, designated A and B. Group A olivine tholeiites, with a maximum Mg-value ($=100 \cdot Mg/Mg+Fe^{+2}$) of ~61, show no significant variation in their chemistry. Group B tholeiites have an Mg-value of ~64. A well-defined trend, possibly the result of low-pressure fractional crystallization originates from these tholeiites and extends with decreasing Mg-values through andesitic (b) to dacitic (bb) compositions. Since Group B tholeiites have higher Mg-values than those of Group A, they cannot be simply related to A

*These analyses were obtained by electron microprobe analysis of glass beads fused from whole-rock powders on an Ir-strip furnace. Because the oxidation state of Fe cannot be determined by the microprobe, $Fe^{+3}/Fe^{+2}+Fe^{+3}$ has been set equal to 0.1.

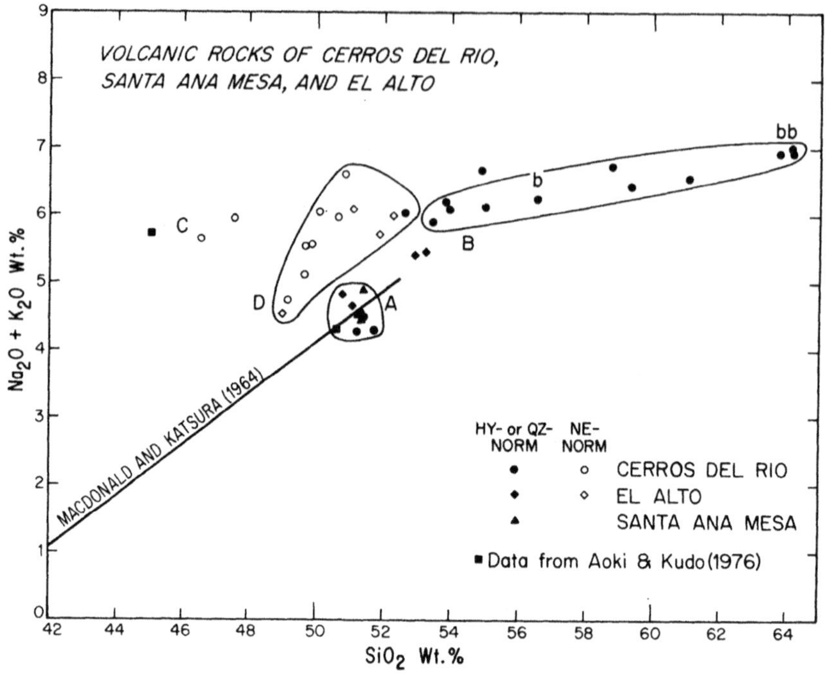

Fig. 2. Alkalis vs. silica diagram for rocks of the Cerros del Rio, Santa Ana Mesa, and El Alto volcanic fields. Line dividing tholeiitic and alkali olivine basalts from Hawaii [20] has been added for reference.

by fractional crystallization. Tholeiitic Groups A and B are therefore considered to be independent parental magmas. (2) At least two separate groups of alkaline rocks, designated C and D, are also present. Group C lavas are basanites (>5% normative nepheline) with Mg-values of ~64. Group D lavas are alkali olivine basalts (<5% normative nepheline) with Mg-values of ~63. A broad trend in compositions originates near D and extends with generally decreasing Mg-values toward higher SiO_2 and (Na_2O+K_2O). These compositions are collected together on Fig. 2 with the basalts of Group D for convenience, but their petrogenetic relationship to either Group D basalts or Group C basanites is presently uncertain. Groups C and D can not be simply derived from each other nor from tholeiites A or B by fractional crystallization, hence are also considered to be independent parental lavas. Lavas in equilibrium with undepleted mantle peridotite have Mg-values of between 66 and 75 [12]. All of the parental lavas of Fig. 2 are slightly less magnesian, hence are not direct partial melts of mantle material and must have undergone some crystal fractionation or other process to alter their chemistries before reaching the surface.

2.2 Volcanic rocks of the northern Albuquerque-Belen basin

For comparison limited data on the volcanic rocks of the northern Albuquerque-Belen basin are shown on Fig. 3. Volcanic rocks of this basin occur as isolated centers along the western edge of the rift. Lavas of the Albuquerque Volcanoes consist exclusively of olivine tholeiites, less alkalic than the olivine tholeiites of the Cerros del Rio, Santa Ana Mesa, and El Alto fields, with only a slight range toward more silicic compositions. No correlation between composition and stratigraphic position within this field has been found. The most magnesian of these lavas has an Mg-value of 64, indicating that it is not in equilibrium with the upper mantle. Analyses of the Wind Mesa and Isleta lavas show that both olivine tholeiites and alkali olivine basalts are present and that a slight range in compositions exists. Tholeiites from the Wind Mesa field have the highest Mg-values (~62) and may approximately represent parental magmas. More precise characterization of the lavas, including a recognition of parental compositions for the alkali olivine basalts, will have to await further data.

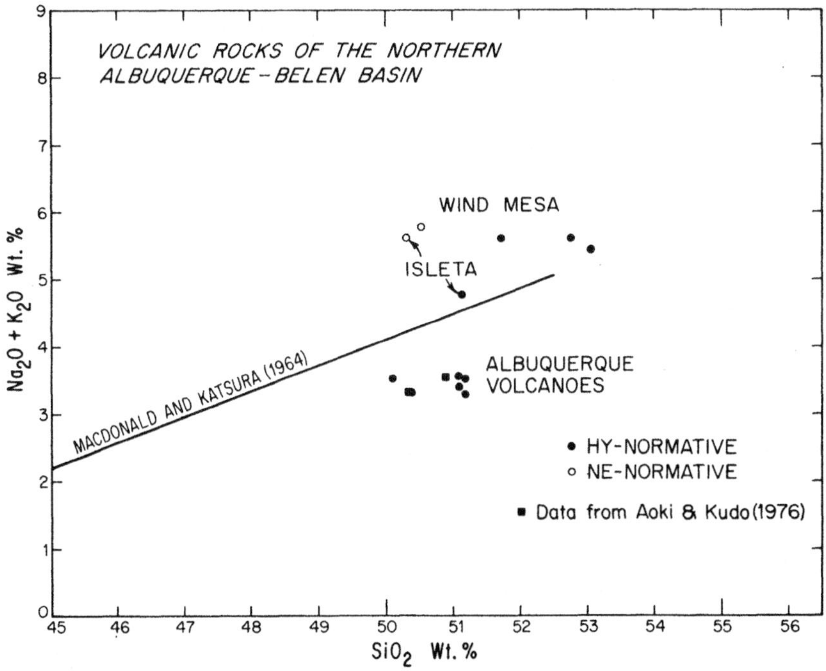

Fig. 3. Alkalis vs. silica diagram for volcanic rocks of the northern Albuquerque-Belen basin. Line dividing tholeiitic and alkali olivine basalts from Hawaii [20] has been added for reference.

3. DISCUSSION

3.1 Transverse shear structure

The Cerros del Rio and Santa Ana Mesa volcanic fields occupy an area of great structural complexity in a direction transverse to the main rift lineation. In this area the southern Española basin shallows and ramps out against the Cerrillos and Ortiz uplifts. The rift is stepped westward a distance of 50-60 km by a series of en echelon faults through the small Santo Domingo basin to the main Albuquerque-Belen basin [1,13,14]. Gravity modelling indicates that the Santo Domingo basin is extremely deep, probably 2-3 km deeper than the adjacent basins, and that very steep gradients (presumably fault zones) separate this basin from the shallower Española and Albuquerque-Belen basins [3]. The Cerros del Rio and Santa Ana Mesa volcanic fields are located directly over the eastern and western boundaries, respectively, of the Santo Domingo basin and it is likely that the fault zones bounding the basin have served as conduits for the lavas. In addition, the southern extension of the Sierra Nacimiento uplift merges with the Albuquerque-Belen basin at the latitude of this transverse zone. That this transverse shear zone has been a zone of crustal weakness for substantially longer than the Rio Grande rift has existed is indicated by the fact that (1) the Sierra Nacimiento is a structure of Laramide age [15], (2) the north and south margins of the Santo Domingo structural basin are aligned in a northeast-southwest direction parallel to Precambrian discontinuities [3], and (3) in Eocene and Oligocene time this zone was the locus of extensive intrusive and extrusive igneous activity [13,16].

3.2 Model for intra-rift chemical variations

Geophysical evidence from several continental rift zones, including the Rio Grande rift, indicates that the lithosphere is thinned beneath rifts due to intrusion into the lithosphere of buoyant, low velocity mantle material [17,18,19]. This diapir is the source of extensive mafic intrusives into the crust [18,19] and of surface volcanism within the rift. However, this (steady state) model is an oversimplification and cannot explain the contemporaneous eruption of fundamentally different parental magmas in close proximity such as occurs in the central Rio Grande rift. Clearly an additional factor is required and this must be the structural setting. Lavas occurring in the offset zone (Santo Domingo basin) are unique chemically in relation to the lavas to the north (on the Taos plateau [7,8]) and south (Albuquerque-Belen basin). The small El Alto field does not occur within this transverse zone. However, it was erupted along fractures trending generally into the Santo Domingo basin. This transverse shear zone may reach deeply into the low-velocity mantle diapir and bleed off partial melts at substantially greater depths (>35 km) than in adjacent "normal" regions (15-30 km), thus resulting in extensive extrusion of alkali olivine basalts and related rocks. Complications within this zone, such as magma genesis from shallower depths (yielding tholeiitic compositions) and possibly contamination of magmas with crustal materials or with themselves, help account for some of the dispersion in composition seen at the surface. This hypothesis raises a number of further questions such as what is the nature of rheologic discontinuities (manifested as cracks at the earth's surface) in the mantle and lower crust, especially in a region capable of supplying small amounts of partial melt; and why are certain regions of the earth's crust/mantle weak over geologically long periods of time. The questions cannot yet be answered, yet are critical to our understanding of magma genesis.

ACKNOWLEDGMENTS

T. R. McGetchin provided much support, particularly for field work, through the facilities of Los Alamos Scientific Laboratory. Discussions with him and with R. J. Bridwell have yielded continuing stimulation. The manuscript was considerably improved by discussions with A. L. Albee.

REFERENCES

1. V.C. Kelly, New Mex. Geol. Soc. Guidebook, 7th Field Conf., 109, 1956.
2. C. E. Chapin, New Mex. Geol. Soc. Guidebook, 22nd Field Conf., 191, 1971.
3. L. Cordell, New Mex. Geol. Soc. Spec. Publ., 6, 62, 1976.
4. C.E. Chapin and W. Seager, New Mex. Geol. Soc. Guidebook, 26th Field Conf., 297, 1975.
5. R.W. Decker. Referenced in: Solid Earth Geosciences Research Activities at LASL, July 1- Dec. 31, 1974, ed. by T.R. McGetchin, Los Alamos Scien. Lab. Progr. Rep. LA-5956-PR, 1975.
6. R.L. Smith, R.A. Bailey and C.S. Ross, U.S. Geol. Surv. Map No. I-571, 1970.
7. K. Aoki, Contr. Min. Petrol., 14, 190, 1967.
8. P.W. Lipman, Geol. Soc. Am. Bull., 80, 1343, 1969.
9. J. Renault, New Mex. State Bur. Mines Miner. Res. Circ., 113, 1970.
10. A.M. Kudo, K. Aoki and D.G. Brookins, Earth Planet Sci., Lett., 13, 200, 1971.
11. K. Aoki and A.M. Kudo, New Mex. Geol. Soc. Spec. Publ., 5, 82 1976.

12. A.J. Irving and D.H. Green, J. Geol. Soc. Australia, 23, 45, 1976.
13. C.E. Stearns, Geol. Soc. Am. Bull., 64, 459, 1953.
14. L.A. Woodward, J.F. Callender and R.E. Zilinski, Geol. Soc. Am. Map and Chart Ser., MC-11, 1975.
15. E.H. Baltz, U.S. Geol. Surv. Prof. Paper, 552, 1967.
16. M. Sun and B. Baldwin, New Mex. State Bur. Mines Miner. Res. Bull., 54, 1958.
17. R.C. Searle, Geophys. J. R. Astr. Soc., 21, 13,1970.
18. R.J. Bridwell, New Mex. Geol. Soc. Guidebook, 27th Field Conf., 283, 1976.
19. I.B. Ramberg, Norges Geol. Unders. 325, 1, 1976.
20. G.A. MacDonald and T. Katsura, J. Petrol., 5, 82, 1964.

RIFTING AND VOLCANISM IN THE NEW MEXICO SEGMENT OF THE BASIN AND RANGE PROVINCE, SOUTHWESTERN USA*

Wolfgang E. Elston

Department of Geology, University of New Mexico,
Albuquerque, N.M. 87131, U.S.A.

ABSTRACT. The Basin and Range province has been the site of extensional orogeny for the past 40 m.y. Its complexities may be caused by the superposition of three tectonic regimes: (a) an Andean-type volcanic arc, which gradually gave way to (b) a spreading ensialic backarc basin and was in turn succeeded by (c) intraplate block faulting and rifting. Total extension may have exceeded 100 percent.

1. INTRODUCTION

The Basin and Range province of western North America covers about 1 million km^2, divided equally between Mexico and the USA. For the past 40 m.y. it has been the scene of an extensional orogeny that differs radically from classic compressional orogenies, as in the Alps or Appalachians. The first half of the Basin and Range orogeny reached its climax in a great "ignimbrite flareup" between 35 and 25 m.y., when the province was inundated by 10^6 km^3 of silicic volcanic rocks, equivalent in volume to major batholiths. The second half was characterized by the development of rifts and block faults which control the present topography. Formation of a basin and range province is not unique to the Cenozoic of North America; somewhat similar events seem to have occurred in the late Paleozoic of Eurasia (1,2).

---

*Research supported by NASA grants NGL-32-004-011 and NGR-32-004-062, U.S. Geological Survey grant 14-08-001-G-255, and grants from the New Mexico Energy Resources Board. I am indebted for ideas to numerous associates and present and former students, especially T.J. Bornhorst, P.J. Coney, E.E. Erb, E.G. Deal, the late R.C. Rhodes, and E.I. Smith.

E.-R. Neumann and I.B. Ramberg (eds.), Petrology and Geochemistry of Continental Rifts, 79–86.

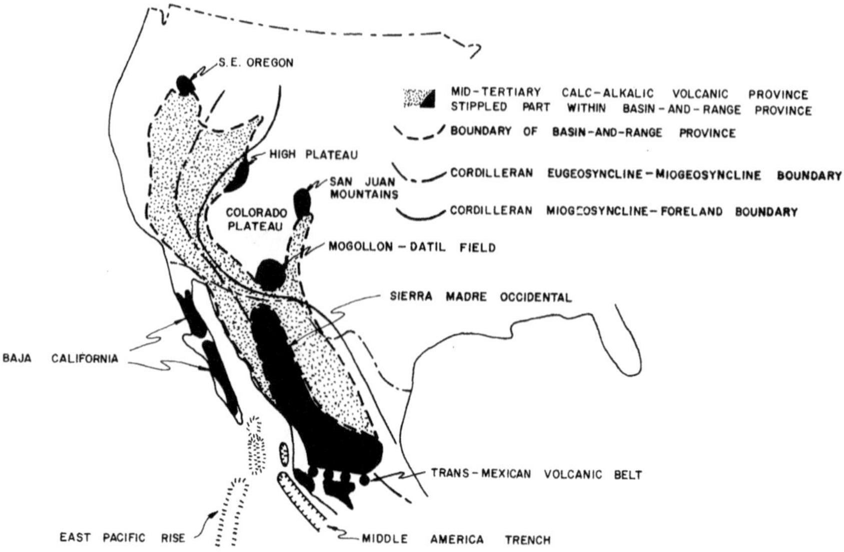

Figure 1. Location and tectonic setting of mid-Tertiary ash-flow tuff fields and of the Basin and Range province, southwestern North America.

This paper is based on work in southwestern New Mexico; details have been published elsewhere (3). Conditions in other parts of the Basin and Range province are broadly similar, although there may be systematic regional variations in the timing of events and the composition of volcanic rocks (4,5).

2. TECTONIC INTERPRETATIONS

The Basin and Range province does not fit easily into geotectonic (geosynclinal-orogenic) concepts. Stille (6) seems to have been puzzled by the association of "alpinotype" volcanism and "germanotype" structures. He designated the area around Bisbee, Arizona, (which presumably includes southwestern New Mexico) as a type example of "subsequent" volcanism, "intercedent" plutonism, and hybrid tectonics ("Zwittertektonik"). Actually, volcanism was not "subsequent" to geosynclinal-orogenic stages. The province overlaps several deformed belts (Laramide, Sevier, Antler, etc.) as well as the boundaries between craton, miogeosyncline, and eugeosyncline, but seems unrelated to all of them (Fig. 1). Over much of the province, there was an "early Cenozoic null" (4) or period of quiescence between Laramide and mid-Tertiary events.

Plate-tectonic interpretations fall into two broad classes: (a) the province was formed actively, in response to forces acting from beneath the plate, such as an overriden East Pacific Rise (7),

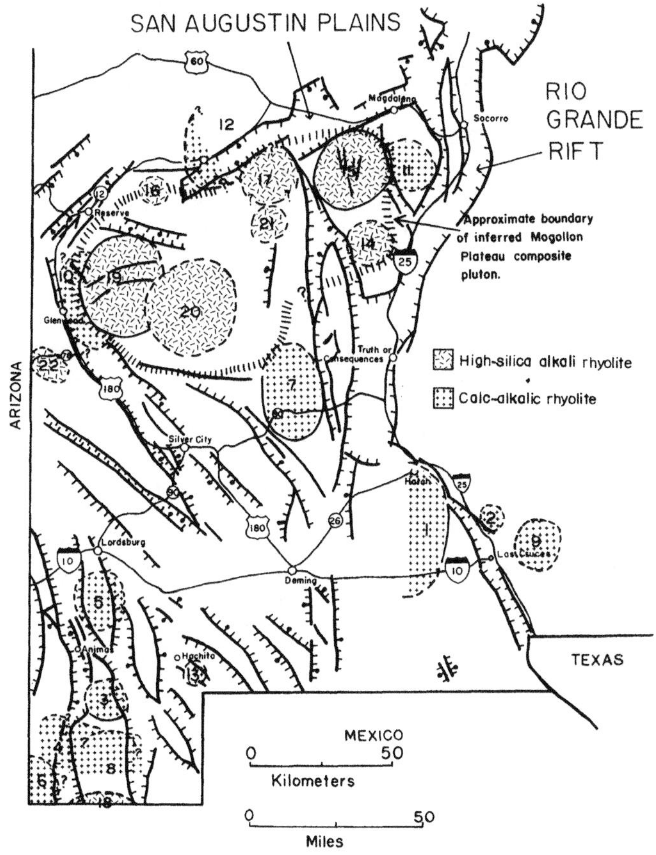

Figure 2. Provisional tectonic sketch map of southwestern New Mexico. Numbered circles outline the inner walls of Oligocene-Miocene ash-flow tuff cauldrons (dashed where covered or imperfectly documented). Hachured lines border basins of the Basin and Range province. The map shows the status of current mapping; none of the cauldrons were documented prior to 1968. Nos. 1, 2, 9 were documented by W.R. Seager; 7 by W.R. Seager, R.E. Clemons, and W.E. Elston; 10 by J.C. Ratté; 11 is being resolved into several cauldrons by C.E. Chapin and associates; 19 and 20 by J.C. Ratté and by W.E. Elston and associates; the rest by W.E. Elston and associates.

thermal diapirs or induced flow, to cause backarc spreading (8,9, 10), convection plumes (11), and thermal runaways (12); (b) the province was formed passively, in response to forces acting at the plate margin, such as subduction of the Farallon plate or transfer of stresses across the San Andreas transform (4,5,13,14). The two classes are not mutually exclusive; active and passive processes tend to interact.

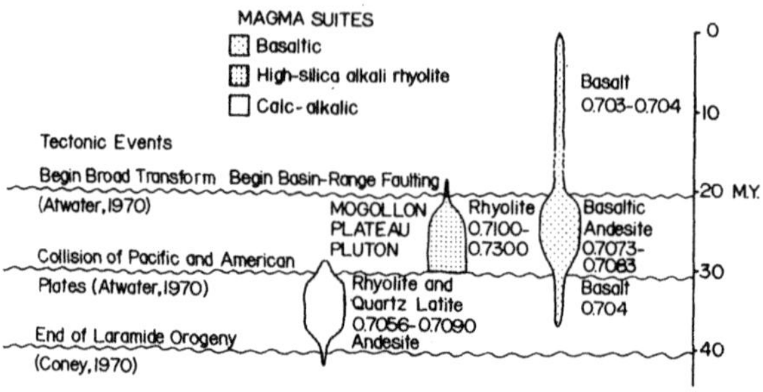

Figure 3. Correlation of volcanic events in southwestern New Mexico with tectonic events at the plate margin. Numbers are $^{87}Sr/^{86}Sr$ ratios (30,31).

3. CHARACTERISTICS OF SOUTHWESTERN NEW MEXICO

Cenozoic volcanic rocks of the Mogollon Datil volcanic field of New Mexico (Figs. 1,2) can be classified into three overlapping suites: (a) calc-alkalic (andesite to rhyolite, including ash-flow tuff, ~40 to 29 m.y.), (b) high-silica rhyolite (mainly ash-flow tuff, ~30 to 21 m.y.), and (c) basaltic (~37 m.y. to the present). Almost all volcanic rocks erupted from central volcanoes. Ash-flow tuff (ignimbrite) sheets spread from large calderas, including resurgent cauldrons of the Valles type (15) up to 40 km in diameter and with up to 2,000 km^3 of ejecta. Cauldrons for calc-alkalic ash-flow tuff probably formed on apices of numerous moderate-sized plutons. High-silica rhyolite cauldrons cluster on the Mogollon Plateau (Fig. 2), interpreted as the surface expression of a composite pluton, 125 km in diameter. As far as known, all plutons were passively emplaced. At a shallow erosion level the province would appear as cluster of nested ring complexes. In many ways the geology of the Mogollon Plateau is a mirror image to that of the better-known San Juan Mountains of southwestern Colorado (16).

Beginning around 21 m.y., the present Basin and Range block-fault topography began to develop. Volcano-tectonic faults associated with cauldrons were reactivated but many regional faults cut across earlier volcano-tectonic structures (Fig. 2). The timing of transitions between volcanic suites and tectonic regimes seems to correspond with timing of events at the plate boundaries (Fig. 3). A possible interpretation of the development of the province is shown in figure 4. Figure 4A, B, and C illustrates active extension; D illustrates passive extension.

By 40 m.y. spreading had slowed at the eastern plate margin (17), relaxing compressional stresses within the plate. At about the same time there seems to have been a change in motion of the Pacific plate, as suggested by the bend in the Hawaii-Emperor

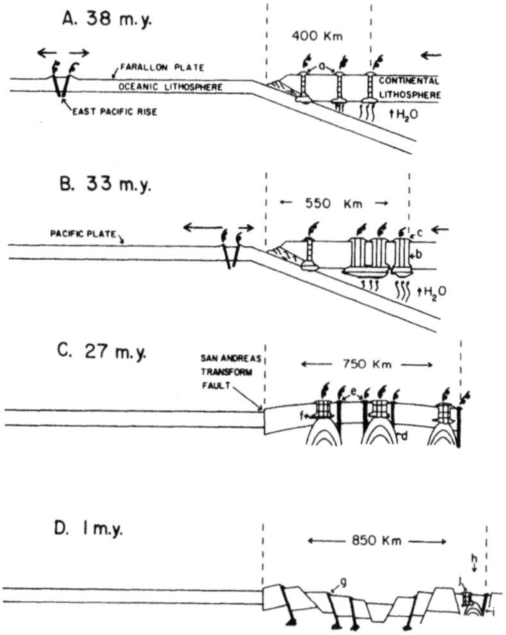

Figure 4. Proposed stages of Cenozoic extensional orogeny, on a line from the north end of Baja California to south-central New Mexico. Details at the plate boundaries and the lithosphere-asthenosphere boundary have been omitted.

chain. At 37 m.y., the New Mexico segment of the Basin and Range province was part of an andesitic volcanic arc (Fig. 4A, a), possibly because partial melting at the base of the continental lithosphere had been triggered by dehydration of a subducted slab of oceanic lithosphere (18,19). By 33 m.y., differentiation of andesitic magma and/or partial melting of the more silicic parts of the lithosphere had resulted in monzonitic plutons (Fig. 4B, b) which rose passively toward the surface. Cauldrons (Fig. 4B, c) formed wherever the roofs of the plutons had blistered, ruptured, and collapsed. Room for the plutons was made by early stages of backarc spreading.

Between 33 and 26 m.y. the Pacific and American plates collided (13); differential motion between them was taken up by the San Andreas transform (Fig. 4C). The plate was no longer confined on either margin and backarc spreading reached its peak. The lithosphere was subjected to increased heat flow and became stretched (10), which allowed diapirs of basaltic material, at the liquidus temperature of basalt, to rise toward the surface (Fig. 4C, d). The diapirs may have caused partial melting of the felsic fraction of the lithosphere. Volcanism was of two types, contaminated basalt (Fig. 4C, e) and granitic plutons (Fig. 4C, f) which broke through to the surface to form cauldrons from which high-silica rhyolite erupted. The transition from the calc-alkalic association to a bimodal association of contaminated basalt and high-silica rhyolite seems to have began relatively early in New Mexico. It spread to other parts of the Basin and Range province as the San Andreas transform lengthened (4,14).

By 20 m.y., volcanism began to wane. Active deformation of a broad mobile plate boundary zone gave way to passive deformation of a cooler and more brittle plate (13,20). The transfer of shearing stresses across the San Andreas transform fault zone caused tensional block faulting and rifting (Fig. 4D). This was Basin and Range faulting in the strict sense, since it led to development of the modern physiographic Basin and Range province. Deep fractures tapped tholeiite and alkali basalt (Fig. 4D, g). In the areas of most intense rifting, such as the Rio Grande rift (Fig. 2; 4D, h) the characteristics of the active extensional regime persist into the present. Seismic soundings, heat-flow measurements, and precise levelling suggest that a basaltic magma body now lies about 18 km below the surface in the Socorro, New Mexico,area (21-23). It may be a modern analog to the basaltic diapirs inferred to have existed during mid-Tertiary backarc spreading. Late Pliocene basalt of the northern Rio Grande rift, near Taos, New Mexico, is locally associated with basaltic andesite (Fig. 4D, i) and high-silica rhyolite that resemble mid-Tertiary rocks (24). The Pleistocene Valles caldera (Fig. 4D, j) 15) may be a modern analog to the cauldrons of the Mogollon Plateau that erupted high-silica rhyolite in mid-Tertiary time.

5. DISCUSSION

The analysis attempted here has much in common with syntheses by other workers (4,5,9,13,14). It differs in emphasizing the distinction between an early (active) volcano-tectonic structural stage (Fig. 4B, C) and a later (passive) Basin and Range stage (strict sense, Fig. 4D). It ascribes the climax of extension to the active stages. On the basis of evidence from experimental petrology (18,19), it is suggested that calc-alkalic magmas are generated near the lithosphere-asthenosphere boundary, not at the surface of a subducted slab. The beginning of Basin and Range faulting (strict sense) is not correlated with the beginning of bimodal basalt - high silica rhyolite volcanism. Calc-alkalic and basaltic volcanism overlap in time and may be difficult to distinguish by chemical criteria. Most important, the estimate of possible extension made in figure 4, over 100 percent, is far greater than the usually-accepted figure of 100 km (5 to 10 percent) or less. Most published estimates of Basin and Range extension (20,25) are based only on the geometry of Basin and Range faults (strict sense). They do not consider the volume occupied by plutons or thinning of the lithosphere during active extension.

The interpretation given here depends on three assumptions: (a) that cauldrons are rooted in the apices of passive steep-sided plutons and that plutons underlie at least one-third of the Basin and Range province (Fig. 2; 4B, b; the cauldrons shown presumably have smaller diameters than underlying plutons and allowance must be made for cauldrons not yet discovered); (b) that the pre-40 m.y. lithosphere under the Basin and Range province had the same thickness as under surrounding regions; and (c) that thinning of the lithosphere is in proportion to observed thinning of the crust, from 40 km or more under the Sierra Nevada, Colorado Plateau, and Great Plains to 25-35 km under the Great Basin (26). These assumptions have not been proved but they are not precluded by present evidence.

By the scheme cartooned in figure 4, maximum extension occurred during spreading of an ensialic backarc basin. Geophysical characteristics of the Basin and Range province (thin crust, high heat flow, and S_n-wave attenuation; (26-28)) are similar to those of oceanic backarc basins (8,10,29). In support of the analogy between oceanic backarc basins and continental ash-flow tuff provinces, the Lau backarc basin of the southwest Pacific seems to strike into the ash-flow tuff province of New Zealand (8,29) and the Andaman backarc basin in the Indian Ocean seems to strike into the ash-flow tuff province of Sumatra.

There are likely to be important differences between oceanic and continental backarc basins. During backarc spreading the thin lithosphere of an ocean basin or continental margin is likely to rupture. The thick continental lithosphere of western North America seems only to have stretched. On the ruptured floor of an oceanic backarc basin, ophiolite may mix with turbidite to form classic eugeosynclinal assemblages. In an ensialic backarc basin, such as the Basin and Range province, contaminated basalt and high-silica rhyolite are associated with continental fanglomerate. Backarc spreading of the Basin and Range province was not followed by orogenic compression, as in many eugeosynclines, because it was not followed by collision between two continental plates. Instead, transform motion along the San Andreas system had taken the place of convergence at the western plate margin and large-scale tensional block faulting and rifting ensued.

Why were the extensional processes of the Basin and Range province orogenic, in the sense of building present-day mountains? Why did the province not subside during backarc spreading to form a mirror image to the Sea of Japan? The highest mountains are great volcanic plateaus, such as the San Juan Mountains and the Mogollon Plateau. They may have been raised to their present height (up to 4,500 m) by the buoyancy of underlying batholiths during regional extension. Their scenery may rival that of the Alps, but they were not formed by classic alpine compressional tectonics.

REFERENCES

1. Lorenz, V. and Nicholls, I.A., The Permocarboniferous Basin and Range province of Europe. An Application of Plate Tectonics, in The Continental Permian of West, Central, and South Europe, ed. by H. Falke, D. Reidel, Dordrecht, 1976.
2. A.A. Mossakovskiy, Geotectonics, 4, 247, 1970.
3. W.E. Elston and S.A. Northrop, editors, Cenozoic Volcanism in Southwestern New Mexico, N. Mex. Geol. Soc. Spec. Publ. 5, 1976.
4. W.S. Snyder, W.R. Dickinson, M.L. Silberman, Earth and Planet. Sci. Letters, 32, 91, 1976.
5. P.W. Lipman, H.J. Prostka, and R.L. Christiansen, Phil. Trans. Roy. Soc., A, 271, 1972.
6. H. Stille, Enführung in den Bau Amerikas, Borntrager, Berlin, 1940.
7. H.W. Menard, Sci. American, 205 (6), 52, 1961.
8. D.E. Karig, Jour. Geophys. Res., 76, 2542, 1971.
9. C.H. Scholz, M. Barazangi, and M.L. Sbar, Geol. Soc. Amer. Bull., 82, 2979, 1971.
10. D.J. Andrews and W.H. Sleep, Geophys. Jour. Roy. Astron. Soc., 38, 237, 1974.
11. W.J. Morgan, Nature, 230, 42, 1971.
12. O.L. Anderson and P.C. Perkins, Jour. Geophys. Res., 79, 2136, 1974.
13. T. Atwater, Geol. Soc. Amer. Bull., 81, 2531, 1970.
14. R.L. Christiansen and P.W. Lipman, Phil. Trans. Roy. Soc., A, 271, 249, 1972.
15. R.L. Smith and R.A. Bailey, Geol. Soc. Amer. Mem., 116, 613, 1968.
16. T.A. Steven and P.W. Lipman, U.S. Geol. Survey Prof. Paper 958, 1976.
17. P.J. Coney, Am. Jour. Sci., 272, 603, 1972.
18. P.J. Wyllie, Tectonophysics, 17, 189, 1973.
19. W.S. Fyfe and A.R. McBirney, Am. Jour. Sci., 275-A, 285, 1975.
20. W. Hamilton and W.B. Myers, Rev. Geophys., 4, 509, 1966.
21. A.R. Sanford, O.S. Alptekin, and T.R. Toppozada, Seismol. Soc. Amer. Bull., 63, 2021, 1973.
22. J.E. Oliver and S. Kaufman, Geotimes, 21 (7), 20, 1976.
23. R. Reilinger and J. Oliver, Geology, 4, 583, 1976.
24. P.W. Lipman and H.H. Mehnert, Geol. Soc. Amer. Mem., 144, 119, 1975.
25. J.H. Stewart, Geol. Soc. Amer. Abs. w. Progr., 7, 1284, 1975.
26. L.C. Pakiser and I. Zietz, Rev. Geophys., 3, 505, 1965.
27. R.F. Roy, E.R. Decker, D.D. Blackwell, and F. Birch, Jour. Geophys. Res., 73, 5207, 1968.
28. P. Molnar and J. Oliver, Jour. Geophys. Res., 74, 2648, 1964.
29. M. Barazangi and B. Isacks, Jour. Geophys. Res., 76, 8493, 1971.
30. J.W. Stinnett, Jr. and A.M. Stueber, Geol. Soc. Amer. Abs. w. Progr., 8, 636, 1976.
31. M. Bikerman, Geol. Soc. Amer. Abs. w. Progr., 8, 569, 1976.

MINERALOGY OF THE TULLU MOJE ACTIVE VOLCANIC AREA (ARUSSI: ETHIOPIAN RIFT VALLEY)*

H. Bizouard and G.M. Di Paola

Laboratoire de Pétrographie-Volcanologie, Bât. 504,
Univ. Paris Sud, 91405 Orsay, France.
Laboratorio di Geocronologia e Geochimica Isotopica,
via C. Maffi 36, C.N.R. Pisa, Italy.

ABSTRACT. Tullu Mojé is one of the most impressive active volcanic areas within the Ethiopian Rift Valley. Late Quaternary volcanic rocks, ranging in composition from mildly alkaline basalts to comendites have been erupted mostly through fissures in a very close time-space association. Major and trace elements geochemistry indicates the existence of two distinct (K_2O-rich and K_2O-poor) magmatic series. Microprobe analyses of the six most representative samples indicate that only the basalt, hawaiite and dark trachyte of the K_2O-rich series can have been derived from one another by crystal fractionation. Olivine is present in all the rocks of this series and shows a progressive increase in Fe with degree of fractionation. Clinopyroxene shows a gradual Fe enrichment and a rather important Ca decrease in dark trachyte. Plagioclase shows a continuous Na enrichment with advancing differentiation. Compositions of coexisting Fe-Ti oxide phenocrysts in hawaiite and dark trachyte indicate that the redox conditions of crystallization were controlled by the QFM buffer. Trachyrhyolite and comendites do not belong to the K_2O-rich series: in trachyrhyolite, orthopyroxene occurs instead of olivine and Fe-Ti oxides equilibrium indicates higher $\underline{f}O_2$ crystallization conditions than those controlled by the QFM buffer. Comendites are either olivine bearing or olivine free: the mineralogy indicates that crystal fractionation is not a process which could relate either the comendites to the trachyrhyolite or one comendite to another. The decrease with time of the K_2O and Rb contents of the basic magma is briefly discussed in terms of plate motions.

---

* This work has been supported by CNRS-France (ATP 1881) and by CNR-Italy (contract 76.00148.05).

E.-R. Neumann and I.B. Ramberg (eds.), Petrology and Geochemistry of Continental Rifts, 87–100.

1. INTRODUCTION

The Ethiopian Rift Valley, although representing one of the most impressive active continental rifts of the world, is still incompletely known. The aim of this paper is to make a contribution concerning the petrology of Quaternary volcanic rocks which are related to the most recent plate movements in this part of East Africa.

The Tullu Mojé (8°00'-8°30'N ; 39°00'-39°15'E) is one of the most representative areas in the Ethiopian Rift Valley, characterized by intense volcanism from middle Pleistocene to Present time, mostly through NNE-SSW trending fissure swarms affecting a plio-Quaternary "basement" of pantelleritic ignimbrites and Tertiary basic lavas [1-2].

The recent activity occurred in two distinct parts of the area (Fig. 1). In the eastern one, a wide variety of different magma types (almost exclusively lavas) were erupted in a very close time-space association. Rocks with completely different compositions have often been erupted from the same fissure in a geologically insignificant time span. This part of the Tullu Mojé area has an asymmetric position with respect to the morphological median axis of the rift: the huge number of recent normal faults and fissures (NNE-SSW) which affects it, indicates that this is the presently active axis of the structure.

The western part (Bora-Bericcio complex) is characterized by abundant pyroclastics and lavas with a monotonous pantelleritic composition and by the existence of some minor transverse (NW-SE) normal faults. These rocks are assumed to have a different genetic significance due to their different tectonic position in the rift, therefore they will not be considered in this paper.

2. GEOLOGICAL SUMMARY

Except for an embryonic volcanic centre (Tullu Mojé sensu strictu pumice cone), all the activity has developed through fissures, starting with eruptions of important volumes of megaplagioclase-rich fluid lavas of hawaiitic compositions, followed by slightly porphyritic glassy flows of trachyrhyolitic lavas. Very important fissure eruptions of mainly aphyric hawaiitic lavas and minor tephra occurred later both North and South of Tullu Mojé s.s. centre. This activity was followed by minor eruptions of dark trachytic lavas and mildly alkaline basalts associated with local extrusions of porphyritic obsidians and perlites of comenditic composition. One of the largest comenditic fissural domes of this area (Giano) was in activity only two centuries ago [2].

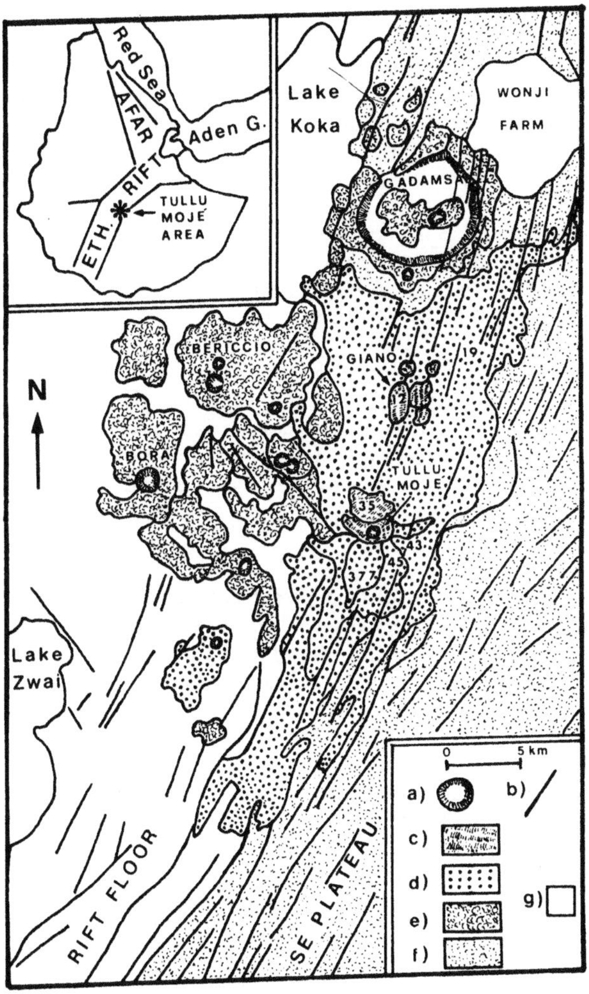

Fig. 1. Sketch map of Tullu Mojé area showing sample locations. a) Caldera and main craters; b) main faults; c) upper Pleistocene to historical comendites; d) Middle Pleistocene to Holocene basic, intermediate and silicic lavas; e) Late Pliocene to Holocene pantelleritic pyroclastics and lavas; f) Plio-Quaternary pantelleritic ignimbrites; g) Middle Pleistocene to Holocene lake sediments.

3. PETROCHEMICAL OUTLINES

In the Tullu Mojé area, basalts are subaphyric or porphyritic rocks with phenocrysts of (in order of appearance) olivine and plagioclase sometimes with augite, set in a microcrystalline groundmass containing the same minerals plus Fe-Ti oxides.

Hawaiites, apart from some plagioclase porphyritic flows, these are aphyric microcrystalline rocks with a mineralogy very similar to that of basalts except for a higher content of plagioclase. In some of these rocks, Fe-Ti oxides crystallized at an early stage.

Megaplagioclase rich hawaiites are characterized by the presence of abundant (up to 25%) plagioclase megacrysts (up to more than 2 cm) associated with minor olivine and augite, abundant Fe-Ti oxides and minor sulfide phenocrysts.

Dark trachytes are slightly porphyritic or aphyric rocks with phenocrysts of olivine, Fe-Ti oxides, minor sulfides and apatite, clinopyroxene and plagioclase sometimes form glomerophyric assemblages. Relics of orthopyroxene phenocrysts, showing large clinopyroxene rims, are sometimes present. The groundmass is fairly glassy with microlites of clinopyroxene, plagioclase, Fe-Ti oxides and minor olivine.

Trachyrhyolites are slightly porphyritic rocks with glomerophyric assemblages of Fe-Ti oxides, sulfides and apatite, minor orthopyroxene, clinopyroxene and plagioclase. The groundmass is abundantly glassy with very small microlites of plagioclase, clinopyroxene and Fe-Ti oxides. The majority of these rocks are related to the activity of Tullu Mojé s.s. centre.

Comendites are rather porphyritic rocks with phenocrysts of ferriferous olivine, Fe-Ti oxides, sulfides and apatite, yellow-green pleochroic clinopyroxene and anorthoclase sometimes showing an oligoclase core. The groundmass is abundantly or completely glassy with variable amounts of alkali feldspar, clinopyroxene, rare Fe-Ti oxides and zircon. In the biggest comenditic dome of the area, located just at the northern base of Tullu Mojé s.s. pumice cone, olivine is absent, while some relics of orthopyroxene phenocrysts with a clinopyroxene rim, are present.

The basalts and most of the aphyric hawaiites are ne-normative, while the megaplagioclase-rich hawaiites and dark trachytes are ol-hy normative or slightly qz normative. The early crystallization of Fe-Ti oxides can account for the modest initial iron enrichment observed in these rocks. The K_2O content of Tullu Mojé peralkaline rhyolites is very similar to that of Pruvost marginal centre in the Afar depression, therefore higher than the equivalent rocks of the Erta Ale axial range [3].

The volcanology and petrography as well as the chemical data on most of the major and on some minor elements (Fig. 2) suggest that the different Tullu Mojé products have a common genetic parentage and that they may be related through one single fractionation process. However, when the behaviour of Na_2O, K_2O and Rb (Fig. 2) is investigated, intermediate rocks appear to be subdivided into two distinct series, one K_2O- and Rb-rich and one K_2O- and Rb-poor. Such a dichotomy is not evident in the most basic types while it exists in some of the more silicic rocks (Fig. 2). It is not clear, however, looking at Fig. 2 whether trachyrhyolites and comendites belong to one of these two series or have an independent origin.

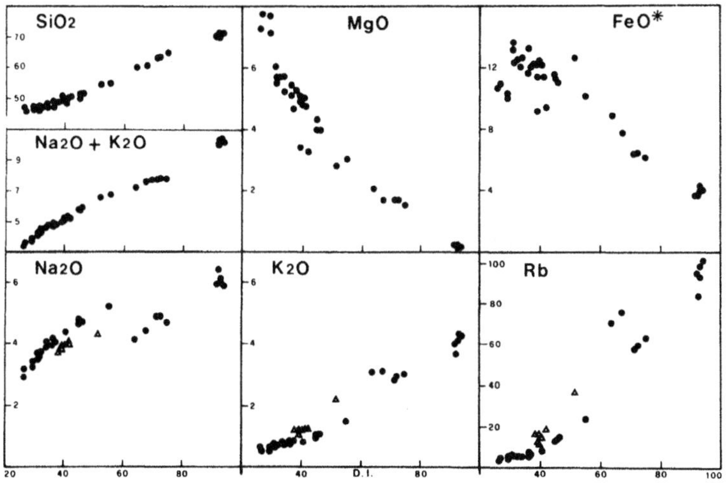

Fig. 2. Some major elements (Wt. %) and Rb (ppm) variation diagrams. D.I. modified for peralkaline rocks by adding Ac + ns.

4. MINERALOGY

The crystalline phases of three representative samples of the K_2O-rich series (basalt RV 19, megaplagioclase-rich hawaiite RV 43, dark trachyte RV 377), those of the most representative trachyrhyolites (orthopyroxene bearing RV 45) and those of two comendites (olivine bearing RV 12 and olivine free RV 35), were analysed by microprobe.

4.1 Olivine (Fig. 3)

This mineral is present from basalt to dark trachyte as the first crystallizing phase. A continuous chemical variation is observed from microphenocrysts (Fo 80) to microlites (Fo 72.5) in basalt, to phenocrysts (Fo 69-48) and microlites (Fo 50-33) in hawaiite and to Fo 45-30 in dark trachyte. MnO increases slightly with FeO (from 0.2 to 1.3%) and the CaO content (0.3%) is typical of volcanic olivine [4]. Olivine is absent in trachyrhyolite while it crystallizes again as Fe-hortonolite (Fo 20-15) in comendite RV 12. This mineral is again absent in the more evolved comendite RV 35.

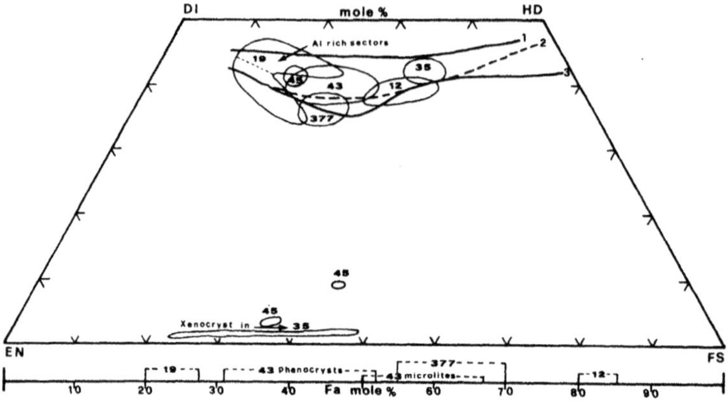

Fig. 3. Pyroxenes variation quadrilateral and olivine variation line. 1) Shiant Isles Sill; 2) Erta Ale; 3) Skaergaard.

4.2 Pyroxenes (Fig. 3)

Ca-clinopyroxenes are present in all the analysed rocks as phenocrysts and microlites. In basalt they crystallize after olivine and plagioclase. They have numerous zoning sectors chemically distinguishable by important differences in the Al content. The Al-rich sectors are enriched in Ti and Cr and slightly in Ca. Fe content increases from phenocrysts to microlites essentially with Ca depletion in low Al sectors and with Mg depletion in high Al sectors. In all sectors, Al decreases with advancing crystallization. The opposite trend is seen for Ti particularly in Al-rich sectors (Fig. 4).

Clinopyroxenes in hawaiite, dark trachyte and trachyrhyolite are richer in Fe than those in basalt. Clinopyroxene phenocrysts of all these three rocks have Fe/Fe + Mg ratios very similar to that of microlites in basalt. This suggests that there is no relation between the clinopyroxene composition and the degree of fractionation of these rocks.

Some significant differences among them can be pointed out (Fig. 3-4):

a) in hawaiite, microphenocryst compositions are very similar in major and minor elements to those of microlites in basalt; as already mentioned, Al continuously decreases from microphenocrysts to microlites; Fe increases and Mg decreases correspondingly with no change in Ca; the slight initial Ti enrichment is followed by a Ti decrease when Fe/Fe + Mg is higher than 0.42.

b) in dark trachyte, neither phenocrysts nor microlites are enriched in Fe relative to those of hawaiite; their Ca content is lower and Ti is drastically lower.

c) in trachyrhyolite, phenocrysts have similar Mg, Fe, Ca contents to those of hawaiite while TiO_2 is definetely lower. In this rock only a few microlites have been analysed because of their small size: they show a subcalcic tendency characteristic of quench pyroxenes [5-6] with a Ti content always higher than that of the phenocrysts.

Ca-clinopyroxenes evolve from En_{30} Fs_{32} Wo_{38} to En_{20} Fs_{40} Wo_{40} in comendite RV 12 and from En_{22} Fs_{35} Wo_{43} to En_{19} Fs_{40} Wo_{41} in comendite RV 35. In the pyroxenes of these two rocks, a Ca enrichment with Fe can be noticed; the Na content begins to increase leading to an Ac content of 5% in the most Na-rich pyroxenes.

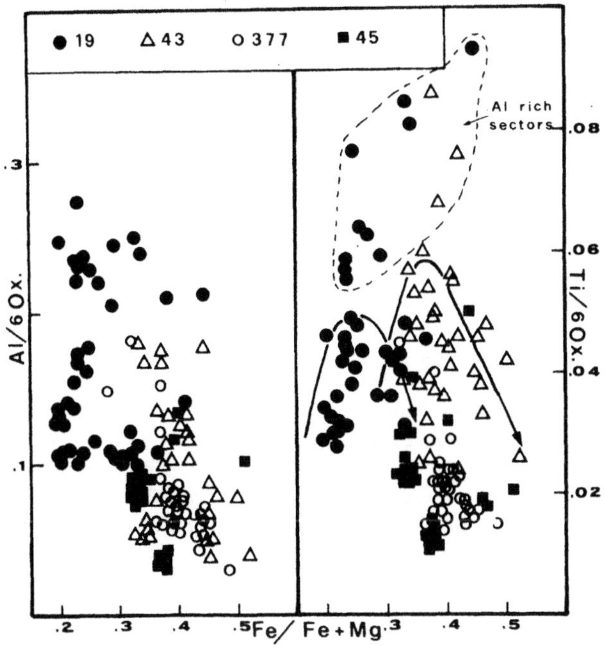

Fig. 4. Al and Ti variations as a function of Fe/Fe + Mg in clinopyroxenes.

Pigeonite (En_{49} Fs_{42} Wo_9) has been found only in trachyrhyolite. It is never abundant and it is a late crystallizing phase together with Ca-clinopyroxene microlites.

Orthopyroxene (En_{62} Fs_{35} Wo_3) is present as a stable phase only in trachyrhyolite. Ca-clinopyroxene, orthopyroxene and pigeonite show the expected tie-lines for such a range of compositions. The trend is similar to that of Skaergaard [7], except for the fact that the field of orthopyroxene is more expanded at Tullu Mojé than at Skaergaard. The compositions of ortho- and clinopyroxene of RV 45 indicate that they are in equilibrium. In dark trachyte, as already mentioned, rare orthopyroxenes with large clinopyroxene rims occur; the orthopyroxene xenocryst analysed in comendite RV 35 has a very large range of composition (from En_{73} $Fs_{25.5}$ $Wo_{1.5}$ to En_{51} $Fs_{47.5}$ $Wo_{1.5}$) and it is clearly not in equilibrium with the Ca-clinopyroxene of this rock. Although orthopyroxene occurs as a stable phase only in trachyrhyolite, the occurrence of orthopyroxene xenocrysts in dark trachyte and in comendite indicates that this mineral crystallized, under intratelluric conditions, from different liquids, as witnessed by the Mg-rich and Ca-poor orthopyroxene xenocryst of comendite RV 35.

4.3 Feldspars (Fig. 5)

Plagioclase crystallized from basalt to olivine bearing comendite with a continuous compositional change from calcic to sodic members; the same trend is observed within each single rock (Fig. 5a). In fact, in basalt, plagioclase composition ranges from An_{86} to An_{37}, in hawaiite from An_{61} to An_{12}, in dark trachyte from An_{49} to An_{27} and in trachyrhyolite from An_{46} to An_{41}. In this last rock, plagioclase has been analysed only as phenocrysts because of the small size of the microlites. In comendite RV 12, oligoclase (An_{18-11})crystallized with a continuous evolution towards alkali feldspar composition (Ab_{74} Or_{23} An_3) while in the most peralkaline comendite (RV 35), anorthoclase (Ab_{74} Or_{24} An_2) is the only salic phase present, apart from apatite.

The plagioclase always has a low K_2O content and as it becomes more sodic two trends can be observed: a K_2O-rich one from basalt to dark trachyte and a relatively K_2O-poor one in comendite RV 12 (Fig. 5b). Plagioclase phenocrysts of trachyrhyolite seem to be related to the low K_2O trend. Over the whole range of rocks the Fe content of plagioclase tends to decrease with increasing Na content, but in basalt and hawaiite the Fe content clearly increased during crystallization, that is to say from phenocrysts to microlites. No clear trend can be established in the remaining rocks (Fig. 6).

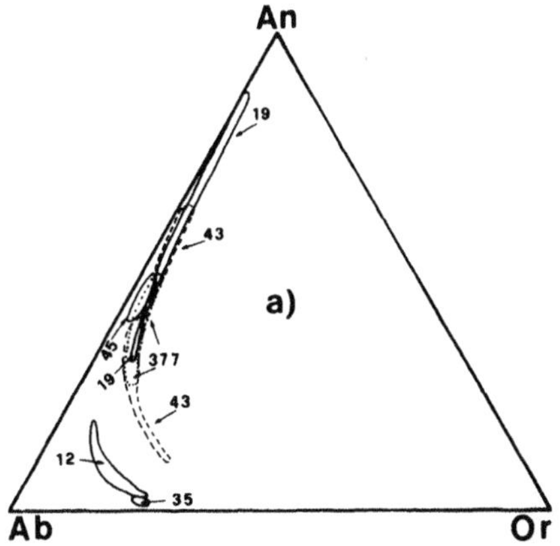

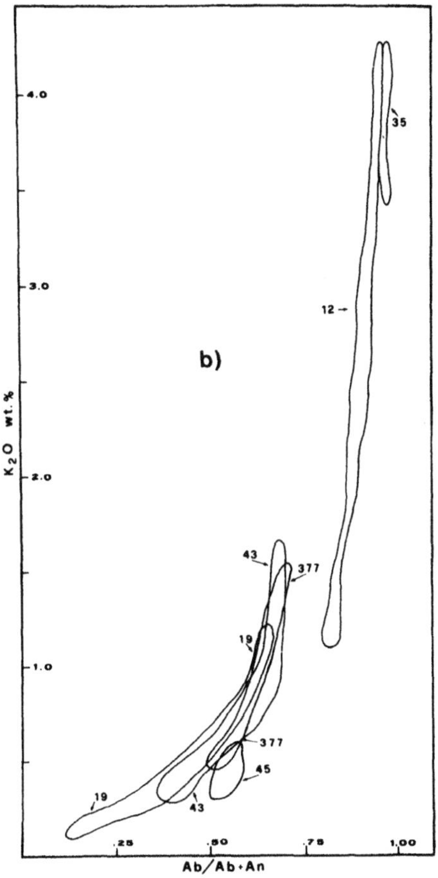

Fig. 5. Feldspar variations illustrated in: a) Ab, An, Or diagram; b) K_2O versus Ab/Ab + An diagram.

4.4 Oxides

The earliest Fe-Ti oxide microphenocrysts occur in megaplagioclase rich hawaiite. Ilmenite crystallized first and its relative abundance decreases in the more evolved rocks. Temperature and $\underline{f}O_2$ decreased continuously during crystallization [8] from dark trachyte (1090 ^{o}C - $10^{-10.2}$) to trachyrhyolite (990 ^{o}C - $10^{-10.7}$; 950 ^{o}C - $10^{-11.5}$) down to comendite RV 12 (870 ^{o}C - $10^{-13.9}$). This indicates a QFM buffer control on crystallization for dark trachyte and comendite where olivine is present. On the contrary, orthopyroxene bearing trachyrhyolite seems to have crystallized at a higher $\underline{f}O_2$ (Fig. 7) close to the "Opx" buffer described by Carmichael [9].

It is well known that during fractionation, Mg and Al decrease both in ilmenite and in magnetite, while Mn increases as a function of the more differentiated character of the liquid. Fe-Ti oxides of

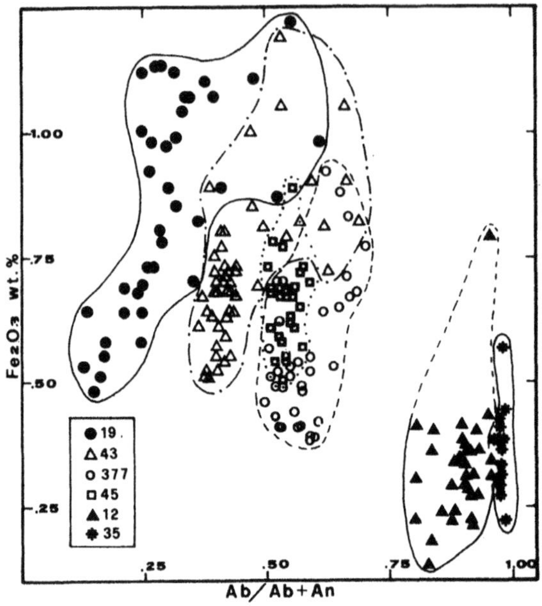

Fig. 6. Fe_2O_3 varation in feldspars as a function of Ab/Ab+An.

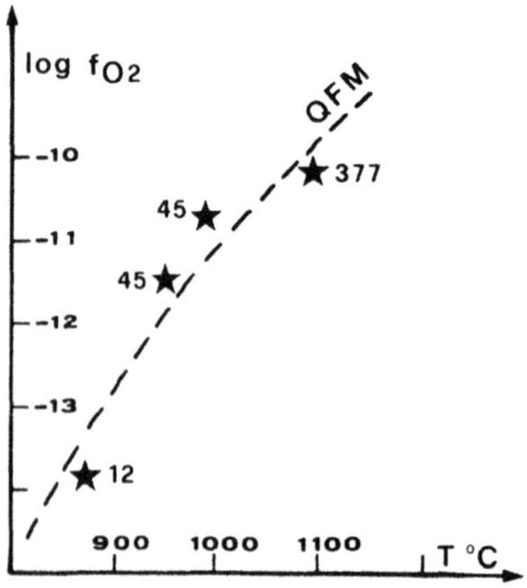

Fig. 7. Log $\underline{f}O_2$ - T °C variation diagram [8].

trachyrhyolite do not show such a behaviour; in fact in this rock, these minerals are enriched in Mg and Al and depleted in Mn, relative to the oxides of the less evolved dark trachyte.

5. DISCUSSION

At Tullu Mojé two distinct magmatic series have been recognized:

a K_2O-rich one in which intermediate types are ol-hy normative, and a K_2O-poor one in which the equivalent types are ne normative. The K_2O-rich series is also characterized by a significantly less important iron enrichment than that of the K_2O-poor one. Both series appear to have an alkaline character intermediate between the alkaline series of Assab [10] and the transitional series of Erta Ale [3]. As already mentioned, complete fractionation trends relating basalts to peralkaline rhyolites are not evident at Tullu Mojé. However, on the basis of the present data, it can be affirmed that some of these rocks are related by crystal fractionation. Samples RV 19, RV 43 and RV 377, which belong to the K_2O-rich series, have petrochemical and mineralogical characteristics which clearly indicate that they are related by crystal fractionation. On the contrary, samples RV 45, RV 12 and RV 35 do not belong to the K_2O-rich series and, at the same time, they cannot be definitely attributed to the K_2O-poor series even if plagioclase and alkali feldspar in these rocks have relatively low K_2O contents. Other evidence discussed below prevents us from considering all these rocks as pertaining to an unique fractionation series.

In RV 19, RV 43 and RV 377, all minerals show a regular evolution from more primitive to more differentiated compositions, reflecting the evolution of the liquids. In basalt, as crystallization proceeds, the liquid is enriched in Ti and Fe and depleted in Cr: this is evident in clinopyroxenes where Cr drops and Ti increases. Al-rich sectors of these clinopyroxenes show a trend similar to that of the alkali suite of the Shiant Isles Sill [11] while Al-poor sectors become slightly depleted in Ca. In hawaiite no drastic changes occurred when the first minerals crystallized, phenocrysts of plagioclase are slightly enriched in Fe as a consequence of the Fe increase in the liquid and clinopyroxene cores are Ti-rich, but the relatively early appearance of Fe-Ti oxides caused a rapid decrease of Ti and Fe in the residual liquid, and the more evolved clinopyroxene microphenocrysts and microlites are Ti-poor. This correlation has been pointed out by Gibb [11] who also found a direct correlation between Al and Ti, not found at Tullu Mojé where Al is always decreasing. This difference is due to the more undersaturated character of the Shiant Isles Sill suite allowing replacement of Si by Al in tetrahedral sites [12-13]. The Wo contents in clinopyroxenes of Tullu Mojé basalt and hawaiite are difficult to compare because these minerals are sector zoned, but a slight decrease of this component seems to occur from basalt to hawaiite. However, the Wo-content in clinopyroxenes of this series is always higher than those of augites of Erta Ale and Boina [6].

In dark trachyte, Ti abruptly decreases because of the extensive crystallization of Fe-Ti oxides at the hawaiite stage while the relative abundance of Fe does not increase (constant Fe/Fe + Mg ratio). Plagioclases are Fe-poor and clinopyroxenes very Ti-poor.

Clinopyroxenes are only slightly enriched in Fe while coexisting olivines become more and more iron rich. This is reflected in an inversion of the slope of the tie-lines connecting coexisting clinopyroxenes and olivines.

Such clinopyroxene-olivine associations in basalt and hawaiite on the one hand, and in dark trachyte on the other, are very similar to those described by Carmichael [14] at Thingmuli for olivine and clinopyroxene in basic rocks, and clinopyroxene and pigeonite in basaltic andesite, suggesting equilibrium crystallization conditions. A certain analogy between the mineral trends of the Tullu Mojé and the Thingmuli series can be found in the fact that, after the crystallization of oxides, the Fe/Fe + Mg ratio in the rock increases only slightly and clinopyroxenes of evolved tholeiites and basaltic andesites have similar Fs contents. Futhermore, immediately after the first crystallization of oxide phenocrysts clinopyroxenes show a small Ca increase followed by a drop of this element close to the minimum of the subcalcic trend. Similar trends of clinopyroxene phenocrysts can be observed in Tullu Mojé and Thingmuli, despite the fact that the latter has a clear tholeiitic affinity. However, it remains unexplained why clinopyroxene microlites of hawaiite do not display the same subcalcic tendency as clinopyroxene phenocrysts of dark trachyte. Anyway, the Tullu Mojé clinopyroxene trend is more calcic than those of Thingmuli [14], Erta Ale and Boina [6] and less calcic than that of Shiant Isles Sill [11] confirming that the alkaline affinity of Tullu Mojé is intermediate between the Erta Ale and Shiant Isles Sill series. The presence of orthopyroxene in trachyrhyolite is consistent with the low Fe/Fe + Mg ratio of this rock. Trachyrhyolite RV 45 could be considered as the residual liquid of dark trachyte RV 377, despite the presence of orthopyroxene instead of olivine. In fact, the presence of orthopyroxene instead of olivine can be explained because the Fe/Fe + Mg ratio of RV 45 is similar to that of RV 377 and at the same time, RV 45 has a higher Qz normative content than RV 377. But, since the amount of oxides crystallized in RV 377 is small, and since all the minerals (plagioclase, pyroxenes, Fe-Ti oxides) of trachyrhyolite RV 45 (a rock with SiO_2 = 63%) have a more primitive composition (*) than those of dark trachyte RV 377 (a rock with SiO = 54%), it is impossible to consider trachyrhyolite RV 45 as derivate by crystal fractionation from dark trachyte RV 377.

Both comendites show a markedly evolved character (high Fs and Wo content in clinopyroxene, low Fe in feldspars). As already mentioned, among Tullu Mojé comendites, RV 35 is the only sample in which olivine is absent, it therefore seems unlikely that this rock originated directly by crystal fractionation from RV 12 or from other more evolved olivine bearing comendites of the area. On the other hand, the anorthoclase of RV 35 is less calcic than that of RV 12 while the clinopyroxene phenocrysts of the former are more Ca-rich than the clinopyroxene microlites of the latter, suggesting a crystal fractionation process. In fact, once the peralkaline field is reached the "plagioclase effect" does not act anymore because the calcium of the liquid is preferentially incorporated in the clinopyroxene [18].

On the basis of the present mineralogical data, the possibility that every single silicic rock of the Tullu Mojé had its own independent origin cannot be excluded. In fact, the gaps existing between the compositions of all the minerals of trachyrhyolite and those of the minerals of both comendites are too large to be explained in terms of equilibrium crystal fractionation in a closed system.

6. CONCLUSIONS

The above discussion shows that the present data are insufficient to give a satisfactory explanation of the Tullu Mojé volcanism. In fact, when this is investigated in relations to the structural significance of the Ethiopian Rift Valley, important problems arise.

1) The absence of olivine in trachyrhyolite and in one comendite as well as their different physico-chemical conditions of crystallization discussed above, could be related to a particular structural environment such as the embryonic Tullu Mojé s.s. centre to which they belong. However, if this assumption is true, why does dark trachyte, which also pertains to the activity of the same centre, show important petrochemical and mineralogical differences?

2) Geological features accounting for the existence of two magmatic series are not evident because of the very short time span between the first and the last volcanic events in the Tullu Mojé area. The only argument seems to be the evidence that the majority of the K_2O-poor ne-normative intermediate lavas are stratigraphically above the equivalent K_2O-rich ol-hy normative products. Does this decrease with time of K_2O and Rb contents in basic magma indicate a tendency towards a more "transitional" character of the magma in response to an acceleration of the extensional process of the Rift as the result of an increase of the Somalian plate motion?

3) Baker [15] emphasized that, except for Socorro and Gran Canaria Islands, where abundant pantellerites occur, peralkaline rhyolites of oceanic islands have a comenditic composition. According to him the different compositions of peralkaline rhyolites seem to be mostly conditioned by the structural environment in which they were generated. Tullu Mojé comendites have been generated within a continental structure, as have the abundant coeval pantellerites already mentioned from the western part of this area. In fact,

geophysical data [16] indicate a Moho depth of about 35 km in this part of the Ethiopian Rift.

This work therefore confirms that both types of peralkaline rhyolites can coexist either in oceanic or in continental environments. Their different compositions probably reflect slight differences in the nature of their respective parent magmas, as pointed out by Barberi et al. [17]. These authors observed that oversaturated peralkaline final liquids generated by crystal fractionation of a transitional basaltic magma, can be either pantellerites or comendites according to the alkali content of the primary basalt. Anyway, as discussed above, it is not yet clear if Tullu Mojé comendites, and even more pantellerites, are generated by crystal fractionation or if they have an independent origin. Further studies are therefore needed to solve the numerous and fascinating problems related to the Ethiopian Rift Valley.

REFERENCES

1. G.M. Di Paola, Bull. Volc., 36, 517, 1972.
2. G.M. Di Paola, Geological map, 1/75 000, of the Tullu Mojé volcanic area (Arussi: Ethiopian Rift Valley), 1976, CNR Pisa.
3. F. Barberi, S. Borsi, G. Ferrara, G. Marinelli & J. Varet, Phil. Trans. Roy. Soc. Lond. A. 267, 293, 1970.
4. J.C. Stormer, Geoch. Cosm. Acta, 37, 1815, 1973.
5. D. Smith and D.H. Lindsley, Am. Min. 56, 225, 1971.
6. H. Bizouard, F. Barberi and J. Varet, Mineralogy of the Erta Ale and Boina series (Afar Rift, Ethiopia), submitted to Journ. Petr., 1977.
7. G.M. Brown, Min. Mag., 31, 511, 1957.
8. A.F. Buddington and D.H. Lindsley, Jour. Petr., 5, 310, 1964.
9. I.S.E. Carmichael, Contr. Min. Petr., 14, 36, 1967.
10. M. De Fino, L. La Volpe and L. Lirer, Bull. Volc., 37, 95, 1973.
11. F.G.F. Gibb, Jour. Petr., 14, 203, 1973.
12. M.J. LeBas, Am. J. Sci., 260, 267, 1962.
13. I. Kushiro, Am. J. Sci., 258, 548, 1960.
14. I.S.E. Carmichael, Am. Min., 52, 1815, 1967.
15. P.E. Baker, Bull. Volc. Spec. Issue, 38, 737, 1974.
16. J. Makris, H. Menzel, J. Zimmerman and P. Gouin, Gravity field and crustal structure of North Ethiopia, in: Afar Depression of Ethiopia ed. by A. Pilger and R. Røsler, V.1, Stuttgart, 1974.
17. F. Barberi, G. Ferrara, R. Santacroce, M. Treuil & J. Varet, Journ. Petr., 16, 22, 1975.
18. I.S.E. Carmichael and W.S. MacKenzie, Am. J. Sci., 261, 382, 1963.

THE VOLCANOLOGICAL DEVELOPMENT OF THE KENYA RIFT

L. A. J. Williams

Department of Environmental Sciences,
University of Lancaster, England.

ABSTRACT. Cainozoic volcanics are exposed along the entire length (750 km) of the sinuous but generally north-south Kenya rift. At various stages in its evolution, the volcanics filled and spread beyond the rift depression to coalesce with the products of eruptions at or far beyond its margins. As a result, Miocene to Recent lavas and pyroclastics have come to be distributed in a meridional belt up to 550 km wide. The volcanics embrace the full compositional range from basalts and nephelinites to trachytes, phonolites and alkali rhyolites. They occupy the graben to depths exceeding 3 km, and accumulations up to 2.5 km thick on the rift shoulders account for most of the highland areas near the equator. The extent to which pre-existing topography controlled the spreads of lavas and ignimbrites is frequently underestimated.

1. INTRODUCTION

This account sets out to summarise the sequence of events in a long and complex history of volcanism in an area of some 375,000 km^2. In many respects, the task is made more difficult because it deals with the best-studied section of the East African rift system. Generalisations and simplified diagrams have become incresingly inadequate in attempts to convey an accurate impression of the history of volcanism in this region. Moreover, it will be appreciated that discussion of volcanological aspects of the evolution of the Kenya rift is very largely concerned with the infilling and periodic overtopping of depressions resulting from faulting and warping, so that the commentary will be meaningless without frequent references to major structural features and events. One

E.-R. Neumann and I.B. Ramberg (eds.), Petrology and Geochemistry of Continental Rifts, 101–121.

of the chief aims of the paper is to demonstrate the importance of topography in controlling spreads of Cainozoic volcanics at various stages in the evolution of the rift valley and wide tracts of flanking ground. For this reason, some aspects of the pre-volcanic history of the region merit brief consideration if only because of their relevance in establishing the form of the Miocene landscape at the time of eruption of the earliest lavas.

Little new information is presented in the paper: indeed, it is not intended to be more than an up-dated and highly selective review based on numerous accounts [1-9] and on observations made over a period of about twenty years. Much of the earlier work in the region was carried out by the Geological Survey of Kenya, and the results are contained in a large number of published reports. The most significant results, however, have come from systematic detailed mapping along the rift by members of the East African Geological Research Unit under the direction of Professor B.C. King. The writer has had the benefit of close association with this project since its inception in 1965, and many of the views expressed in this account stem from knowledge of a great deal of unpublished detail.

In this brief summary of the volcanic and structural evolution of a large part of Kenya, it will be impossible to quote the source or indicate the reliability of much of the data. In general, Survey reconnaissance mapping of areas outside the rift has yet to be superseded by more detailed studies.

2. TOPOGRAPHY AND MAJOR RIFT STRUCTURES

The central section of the Kenya rift is readily identifiable on the simplest of topographic maps (Fig. 1) because it traverses an impressive highland region in which mountains and ranges of hills rise to altitudes of 3000 to 4000 m. The rift is a sinuous but generally north-south feature with the highest part of the floor standing at some 2000 m near the equator. In the north, the floor of the most recently formed NNE-trending narrow trough (Suguta valley) descends to less than 300 m near Lake Turkana (formerly Lake Rudolf), but earlier structures splay out over a wide area so that the connection with the Ethiopian rift is topographically indistinct. At the Tanzanian border in the south, the rift floor stands at 600 m. The trough is well defined here but, across the border, the rift soon loses its identity in a much wider region of block tilting.

It will be apparent from a comparison of Figs. 1 and 2, and from the sections (Fig. 3), that most of the highland areas in Kenya and eastern Uganda consist of Cainozoic volcanics. Some accumulations represent exceptionally large central volcanoes, of which Mt. Kenya (5200 m; Fig. 3, section D) and Mt. Elgon (4300 m; section F) are striking examples. Farther south in Tanzania is the massive triple-centre complex of Kilimanjaro (5900 m; Fig. 1) from which flows extend across the border into Kenya. Other central volcanic complexes are situated closer to the margins of the rift: examples are to be found in the Aberdare range (4000 m; section C) and Timboroa (2900 m; section D). Some important topographic features such as the Mau range (3100 m; section C), the Uasin Gishu plateau (sloping westwards from 2700 m to 1800 m; section E), Bahati (up to 2900 m; section D) and the Kinangop (2500-2700 m; section C) are areas of flood lavas, ash flows and air-fall pyroclastics for which sources are less easily recognisable.

In few areas in the Kenya rift is there any approximation to a simple graben structure. Addition of little more topographic detail to that presented in Fig. 1 would serve to illustrate structural complexity within even those regions in which the rift valley has been seen to be best defined. Among the more obvious features are inner troughs some 25-25 km wide, lying within a broader depression 60-70 km across (Fig. 4). These inner structures are not situated axially with respect to the main rift. They tend to lie closer to the margin determined more by downwarps than single major faults, and to be separated from spectacular boundary fault scarps by a step-fault platform (south of the equator; Fig. 3, section D) or an uparched horst block (equator to 1°N; sections E and F). Single major faults assume less importance in determining either margin of the rift north of 1°N, but the Suguta trough runs close to a severely downwarped flank. Near Lake Turkana, the Suguta is bounded to the west by a fault scarp marking the edge of the Loriu platform (section H), but farther south tilting and warping are more characteristic features of its western margin. Only near Magadi, at the southern end of the Kenya rift, does the most recently active trough run more or less centrally along the main graben (Fig. 4).

Near the equator, a WSW-trending trough (the Kavirondo rift) some 20-30 km wide branches from the main rift towards Lake Victoria. It differs from the main (Gregory) rift valley in showing no uplift of shoulders [5,10] and no Quaternary volcanism along the floor.

The altitude and form of the sub-volcanic surface are of obvious importance, not only in assessing the thickness of the volcanics and displacements along major faults, but also in understanding the roles of episodes of downwarping and uplift through the late Mesozoic and Cainozoic history of the region. Few boreholes have been deep enough to penetrate the Precambrian basement underlying the volcanics, and exposure of basement rocks in scarps at the rift margins tends to be sporadic: indeed, it is exceptional along the eastern margin. Consequently, reconstruction of the

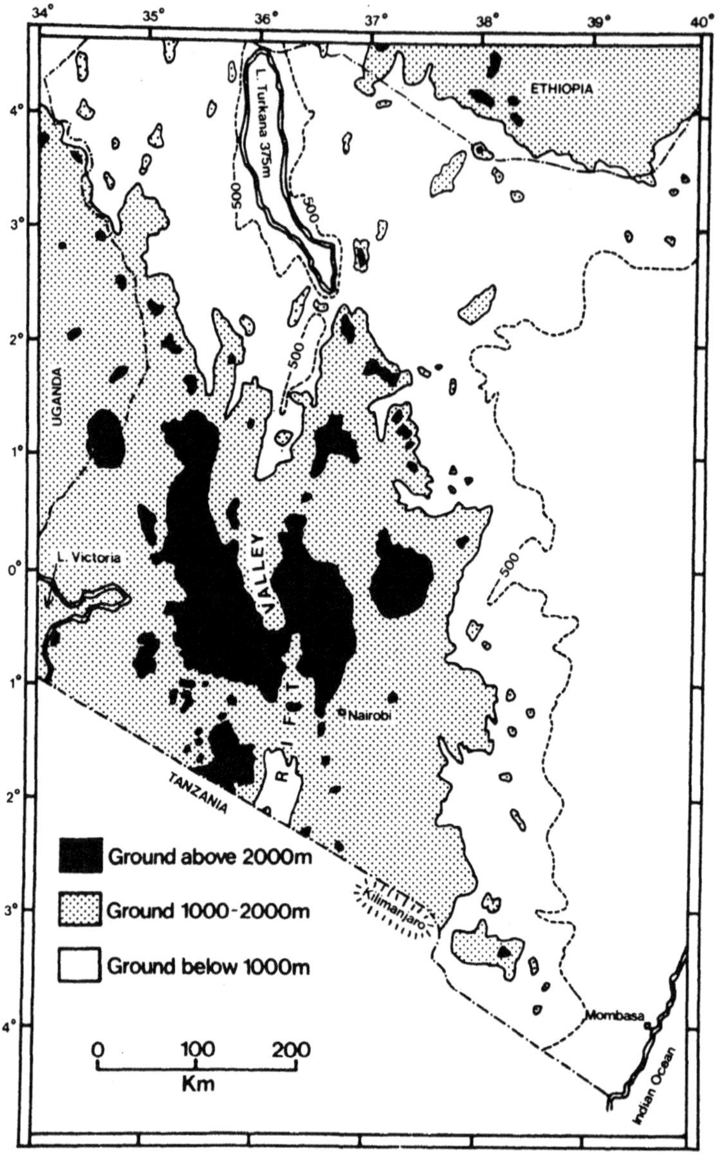

Fig. 1 Present topography in Kenya, eastern Uganda and southern Ethiopia.

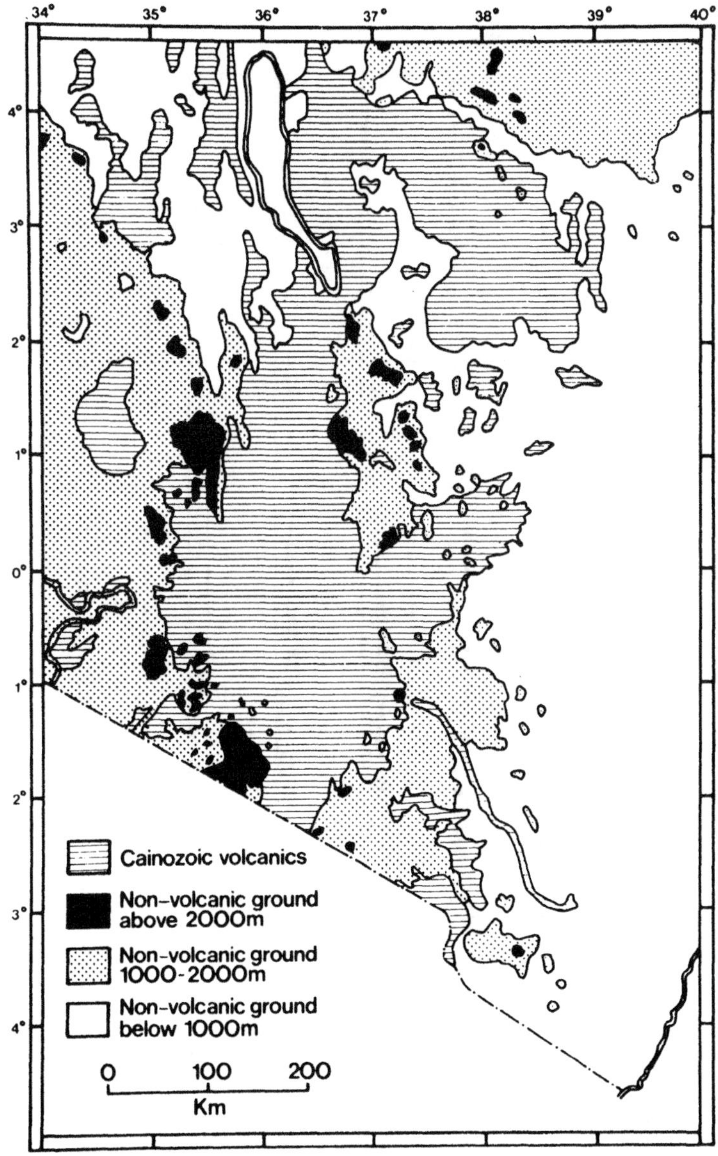

Fig. 2 Non-volcanic topographic features in the region covered by Fig. 1.

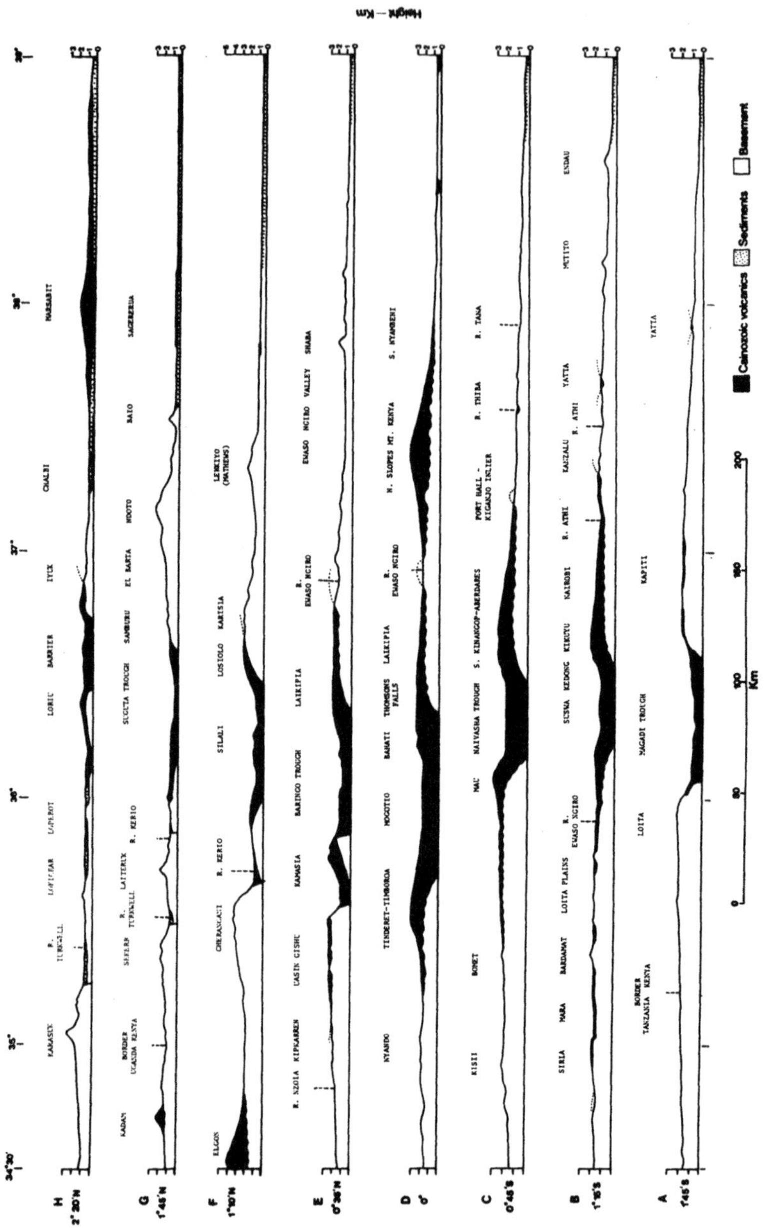

Fig. 3 Sections across the Kenya rift. Lines are indicated in Fig. 4.

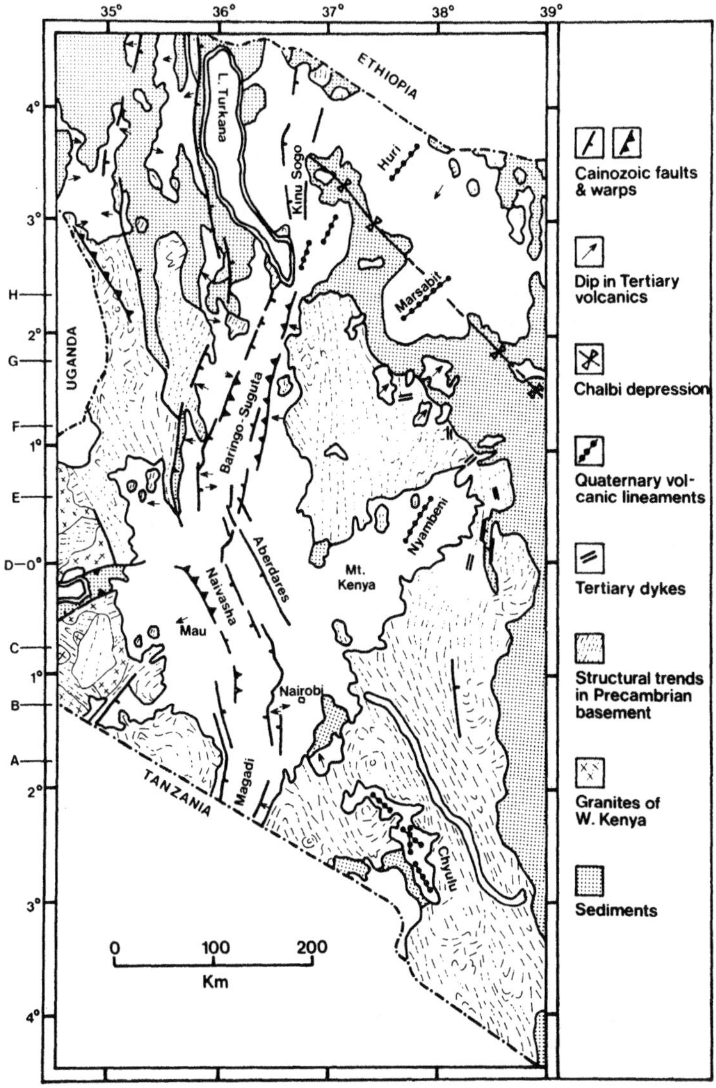

Fig. 4 Structural map of the Kenya rift and flanking areas.

sub-volcanic surface depends heavily on observations along the edges of the volcanic cover, and an appreciation of the ages and significance of erosion surfaces and sediments.

Early investigations revealed widely scattered occurrences of sediments, locally fossiliferous and dated as Miocene, beneath volcanics which rest directly on the basement over large areas. At some localities, Cainozoic volcanics and/or sediments rest on what are evidently dissected remnants of a planar basement surface, but in other cases reconstructions based merely on observations around the margins of lava sheets gave misleading impressions of regularity of the buried basement topography. Misconceptions arose particularly in regions where extremely fluid phonolites or basalts infilled, and were ultimately confined by, rugged terrain. The nature of the pre-existing topography can sometimes be judged by inliers representing remnants of basement hills, either projecting above the volcanics or planed down to the level of the lava surface. Some of these inliers are shown diagrammatically in Fig. 4, but many are not indicated. Not uncommonly and without justification, the sub-volcanic surface between the inliers is assumed to have a planar character.

In some instances, the role of differential erosion was not fully appreciated so that the possibility of original confinement of the volcanics by some of the less resistant basement formations was overlooked or underestimated simply because schists and gneisses now stand at or below the base of the volcanics. The consequences of incorrect interpretation of this situation are most serious in those areas in which the reduced basement surface is now approximately planar, because of the increased liklihood of misidentification of erosion bevels.

These comments assume particular relevance in considering what became known in Kenya as the sub-Miocene peneplain or erosion bevel [10-12]. The recognition and dating of this surface depend to a large extent on tracing it into areas where remnants are preserved beneath volcanics of known age; yet, as explained above, this is the very situation in which it becomes hazardous to identify planar elements in the submerged landscape. It is sometimes forgotten that Pulfrey [11] was well aware of difficulties which arise because the volcanics frequently bury a landscape of considerable relief. Indeed, he recognised that very extensive areas along the flanks of the present rift valley must have stood as hills and mountains above the proposed bevel. It was clear that any planar sub-volcanic surface must have been extremely limited within about 150 km of the rift, and that it would be virtually impossible to trace the bevel into the rift because of complications arising from faulting and general lack of exposure of basement rocks.

Understandably, more detailed work has led to revision of earlier notions concerning the geomorphological evolution of the rift zone, one suggestion being that no widespread planar mid-Tertiary surface can be recognised beyond a region in eastern Kenya [4]: and, as later comments will show, this includes a large and critical area in which the history of volcanism and sedimentation remains poorly documented.

4.1 THE KENYA DOME

On the assumption that originally planar erosion surfaces (described as end-Cretaceous, sub-Miocene and end-Tertiary) had seaward slopes of about one foot per mile before deformation, it was concluded [8,12] that late Mesozoic uplift of at least 400 m in central Kenya was followed by a further 300 m in the Miocene, and some 1000-1400 m in the late Tertiary or Pleistocene. From these claims, and the earlier work of Pulfrey [11], arose the notion of a 'Kenya dome' (meaning a region of domal uplift) separated from a similar feature in Ethiopia by the Chalibi-Habaswein depression (Fig. 4). Reconsideration of all the evidence now leaves little doubt that estimates of the magnitude and timing of uplift require revision.

Recent work has concentrated on tracing two lateritised surfaces (late Mesozoic and 'end-Tertiary') across the rift zone, and on detailed analyses of the evolution of drainage systems. These approaches have been extremely rewarding. King [5], for instance, concludes that the Kenya rift developed along a region of earlier uplift which determined the position of the African continental watershed after the break-up of Gondwanaland, and shows that gentle arching of this watershed region occurred in early Tertiary times before the first rift depressions were formed by downwarping in northern Kenya. Thus, whilst recognising the importance of late uplift of the rift shoulders, he maintains that much of the present elevation of the basement (1400-2000 m) along the margins of the trough has been inherited from earlier, essentially pre-rift, uparching.

Very thick accumulations of volcanics on the basement account for a great deal of the highlands flanking the central part of the rift (Figs. 1 and 2). Because of differences of opinion about the timing of the uplift and in order to emphasise the constructional role of volcanism, in recent years the term 'Kenya dome' has come to refer more to a topographic than a structural feature. It is of interest to note in passing that the term 'domal uplift' is no longer considered appropriate to describe complex block uplifts in the Horn of Africa because no simple culmination exists on the basement surface [13].

The reader not familiar with the history of research in Kenya will perhaps feel that many of these comments on topography and erosion surfaces have doubtful relevance in a discussion of the volcanological development of the rift. If any justification is required for their inclusion here, it is only necessary to examine recent accounts, even those by experienced workers, to find that considerable confusion exists about the nature and significance of the 'Kenya dome'; and indeed about some aspects of the present topography. For instance, in one paper concerned with the interpretation of gravity data [14] contours illustrating the form of a sub-Miocene erosion bevel are taken to represent the topography as it would be after removal of all Miocene and later deposits: as a result, present high-standing areas of basement rocks go unrecognised. Fundamental misconceptions of this kind have prompted the writer to indicate in diagrams (Figs. 5-7), not only the distribution of the volcanics at various stages in the evolution of the rift, but also to show in a very generalised fashion areas of 'higher ground'. It will be appreciated that the elevations of these highland regions cannot be specified with any precision in diagrams referring to Tertiary events, but it is hoped that this approach will convey some impression of the palaeogeography and serve to illustrate the long-standing influence of topographic barriers during the history of volcanic activity.

4. HISTORY OF VOLCANISM

The sucessions exposed in rift boundary scarps and across the flanking ground obviously represent either phases when the volcanics filled and overtopped a developing trough, or activity at eruptive centers located on the rift shoulders. In the former case, there may be some expectation of correlation right across the rift regardless of the nature or thickness of later infill resulting from concentration of tectonic and volcanic activity into inner troughs, but this presupposes that very fluid flows were erupted in sufficient volume and that no constraints existed on flow across the axial region of the rift. It has proved surprisingly difficult to make convincing detailed lithological correlations of this kind across the rift [5]. In the latter case, opposite sides of the rift can be expected to display evidence of volcanism of quite different character, despite broad contemporaneity as indicated by isotopic age determinations.

In the absence of age determinations, or supported by sporadic dates, early attempts were made to subdivide the volcanics on the basis of tectonic events, or to relate episodes of volcanism, domal uplift, faulting and downwarping [8,9,15]. These efforts tended to be based on the notion that a few widespread phases of rift faulting could be confidently identified. Recent reviews [5,16] have criticised, however, the frequently mistaken assumption that it was volcanism rather than tectonic activity which was nearly continuous during the evolution of the rift, and they make the point that this goes a long way to explaining why some apparently convincing examples of neat subdivision into volcanic and tectonic events can be misleading in areas where long periods are simply not represented by any volcanic activity. Moreover, additional problems arise in regions which have been subjected to tilting and warping, for the precise structural history is much more difficult to unravel than even the complex sequence of events associated with rejuvenation of faults.

Nevertheless, detailed work has shown that phases of more intense faulting can be distinguished in records of prolonged tectonic activity in widely separated areas. Along the length of the rift, some of the most significant movements can be assigned to one of two periods: an earlier at about 7 m.y., and a more recent episode around 1.5 to 2 m.y. which largely determined the present morphology [7]. The floor of the rift was disrupted by numerous closely-spaced so-called 'grid' faults at about 0.5 m.y. Throws on individual 'grid' faults are trivial, but locally the movements were accompanied by rejuvenation of boundary faults during late depression of the floor and/or rise of the shoulders of the rift.

In several areas on the western side of the rift, the episode of major faulting which occurred around 7 m.y. ago coincided with a change from essentially basaltic and phonolitic to basaltic and trachytic volcanism [6,17,18,19]. Although still largely unexplained but possibly associated with the rapid evacuation of large magma chambers [8,20], this example of a relationship between volcanic and tectonic activity continues to provide the most satisfactory basis for initial subdivision of the volcanic events in the region. Another approach involves consideration of the timing of periods of maximum overspill of volcanics from the developing trough. Two phases are recognised: the first lead to extensive overtopping of the rift depression by phonolites around 12-15 m.y., and the second by trachytic flows about 2-6 m.y. ago.

In the commentary that follows, the early stage (12-30 m.y.) of volcanism is referred to as Miocene, that spanning the period 2-12 m.y. as Pliocene, and all activity within the last 2 m.y. as Quaternary. No apology is offered for repetition of many comments provided in a recent review [7] in which it was not possible to incorporate diagrams showing the distribution of volcanics.

3.1 Miocene volcanism (Fig. 5)

The oldest volcanics (15-30 m.y.) are basalts which are exposed from the equator through north-western Kenya, and representatives of nephelinite-phonolite-carbonatite suites forming giant cones and intrusive complexes [21] in the vicinity of the Kenya-Uganda border (Fig. 5a).

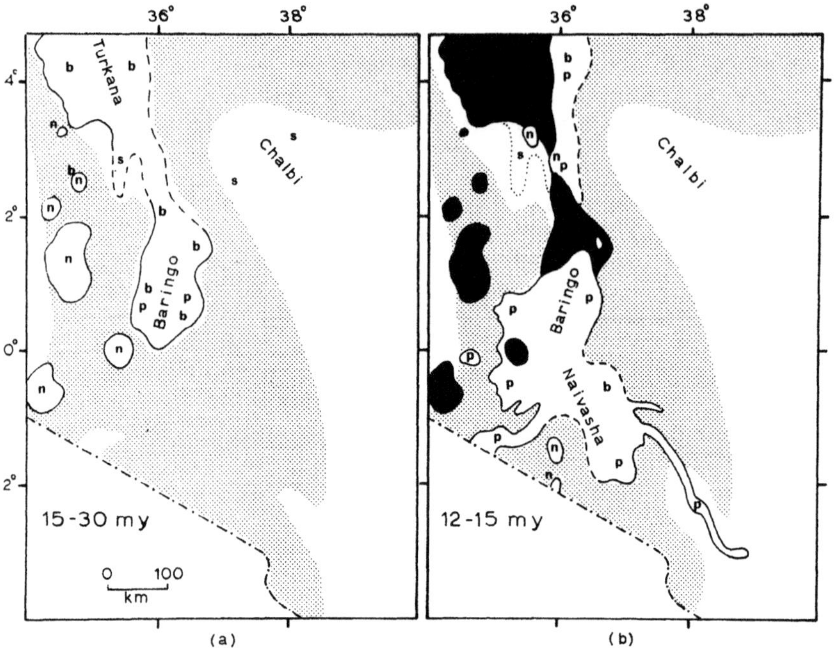

Fig. 5. Distribution of Miocene depressions, volcanics and sediments. b = basalt, n = nephelinite, p = phonolite, s = sediments, black = older volcanics. Stippled areas represent schematically regions of higher ground.

The basalts were erupted in broad downwarps along the crest of the gently uparched continental watershed [5]. The western margin of the Turkana depression was defined by a major flexure, now eroded back to form a prominent NNW-trending scarp, but the location and character of the eastern margin and the distribution of early basalts in that direction are matters for speculation. The Turkana basalts (15-30 m.y.) rest on substantial thicknesses of sediments derived from the basement rocks. Occurrences of similar sediments in the Chalbi region, and the preservation of Mesozoic limestones, sandstones and shales about 200 km to the SE, suggest that the Chalbi depression existed at this time. It is not unlikely, therefore, that some basaltic eruptions in this part of Kenya date back to the Miocene, but no systematic investigations have been carried out.

The limits and orientation of the Baringo depressions are far more reliably known from detailed observations on both sides of the trough [5,6]. During the period 15-20 m.y. basalts, basanites, phonolites and sediments largely infilled this early NNE-trending downwarp, but flanking high ground effectively contained the volcanics. No direct connection existed between the Baringo and Chalbi depressions.

Miocene basalts near the equator had been subjected to faulting and erosion before eruption of the bulk of the overlying 'plateau' phonolites (Fig. 5b), but several dissected phonolitic/ syenitic complexes emplaced in basalts testify to localised episodes of salic volcanism earlier than the main phonolites [6]. Nephelinites and phonolites resting on or emplaced in basalts in Turkana indicate phases of strongly alkaline volcanism far removed from the older chain of nephelinitic volcanoes along the Kenya-Uganda border [2]. Melanephelinites (12-15 m.y.) in southern Kenya were the earliest manifestations of volcanic activity along what later became the western margin of the Magadi trough [19].

Deepening of the Baringo depression around 12-15 m.y. ago probably extended the rift to the SSE to form the Naivasha trough (Fig. 5b): both depressions were filled and overtopped by phonolites. Recognition of intrusive complexes in the Baringo region show that at least some of these 'plateau' phonolites came from large shield volcanoes [6,22]. In many areas, it can be demonstrated that the lavas spread across and largely submerged extremely irregular basement terrain flanking the main troughs. This is clearly illustrated west of the rift by basement inliers across the Uasin Gishu plateau (Fig. 3, section E), at Bomet (section C), and by deflection of the flows around basement hills in the Mara region (section B) where the phonolites probably occupied a broad valley. On the eastern side of the rift, confinement of the lavas by formerly near-continuous high ground extending from Karisia (section F), through the Fort Hall-Kiganjo inlier (section C) to Kanzalu (section B) is just as striking a feature as the much better known valley flow which broke through this barrier, and which is now preserved as a thin lava capping on the Yatta plateau following inversion of the topography.

The phonolites were evidently also confined by high ground at the southern end of the Naivasha trough. The flows now terminated in an erosional scarp probably cut during end-Tertiary dissection, and it is not unlikely that the phonolites are at least as thick here as in areas closer to Nairobi. Flow-tops cannot be recognised between Kapiti and Nairobi, but elevations of the phonolite outcrops indicate northward back-tilting of the lavas (Fig. 4); this could well have been produced by a combination of downwarping in the Nairobi region and late uplift of the rift shoulders around Kapiti (Fig. 3, sections A and B). Elsewhere, the extremities of the phonolite sheets are frequently marked by a reduction in the number of flows, and thinning of individual units [22]. Lack of terminal ponding of these extremely fluid lavas is understandable since the flows were essentially horizontal at the time of emplacement: the present dips provide, therefore, a reliable insight into the nature of post-Miocene movements of the rift shoulders [20].

4.2 Pliocene volcanism (Fig. 6)

Detailed investigations of the 'plateau' phonolites and other volcanics over a large area north of the equator now allows comparison of successions within the rift with those on the shoulders. A relatively up-faulted or up-arched block in the Baringo region (Fig. 3, sections E and F) exposes a well-calibrated succession of volcanics and sediments some 3 km thick, resting on basement gneisses [6]. The demonstrated thickness of phonolites (1600 m) is far in excess of that observed elsewhere in the rift, and the lavas (7-16 m.y.) were erupted over a much longer time range than flows which spread beyond the trough. Phonolites west of the rift (Uasin Gishu phonolites) are mostly 12-13.5 m.y. old (described here as Miocene), but those on the eastern side range in age from 15 up to 10 m.y. The existence of phonolites at Laikipia younger than the Uasin Gishu flows had been indicated by isotopic dating [23], but complexity arising from intercalated basalts, trachytes and tuffs, as well as from overstep of older formations [5], could not have been anticipated from the results of reconnaissance mapping.

The youngest flows (Rumuruti phonolites) covered much of Laikipia before encountering the Karisia-Kanzalu chain of hills, which was mentioned earlier (see also Figs. 3 and 6a). Part of this barrier survives as high ground in the north, but much of it has been destroyed by erosion so that the basement rocks along the upper reaches of the Ewaso Ngiro valley now stand well below the level of the more resistant phonolites.

The nature and significance of erosional scarps marking the edge of phonolite sheets have largely escaped comment in review articles though Shackleton [24], for instance, was well-aware of the implications in the Laikipia area where he also mapped remnants of a valley flow now detached from the main phonolite outcrop as a result of considerable erosion. Pulfrey [11], too, appreciated the fact that a long-standing major barrier must have existed southwards from the Karisia hills, but he did not extend the hills and mountains on his sub-Miocene bevel as continuous features to link up with the Machakos hills SE of Nairobi. The area south of the Forth Hall-Kiganjo inlier also proved troublesome in later work on erosion surfaces [12], and it became necessary to conclude that the end-Tertiary erosion cycle was far advanced before eruption of lavas correlated with the Kapiti and Yatta phonolites (13-13.5 m.y.) since locally flows descend some 240 m below the level of a postulated sub-Miocene surface.

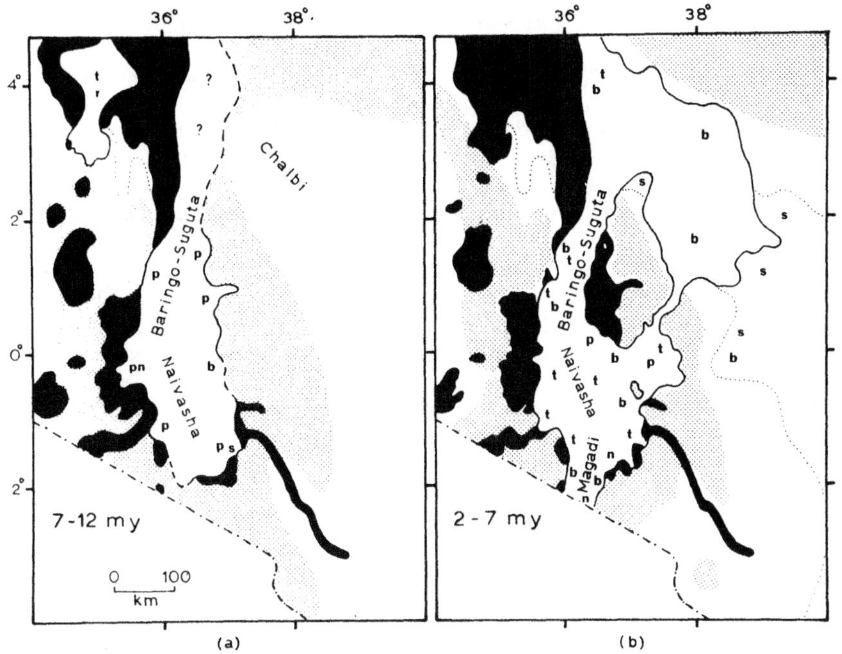

Fig. 6 Distribution of Pliocene volcanics, sediments and main depressions. b = basalt, p = phonolite, n = nephelinite, pn = phonolitic nephelinite, t = trachytic volcanics, r = rhyolitic flows, s = sediments, black = older volcanics, stipple = higher ground.

One further comment can be made about the approximately linear nature of the eastern margin of volcanics older than about 7 m.y. (Fig. 6a), lest it be taken to represent tectonic rather than topographic control over their distribution. The structural map (Fig. 4) shows that the proposed chain of hills follows closely the structural grain of the basement, and that the NNW line is not widely divergent from some rift trends (Naivasha trough; Uganda escarpment bordering the Turkana depression); but of greater significance is the fact that the Karisia-Kanzalu line converges on and is truncated by the Baringo-Suguta trough.

Major faulting which took place after eruption of the 'plateau' phonolites coincided with marked compositional changes in the volcanics. Trends in both mafic and felsic rocks in areas north of the equator were towards increasing silica saturation, so that post-Miocene volcanics belong to basalt-mugearite-trachyte suites (6). Phonolitic varieties are present, but are never as undersaturated as the earlier 'plateau' lavas. Some basaltic and trachytic formations were evidently derived from fissures or multicentre sources, but over a large area major trachytic shield volcanoes (2.5-6 m.y.) built up on a foundation of faulted and tilted flood lavas [25]; these shields are now deeply dissected and the flanks frequently overlap. Despite resulting complexity, it can be shown that the older volcanoes are more strongly tilted (inwards towards the centre of the rift) than the younger examples which are situated axially [5].

The junction between the main Kenya rift and the transverse Kavirondo trough is obscured by great thicknesses of nephelinitic and phonolitic volcanics (7-20 m.y.) belonging to overlapping major centres [17]. Miocene sediments and 'plateau' phonolites are intercalated in the successions. The major faulting episode around 7 m.y. was followed by eruptions of basalts, basanites and mugearites, as well as shield-forming trachytic volcanics; ignimbrites (4-6 m.y.) are prominent among flows which spread beyond the rift depression [17,26]. Farther south towards the Tanzanian border, phonolitic activity (7-10 m.y.) was superseded by eruptions of trachytic lavas and ignimbrites (from 7 m.y. onwards) and basalts (2.5-3 m.y.). Occasional melanephelinites at the western margin of the rift [19], and nephelinite-phonolite suites at many of the eroded remnants of Pliocene (2-6 m.y.) cones between the border and Nairobi, all indicate proximity to a strongly alkaline province in northern Tanzania.

Pliocene successions on the flanks of the rift at the latitude of Nairobi are dominated by trachytic lavas and ignimbrites (2-3.5 m.y.) which rest locally on trachyphonolites. Problem in dating the phonolitic rocks have yet to be resolved, but they probably range back to 10 m.y. [27]. The distribution of the most extensive ash-flows on both sides of the rift (3-6 m.y.) shows that the trough had already been largely infilled by volcanics and sediments by that time. A basaltic shield forms much of the highland area (the Aberdares) from the equator to within 50 km of Nairobi. Some of the older formations there may be of Miocene age for they certainly underlie phonolitic flows dated at 6.5 m.y. [28], but an extended range of activity is indicated by relationships on the eastern side of the complex where younger basalts are ultimately overlain by volcanics of the adjacent Mt. Kenya phonolitic volcano (2.5-3.5 m.y.). Activity in this central complex, which culminated in eruptions of trachytes and basalts, carried volcanism into the region east of the Karisia-Kanzalu chain of hills (Fig. 6b), but basalts on the northern flanks of the mountain flowed across a basement surface of considerable relief before spreading out on the plains beyond [24]. There can be no doubt, therefore, that the Mt. Kenya volcano built up on very irregular terrain: possibly between two parallel chains of hills (see Fig. 7).

Trachytic to rhyolitic lavas and ignimbrites in NW Kenya remain undated, they are provisionally assigned to the period 7-12 m.y. (Fig. 6a).

A vast plain in NE Kenya is generally ascribed to 'end-Tertiary' planation [29], though it has been pointed out that part of it is sedimentary rather than erosional in origin [12]. Between the Nyambeni range and Marsabit (Fig. 4), however, differences of opinion have arisen over precisely which features represent the end-Tertiary surface, particularly in those areas in which dissected basalt plateaus dominate the landscape. Thus, in one account [12], the bases of the lava sheets were taken as the erosion surface, although it was clear that the lavas rest on basement pediments in some areas, and on unfossiliferous conglomeratic sediments in others. The assumption that the sub-volcanic surface represents the end-Tertiary ignores the fact that a striking feature of the region is the widespread occurrence of thick lateritic soils on both the plateaus and intervening plains [30]. In this southern part of this region, remnants of the end-Tertiary (characterised by red soils and sheets of basement pebbles) occur on interfluves near the Ewaso Ngiro river, which was incised into the surface before eruptions of the Quaternary Nyambeni basalts. Older basalts capping one of the plateaus nearby rest on sediments and, locally, on the tops of basement hills buried in the sediments. These basement relics, and other residual hills standing above the surrounding plains, were considered to represent considerably dissected remnants of a sub-Miocene erosion bevel [31]. The sediments, plateau basalts and feeder-dykes (Fig. 4) are all assigned to the Pliocene [28,30]. It is of interest to note that some of the dykes follow the N to NNW basement grain, but many display E to NE trends which are parallel to structural culminations and depressions in the basement. Different views on the interpretation of the evidence largely account for lack of agreement about the age of the plateau basalts which probably covered a very large area in northern Kenya (Fig. 6b). A single age determination gave a result around 28 m.y. and does little to solve the problem.

There is still lack of agreement about the time of maturation of the 'end-Tertiary' surface. A mid-Pliocene age was considered plausible [12] on the basis of what the writer regards as controversial evidence from the coastal region. A lateritised surface can be dated more precisely across the volcanics near the rift, where deep red weathering must have developed less than 2 m.y. ago but before the most recent faulting episode [5].

4.3 Quaternary volcanism (Fig. 7)

The reader is referred to a review [30] which provides more detailed discussion of some of the issues mentioned above, and which goes on to indicate their bearing on the character of Quaternary volcanism. The most recent episodes of volcanic activity are, therefore, dealt with very briefly in this account.

It is generally true that Quaternary volcanism within the Kenya rift was predominantly trachytic, whereas activity to the east produced mainly basalts [2,9,23], but this oversimplified statement fails to indicate the increasingly important role of basalts in northern parts of the rift floor.

The most spectacular volcanism within the rift valley was associated with the evolution of a chain of trachytic caldera volcanoes along the floors of the young inner troughs (Figs. 4 and 7), but massive eruptions of trachytic lavas and pyroclastics (locally at least 1 km thick) and trachyphonolites occurred in southern and central Kenya. Some of the flows (2 m.y.) can be traced over the rift margin eastwards to Nairobi, and these are generally regarded as flood trachytes: but it could well be that the entire Plio-Pleistocene trachytic succession in the Nairobi area [27,32,33] represents the eastern flanks of a giant shield which was truncated by major faulting [30].

The caldera volcanoes are unevenly distributed [34]. Centres south of the equator have calderas 10 km or more in diameter, and they provide likely sources for the youngest air-fall pyroclastics which thickly mantle areas up to 100 km west of the axial line of the rift. In contrast, the volcanoes in a northern group are surmounted by calderas which seldom exceed 5 km across. They gave rise to much more restricted pyroclastics, but all are characterised by basalts that were mostly erupted from fissures aligned parallel to the rift faults. The basalts are not confined to the volcanoes, and they now cover many intervening areas of the rift floor. The youngest eruptions at one of the northern caldera volcanoes (Emuruangogolak) took place not more than 250 years ago [30]; and ash emission from a cone on the flanks of the 'Barrier' shield, at the southern end of Lake Turkana (Rudolf), was reported late last century.

At some of these trachytic shields, more than one stage of caldera collapse is clearly recognisable (for instance, at Longonot and Suswa volcanoes in the southern part of the rift), but at others early structures are largely obscured by continued volcanism and by late-stage caldera formation. In one or two examples, the existence of a large early caldera can be inferred from, or its full extent traced by, geomorphological evidence. This is the case at Emuruangogolak, where the most obvious feature is a shallow summit caldera 4.5 x 3 km across. An earlier structure about 9.5 x 7.5 km went unrecognised during initial studies and it was not reported previously [30,35]. The writer has now assembled evidence to show that erosion of friable tuffs, which formed a large part of the rim of the first caldera and which impounded substantial lava infill, led to inversion of the topography. A marked break in slope on the flanks of the shield had been a puzzling feature during fieldwork, when it was attributed to late doming of the summit area. It is not unlikely that similar complexity will be found at the 'Barrier' when that caldera volcano is mapped in detail.

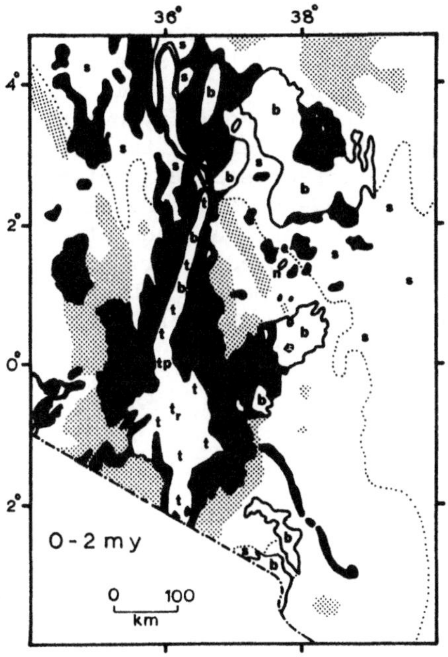

Fig. 7. Distribution of Quaternary volcanics and sediments.
b = basalt,
t = trachyte,
tp = trachyphonolite,
r = rhyolite,
n = nephelinite,
s= sediments,
black = older volcanics,
stipple = higher ground now above 1500 m.

The Tertiary plateau basalts described in a previous section were early manifestations of activity that culminated in construction of several Quaternary linear multicentre ranges east of the rift. Each of these basaltic chains consists of several hundred cinder cones and/or maars surmounting basalt piles up to 1500 m thick, and they are unlike any landforms preserved within the rift valley.

The orientations of fissures which must have controlled the trends of these multicentre ranges are shown as major volcanic lineaments in Fig. 4. Some are parallel to the local basement grain; others cut across the more obvious older structural lines and in a manner yet to be confirmed or explained, they evidently follow more subtle structural features such as culminations and depressions in the metamorphic basement. This is perhaps a problem that could be profitably studied before entering into further speculation about the extent to which basement trends controlled major rift structures.

ACKNOWLEDGEMENTS. Over the last 12 years, the writer has been deeply involved in all activities of the East African Geological Research Unit (supported by the Ministry of Overseas Development, the Natural Environment Research Council and the Kenya Ministry of Natural Resources), and he is grateful to all his colleagues for numerous discussions on rift problems. Particular thanks go to the director of the project, Professor Basil King, for unstinting advice and encouragement over many years: one lesson that will not be forgotten had as its theme the fact that rivers do not, and never did, flow uphill. Little has been said in this account about the history of sedimentation in the rift, but these acknowledgements will be incomplete without reference to the fine work and friendly cooperation of the late Professor Bill Bishop. The writer's earliest observations were made during service with the Geological Survey of Kenya under the expert guidance of Dr. William Pulfrey, to whom must go much of the credit for the success of a long programme of reconnaissance mapping which provided the basis for detailed studies. Finally, support from the University of Nairobi and the University of Lancaster is gratefully acknowledged.

REFERENCES

1. L.A.J. Williams, Nature, 224, 61, 1969.
2. L.A.J. Williams, Bull. volc., 34, 439, 1970.
3. L.A.J. Williams, Tectonophysics, 15, 83, 1972.
4. B.C. King, in: African Magmatism and Tectonics, ed. by T.N. Clifford & I.G. Gass, Oliver & Boyd, Edinburgh, 263, 1970.
5. B.C. King, Structural and volcanic evolution of the Gregory Rift Valley, J. Geol. Soc. (in press).
6. B.C. King & G.R. Chapman, Phil. Trans. R. Soc. Lond., A 271, 185, 1972.
7. B.C. King & L.A.J. Williams, in: Geodymanics: Progress and Prospects, ed. by C.L. Drake, American Geophysical Union, Washington, 63, 1976.
8. B.H. Baker & J. Wohlenberg, Nature, 229, 538, 1971.
9. B.H. Baker, P.A. Mohr & L.A.J. Williams, Geol. Soc. Am. Special Paper, 136, 1972.
10. R.M. Shackleton, Q.J. Geol. Soc. Lond., 106, 345, 1950.
11. W. Pulfrey, Bull. Geol. Surv. Kenya, 3, 1960.
12. E.P. Saggerson & B.H. Baker, Q. J. Geol. Soc. Lond., 121, 51, 1965.
13. P.A. Mohr & P. Gouin, in: Geodynamics: Progress and Prospects, ed. by C.L. Drake, American Geophysical Union, Washington, 81, 1976.
14. J.D. Fairhead, Tectonophysics, 30, 269, 1976.
15. B.H. Baker, East African Rift System: Upper Mantle Committee-UNESCO Seminar, Nairobi, April 1965, University College Nairobi, Pt. 2, 1, 1965.

16. G.R. Chapman & M. Brooks, Isotopic ages and magnetostratigraphy of the Baringo basin, Kenya, J. Geol. Soc. (in press).
17. W.B. Jones, Ph.D. Thesis, Univ. Lond. 1975.
18. R.M. Knight, Ph.D. Thesis, Univ. Lond., 1976.
19. R. Corssley, Ph. D. Thesis, Univ. Lancaster, 1976.
20. G.R. Chapman, S.J. Lippard & J.E. Martyn, The structure and structural evolution of the Kamasia Range and Elgeyo Escarpment, Kenya Rift Valley (submitted for publication).
21. B.C. King, M.J. LeBas & D.S. Sutherland, J. Geol. Soc., **128**, 173, 1972.
22. S.J. Lippard, Geol. Mag., **110**, 543, 1973.
23. B.H. Baker, L.A.J. Williams, J.A. Miller & F.J. Fitch, Tectonophysics, **11**, 191, 1971.
24. R.M. Shackleton, Rep. Geol. Surv. Kenya, **11**, 1946.
25. P.K. Webb & S.D. Weaver, Bull. volc., **39**, 294, 1975.
26. L.A.J. Williams, Rep. Geol. Surv. Kenya, **96**, (in press).
27. B.H. Baker & J.G. Mitchell, J. Geol. Soc., **132**, 467, 1976.
28. L.A.J. Williams, in: Rifting Problems ed. by: N.A. Logatchev & N.A. Florensov, U.S.S.R. Acad. Sci., Irkutsk (in press).
29. F. Dixey, Rep. Geol. Surv. Kenya, **15**, 1948.
30. L.A.J. Williams, Character of Quaternary volcanism in the Gregory Rift Valley, J. Geol. Soc. (in press).
31. L.A.J. Williams, Rep. Geol. Soc. Kenya, **75**, 1966.
32. E.P. Saggerson, Bull. volc., **34**, 38, 1970.
33. L.A.J. Williams, in: Nairobi: City and Region, ed. by W.T.W. Morgan, Oxford Univ. Pres, Nairobi, 1, 1967.
34. P.A. Mohr & C.A. Wood, Earth Planet. Sci. Lett., **33**, 126, 1976.
35. G.R. Chapman, S.D. Weaver & L.A.J. Williams, J. Geol. Soc., **130**, 179, 1974.

SPATIAL AND TEMPORAL VARIATIONS IN BASALT GEOCHEMISTRY IN THE N. KENYA RIFT*

S. J. Lippard and P. H. Truckle

East African Geological Research Unit, Department of Geology, Bedford College, University of London

ABSTRACT. Major and trace element studies of basic lavas from the northern part of the Kenya Rift show that the lavas erupted along the axial zone of the Rift have become progressively less undersaturated with time from basanites and alkali basalts in the Miocene to transitional basalts in the Quaternary. The lavas erupted outside the Rift and in the subsidiary Kavirondo Rift are more undersaturated types.

1. INTRODUCTION

Fig. 1 is a simplified geological map of the Kenya Rift between the Equator and 2°N showing the distribution of Miocene (20 - 7 Ma), Pliocene (7 - 2 Ma) and Quaternary (<2 Ma) volcanics. Within this sector, although the volcanics form reasonably well defined belts, the structures are complex. The Rift changes from a fairly well defined, although markedly asymmetric, faulted graben in the south to a broader depression bounded by faulted downwarps in the north [1]. Basement gneisses are exposed on the flanks and, in the Kamasia and Tiati areas, locally within the Rift. The oldest volcanics are Miocene (∿ 20 Ma) in age. Pliocene (7 - 2 Ma) volcanics are well developed on the western side, but are only poorly represented to the east where locally Quaternary lavas rest

* The East African Geological Research Unit work, directed by Professor B. C. King, has been supported by the Ministry of Overseas Development (O.D.M.), the Government of the Republic of Kenya, the Natural Environment Research Council (N.E.R.C.) and the Central Research Fund of the University of London

E.-R. Neumann and I.B. Ramberg (eds.), Petrology and Geochemistry of Continental Rifts, 123–131.

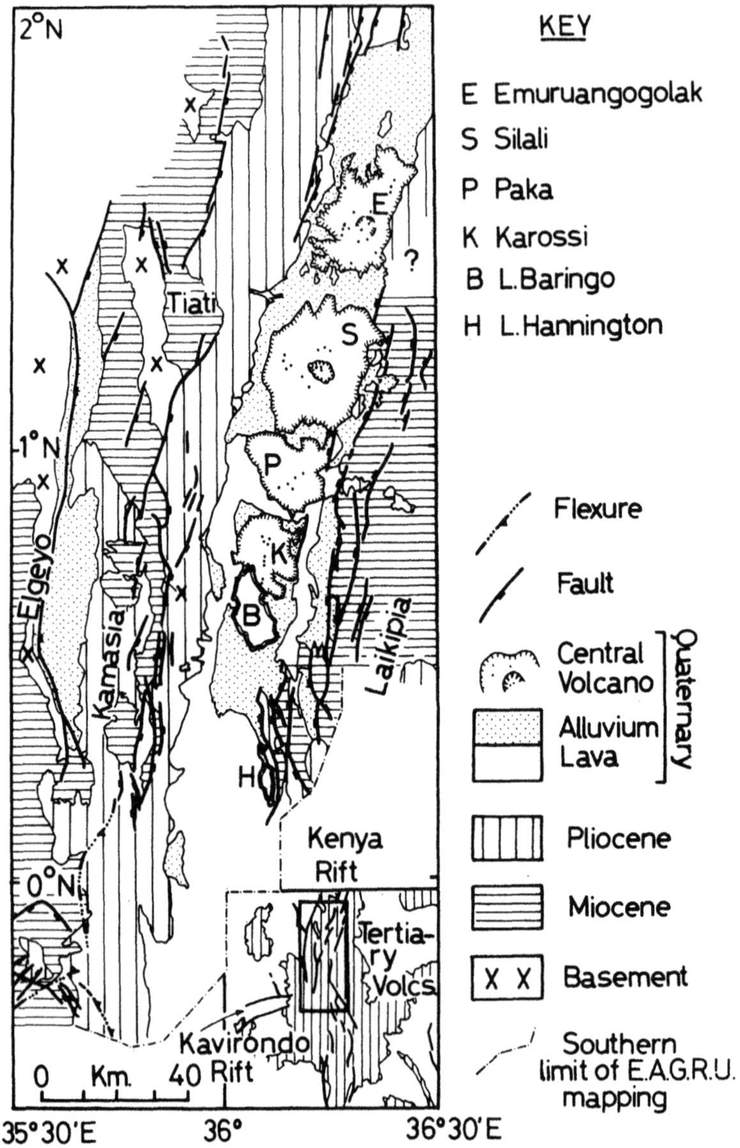

Fig. 1 Geological Map of the North Kenya Rift

unconformably on lower - middle Miocene rocks (e.g. E. of L. Baringo). The Quaternary volcanics occupy a well-defined trough some 20 - 30 km broad. This structure is not axial to the Rift as a whole but displaced to its eastern side.

An earlier account of the volcanism of the northern part of the Kenya Rift was given by King & Chapman [2].

1.1 Techniques and nomenclature of the lavas

The analyses were carried out by wet chemical methods (major elements) and by XRF on a Phillips Automatic PW1212 Spectrometer (major and trace elements) at Bedford College, University of London.

The major element analyses were recalculated to 100% H_2O and CO_2-free and with Fe_2O_3/FeO standardised to 0.20. The CIPW Norm was calculated on these data and those rocks with >30% normative (ab + an) are plotted on Fig. 2. Most analyses (72% of the samples) fall within the fields of basalts and basaltic hawaiites. Only those rocks with 40<DI<25 (94 in all) are plotted on the remaining variation diagrams.

Nephelinites and analcitites are identified petrographically by the lack of modal plagioclase. The division between alkali basalt and basanite is taken at 5% ne. Those rocks with small amounts of hy (up to 5%) in the norm, but still showing alkali basalt affinities (groundmass olivine, absence of Ca-poor pyroxene) are called transitional basalts.

2. DISTRIBUTION OF TYPES

The total volume of Cenozoic volcanics in N. Kenya is of the order of 250,000 - 300,000 cu km of which about 40 - 50% are basalts and related types. The majority of the remainder are salic volcanics, mainly phonolites and trachytes [1].

2.1 Miocene lavas

The oldest formations are the Samburu basalts which have given K/Ar ages in the range 15 - 20 Ma [3] and consist largely of alkali basalts and basanites containing phenocrysts of olivine and titan-augite. Plagioclase-phyric types occur less frequently. Peridotite (lherzolite and harzburgite) nodules of probable mantle origin are locally found. Resorbed kaersutite phenocrysts are occasionally present. Ankaramites are abundant but have generally been excluded from this study for their cumulate origin. The Saimo and Elgeyo

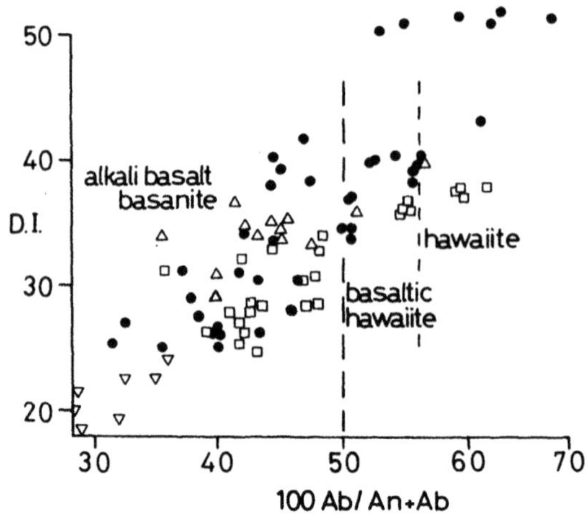

Fig. 2 Normative plagioclase versus Thornton-Tuttle differentiation plot for basic lavas from the North Kenya Rift

Triangles - Miocene, Inverted triangles - cumulate rich
Circles - Pliocene
Squares - Quaternary

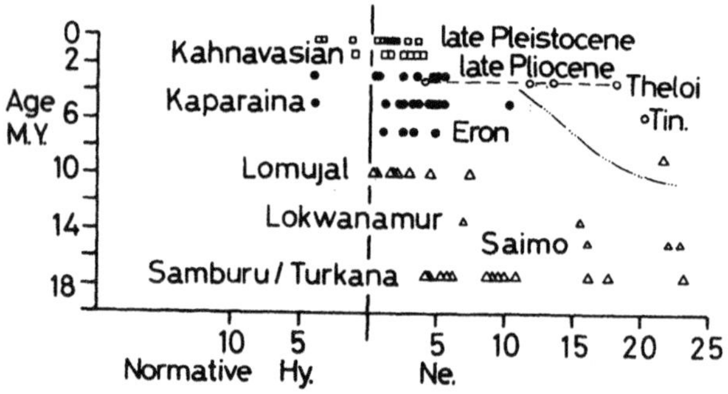

Fig. 3 Symbols as in Fig. 2 (open circles, Kavirondo Rift)

formations in the Kamasia-Elgeyo area consist largely of more undersaturated plagioclase-free limburgites and analcitites. These have been dated radiometrically and stratigraphically at 15 - 16 Ma.

In the southern half of the area the period 16 - 7 Ma was dominated by voluminous eruptions of "plateau" phonolite lavas [1] [4] to the almost total exclusion of more basic types. However, in the northern part thick basalt formations occur that are equivalent in age to the phonolites. Around $2^{o}N$ the Miocene (Turkana basalt) succession consists of a basal basalt formation (17 - 16 Ma), the Lokwanamur basalts (14 - 13 Ma), a thin phonolite sequence (11 Ma) and the Lomujal basalts (ca. 10 - 9 Ma).

2.2 Pliocene lavas

Pliocene basalt formations occur extensively along the western margin of the Rift throughout its entire length north of the Equator. They are well developed in the Kamasia area, particularly on the eastern flanks of the Tugen Hills, and in S. Turkana. The basalts are separated by intervening trachyte formations, often constituting recognisable central, shield-like volcanoes [5]. The Pliocene basalts of the rift zone are mildly to moderately undersaturated alkali basalts. Several petrographic types are found; olivine and titanaugite-phyric basalts tending towards ankaramites, strongly feldspar-phyric types, often basaltic hawaiite or hawaiite in composition; and olivine and plagioclase microporphyritic basalts. Mugearites are represented in most sequences.

Harzburgite nodules have been found in one locality in Pliocene basalt where their mineralogy/chemistry indicates that they are residual mantle material (P.H.T. in prep.).

2.3 Quaternary lavas

The Quaternary basalts within the Rift are confined to the central trough. They form the foundation of and are associated with a series of trachyte and trachyte-basalt central volcanoes (Fig. 1). South of Lake Baringo an extensive series of trachyphonolite flood lavas, the Hannington Trachyphonolites, occupy most of the belt. Another central volcano, Menengei, lies just south of Fig. 1.

Petrographically the Quaternary basalts are not readily distinguished from the Pliocene, except for one rather coarse grained diktytaxitic type which seems to be characteristic, but they contain a higher proportion of transition basalts recognised by the presence of small amounts of hy in the norm (Fig. 3). Gabbro xenoliths are commonly found in flows and tuff deposits and

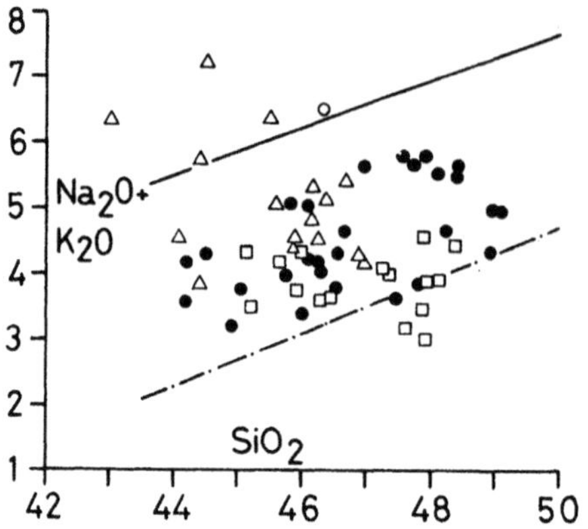

Fig. 4 Symbols as in Fig. 3
Solid Line - Saggerson/
Williams Line
Dashed Line - Irvine/
Baragar Line

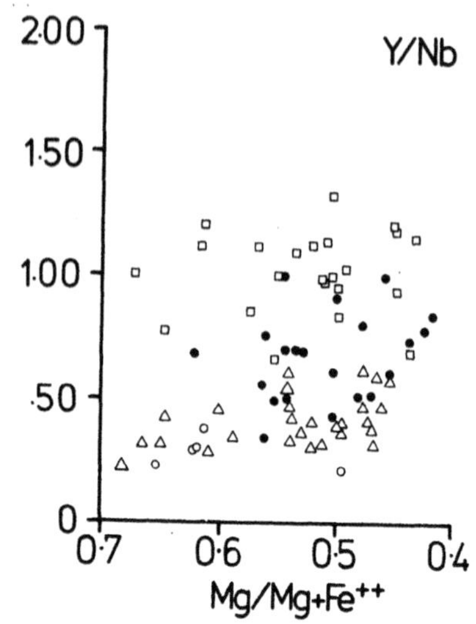

Fig. 5 Symbols as in Fig. 3

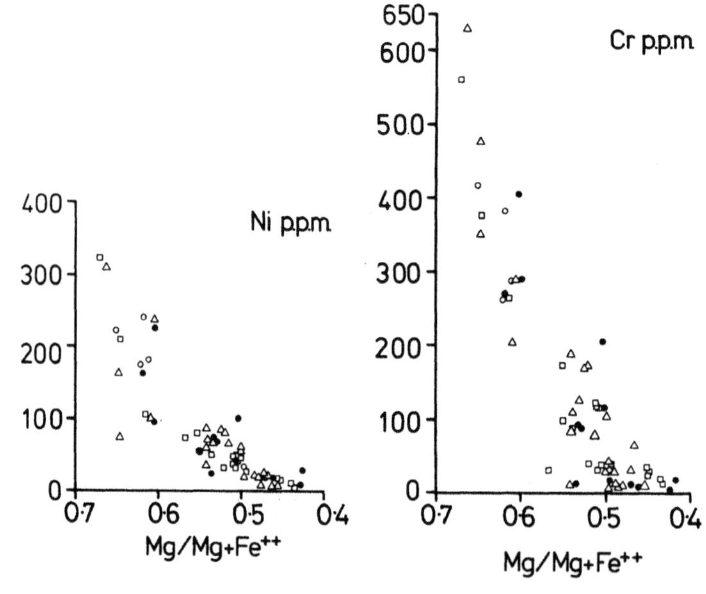

Fig. 6 Symbols as in Fig. 3

probably represent cumulates from high-level magma chambers.

2.4 Contemporaneous volcanics outside the Rift

In the Tinderet-Timboroa area at the eastern end of the Kavirondo Rift eruptions of strongly undersaturated nephelinites occured sporadically throughout the Miocene [6] [7]. In the Pliocene (5 - 6 Ma) analcite basanites were erupted in the Tinderet and Londiani areas, analyses of some of these rocks are included in this study.

Late Pliocene-Quaternary basalts occur east of the Rift in the Nyambeni and Marsabit areas [1]. Although chemical data are sparse, these seem to be more undersaturated types than those of equivalent age in the Rift [8].

3. GEOCHEMISTRY

The alkali/silica plot (Fig. 4) shows that most of the basalts from the Rift fall within the field of alkali basalts with some of the Quaternary lavas plotting closest to the alkali basalt/tholeiite line. The Y/Nb vs Mg/Mg + Fe plot (Fig. 5) shows that rocks belonging to each age group span a wide range of compositions, in terms of differentiation index, while Y/Nb increases with decreasing age. (According to the data compiled by Pearce & Cann (1973) this ratio is an index of the degree of alkalinity.) This plot also emphasises that the Pliocene lavas of the Kavirondo Rift are distinctly more alkaline than those of the main Rift.

The majority of the basalts are rather highly differentiated, few of the aphyric lavas have Ni and Cr contents (Fig. 6) and Mg/Mg + Fe values appropriate to magmas in equilibrium with the mantle [9]. There are some exceptions however, notably 13/706 and 13/333 which are olivine-phyric Quaternary basalts from south of Lake Baringo and 10/749 which is a Miocene basalt containing peridotite nodules (removed before the analysis was carried out). Most of the basic lavas, with Ni < 100 ppm, Cr < 200 ppm and Mg/Mg + Fe < 0.57 have probably undergone at least 20% separation of olivine and clinopyroxene prior to eruption.

4. DISCUSSION AND CONCLUSIONS

In N. Kenya we have presented evidence for two trends:

(1) Increasing undersaturation at any one time away from the rift axis, notably along the subsidiary Kavirondo graben.

(2) Decreasing undersaturation with time leading to the development of transitional basalts in the Quaternary along the axis of

the rift.

Similar patterns have been recognized in other rift systems [10] [11] [12]. Furthermore there appears to be a complete gradation in magma types from basanites, through alkali basalts to transitional basalts. There is at present incomplete data to extend this to more undersaturated compositions. Wright [13] and Saggerson [14] in the central Kenya rift area recognized two distinct series, one mildly undersaturated, alkali basaltic, the other strongly undersaturated, nephelinitic. Wright (op. cit.) also noted that the former series is largely confined to the rift zone.

Kay & Gast [15] concluded that, on the basis of trace element data, nephelinites and alkali basalts are generated by less than 3% partial melting of garnet-bearing lherzolitic mantle at depths greater than 60 km. These figures agree broadly with the results of experimental studies [16]. Transitional basalts are believed to be generated at either shallower depths or by greater degrees of melting [9]. The tectono-magmatic development of the N. Kenya Rift can be interpreted in terms of decreasing depth and/or greater degrees of melting during the period from 20 Ma to the present in a lherzolite source region some 50-100 km beneath the Rift. Most of the magma is subsequently trapped at various levels, probably mainly within the crust and uppermost mantle, where it has undergone extensive crystal fractionation to produce the range of lava types seen at the surface. The positive gravity anomaly located along the rift axis in N. Kenya [17] can be interpreted in terms of a body of ultrabasic to basic cumulate material and lends support to this hypothesis. The Kenya Rift, although it is about 20 Ma since it was initiated, still seems to be in its development or waxing stage from the point of view of its magmatic evolution.

REFERENCES

1. B.H. Baker, P.A. Mohr and L.A.J. Williams, Geol. Soc. Am. Sp. Paper, **136**, 1972.
2. B.C. King and G.R. Chapman, Phil. Trans. Roy. Soc. Lond. A271, **185**, 1972.
3. B.H. Baker, L.A.J. Williams, J.A. Miller and F.J. Fitch, Tectonophysics, **11**, 191, 1971.
4. S.J. Lippard, Lithos, **6**, 217, 1973.
5. P.K. Webb and S.D. Weaver, Bull. Volc. **39**, 1, 1975.
6. R.M. Shackleton, Quart. Jour. Geol. Soc. Lond., **106**, 345, 1950.
7. B.C. King, M.J. Le Bas and D.S. Sutherland, Jour. Geol. Soc. Lond., **128**, 173, 1972.
8. L.A.J. Williams, Tectonophysics, **15**, (1/2), 83, 1972.
9. A.E. Ringwood, Composition and petrology of the Earth's Mantle, McGraw Hill, 1975.

10. P.W. Lipman, Geol. Soc. America Bull., 80, 1343, 1969.
11. C.K. Brooks and J.C. Rucklidge, Lithos, 7, 239, 1974.
12. P.A. Mohr, Bull. Volc., 34, 141, 1970.
13. J.B. Wright, Geol. Mag., 102, 541, 1965.
14. E.P. Saggerson, Bull. Volc., 34, 38, 1970.
15. R.W. Kay and P. Gast, Jour. Geol., 81, 653, 1973.
16. R.J. Bultitude and D.H. Green, Jour. Petrology, 12, 121, 1971.
17. M.A. Khan and J. Mansfield, Nature, 229, 72, 1971.

RIFT-ASSOCIATED IGNEOUS ACTIVITY IN EASTERN NORTH AMERICA

A. R. Philpotts

Department of Geology and Geophysics, and Institute of Materials Science, University of Connecticut, Storrs, Ct. 06268, USA

ABSTRACT. Although many of the prominent features of the eastern North American rift systems were developed in Triassic to Cretaceous time, much of the fracture system was formed in the late Precambrian. The associated igneous activity was episodic, beginning in the Proterozoic with the intrusion of a diabase dike swarm along the north side of the Ottawa graben. This was followed in the Cambrian by the emplacement of an anorthosite body and several alkali syenite and granite complexes along the northern side of the St. Lawrence rift valley. The most widespread activity came in early Jurassic time with the extrusion of tholeiitic basalts which are now preserved in the down-faulted Triassic-Jurassic basins along the eastern seaboard. This activity has been correlated with the early opening of the North Atlantic. The White Mountain complexes, which began forming during the final subsidence of the down-faulted basins, formed a belt of intrusions with generally decreasing ages toward the southeast. These may have formed as the North American plate moved over a hot spot, which later formed the New England seamounts [20]. Finally, in Cretaceous time strongly alkaline, critically undersaturated magmas were intruded at the intersection of the Ottawa and St. Lawrence rift valleys to form the Monteregian intrusions.

The igneous rocks associated with the rift system have, throughout time, shown a marked bimodal distribution in compositions - basic and felsic. This distribution can be attributed, at least in part, to liquid immiscibility. Many of the tholeiitic Jurassic basalts contain evidence of immiscibility, which in unaltered samples consists of iron-rich droplets of glass enclosed in silica-rich glass. Similar immiscible droplets are found in basalts from the mid-Atlantic ridge. Many fine-grained basic

E.-R. Neumann and I.B. Ramberg (eds.), Petrology and Geochemistry of Continental Rifts, 133–154.

alkaline rocks associated with the rift systems contain ocelli, small spherical bodies of felsic material which also are interpreted as having formed as immiscible droplets.

The repeated occurrence of iron-rich tholeiite and quartz syenite with rather restricted compositions is thought to be due to their formation as conjugate pairs of immiscible liquids produced during partial melting in the source region. The iron-rich tholeiitic magma, being rather dense, is intruded mainly when there are readily available tension fractures, and then intrusion occurs sufficiently rapidly so that little differentiation takes place during emplacement.

The generation of felsic and basic rocks through immiscible splitting of mantle-derived magmas beneath rift systems, is proposed as an important mechanism by which continental type material has been generated at spreading plate boundaries.

1. INTRODUCTION

The igneous rocks associated with the rift systems in eastern North America are of two general types. Tholeiitic rocks have formed in great abundance along those rifts where considerable crustal extension has occurred, whereas alkaline rocks have formed on a more limited scale where rifting has resulted mainly in down-faulting. The tholeiitic rocks were intruded rapidly along dikes with little differentiation taking place, whereas the alkaline rocks were intruded mainly in pipe-like bodies where differentiation was extreme.

During the last one billion years several episodes of igneous activity occurred along the rift systems of eastern North America, and while each had its own distinctive features, certain major rock types have occurred repeatedly. Such rocks must be manifestations of some fundamental mechanism of magma generation beneath the rifts.

Both the tholeiitic and the alkaline rocks associated with these rift systems have recently been shown through textural and experimental investigations to exhibit abundant evidence of liquid immiscibility [26]. Many of the trends of differentiation, in particular of the alkaline rocks, can now be explained using this mechanism. If liquid immiscibility occurs during the cooling and crystallization of a magma, it may also have occurred during the initial melting and formation of that magma. Although high pressure experimental data are not available for the magmas dealt with here, this paper will speculate about the possible role immiscibility might play in the formation of magmas beneath these rift systems.

2. DISTRIBUTION AND AGES OF EASTERN NORTH AMERICAN RIFT SYSTEMS

Although several different ages of rift systems exist in eastern North America, the Triassic-Jurassic-aged ones that extend along the eastern seaboard are the most extensively preserved [31]. This rifting produced a series of elongated fault basins which are filled mainly with reddish-brown arkose and shale and extensive multiple flood basalts and sills. The largest of these basins are, from north to south, the Fundy (Fig. 1, A), Connecticut (B), Newark (C), Gettysburg (D), Culpeper (E), and Durham-Deep River (F) basins. Other basins occur to the south and to the east on the continental shelf but are covered by later Mesozoic and Cenozoic sediments. The basins are located within the Appalachian orogenic belt, the tectonic grain of which appears to have played a role in determining their orientations.

An equally important but less clearly defined system has been named the St. Lawrence valley rift system by Kumarapeli and Saull [18]. The St. Lawrence valley, which is underlain mainly by flat-lying Lower Paleozoic rocks, is an elongated depression bounded on its northwest side by Precambrian rocks of the Canadian Shield and on its southeast side by the Appalachian fold belt (Fig. 1). Also forming part of the rift system are the Saguenay, Ottawa, and Champlain valleys, all of which are occupied by grabens that branch off from the main St. Lawrence rift valley. Faulting along this system has allowed the development of locally thick accumulations of Lower Paleozoic sediments and has provided conduits for magma to rise along and form characteristic alkaline igneous complexes.

While the main igneous activity associated with the St. Lawrence rift system occurred in Cretaceous time, there was activity during several episodes dating back to the Late Precambrian. This rift system, therefore, has had a long history and, indeed, the high frequency of seismic activity along the system indicates it may still be active.

An extensive diabase dike swarm extending along the northern side of the Ottawa graben (Fig. 1), may provide the earliest evidence of activity along the St. Lawrence rift system. The dikes, which have an age of approximately 790 million years [12,21], intrude Grenville-aged (1 billion year) anorthosites and granulite facies metamorphic rocks. Exposures of the dikes are limited to the northern side of the graben, the southern side being covered with a thin veneer of Lower Paleozoic rocks. The dikes, which are mostly from 40 to 50 meters thick and up to 3 kilometers long, parallel the graben and increase in abundance toward it [27]. Large numbers of irregular dikes only centimeters in thickness occur in the immediate vicinity of the fault that forms the northern boundary of the graben. Some of these intrude along fault zones, while others are offset by faults. Clearly the

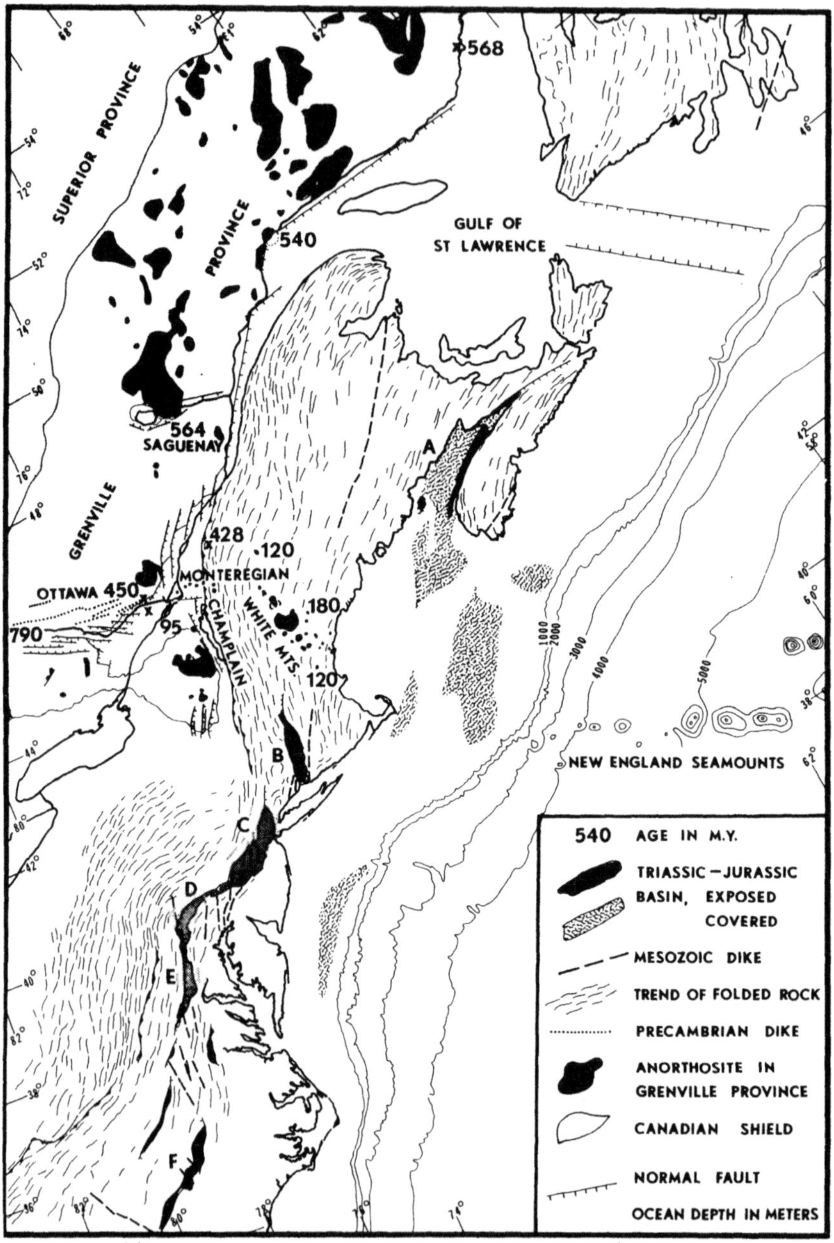

Figure 1. Rift systems and associated igneous rocks of northeastern North America. Compiled from: Tectonic map of North America (King, 1969), Fahrig and Wanless (1963), Kumarapeli and Saull (1966), Gold (1967), Doig and Barton (1968), Foland et al., (1971), Papezik et al., (1975), Van Houten (1977).

northern boundary of the Ottawa graben existed at the time of this diabase intrusion.

Activity along the entire length of the St. Lawrence valley rift system in Cambrian time produced numerous alkaline intrusions which have potassium-argon ages of 510 to 580 m.y. [11]. Considerable relief may have existed on the rift valley at this time, as indicated by the highly variable thickness of the Late Cambrian Potsdam sandstone, the lowest member of the Paleozoic sequence of sedimentary rocks to be deposited in the rift system. In the vicinity of Montreal, where the St. Lawrence system bifurcates into the Ottawa and Champlain grabens, the thickness of sandstone, as intersected in deep drill-holes, indicates that the topographic relief was approximately 1 kilometer [7].

Renewed alkaline igneous activity in Ordovician time resulted in the formation of three known igneous complexes. Immediately west of Montreal, intrusion along the northern and southern boundary faults of the Ottawa graben produced the Chatham-Grenville and Rigaud stocks, both of which have absolute ages of 450 m.y. [11]. Another complex, which is situated near the village of Bon Conseil on the southeastern side of the St. Lawrence rift valley, has an age of 428 m.y. [11]. This Ordovician activity may have resulted from adjustments along the rift system brought about by the Taconic orogeny in the Appalachian rocks immediately to the east.

The main period of igneous activity in the St. Lawrence rift system came in Cretaceous time with the formation of the Monteregian intrusions which form a series of hills along the eastward extension of the Ottawa graben across the St. Lawrence valley [25]. These intrusions range in age from 125 to 90 m.y [11] with the oldest occurring toward the east where they may connect with the northern members of the White Mountain magma series of New Hampshire (see below).

The time of development of the Triassic-Jurassic rift system has been carefully worked out by palynological dating of the sediments that fill the fault basin [8,9]. Deposition began in late Triassic time (Carnian and Norian) with mainly coarse sediments, including fanglomerates which spread out from fault escarpments [31]. Extrusive activity within the basins lasted for only a short period in the Lower Jurassic, from approximately 195 to 185 m.y. [8], but intrusion of diabase dikes may have continued after this. Following the extrusive period, there was further subsidence of the basins with deposition lasting until the Lower to Middle Jurassic.

As the extrusive activity within the basins was coming to an end, a major period of intrusion began along a belt to the northeast of the Connecticut basin. Here, a series of ring dikes, stocks, and batholiths of the White Mountain magma series were intruded mainly at about 180 m.y. [13], a time that coincides with the final subsidence of the basins [8]. Igneous activity continued after this, however, reaching another peak between 120 and 100 m.y ago [13]. Although this activity was limited mainly to the southern part of the belt, these ages correspond with those of the Monteregian intrusions to the north with which the White Mountain rocks appear to be petrogenetically and geographically related. This period of igneous activity, therefore, may have begun in northern New Hampshire and then migrated simultaneously to the south and northwest.

The Triassic-Jurassic rifting is thought to have been related to the break-up of Pangaea and the early opening of the North Atlantic [19,31]. The associated igneous activity is, therefore, of importance in revealing the types of magma present during the initial stages of spreading, which in turn can be compared with those presently being erupted along the mid-Atlantic ridge.

3. SUMMARY OF RIFT-ASSOCIATED IGNEOUS ACTIVITY IN EASTERN NORTH AMERICA

The Precambrian diabase dike swarm along the northern side of the Ottawa graben is typical of the numerous dike swarms in the Canadian Shield. The cumulative thickness of the dikes is sufficient to clearly indicate a period of crustal extension during their intrusion, and as is found in many other parts of the world the associated magma is of tholeiitic character (Table 1). Serpentinized olivine phenocrysts can be found in the chilled margins of some of the dikes, but in more slowly cooled parts, these are removed by reaction with the magma and the rock is a normal quartz diabase.

The igneous activity associated with the St. Lawrence valley rift system in Cambrian time was quite different, and instead of dikes, widely scattered stock-like bodies were intruded at various localities along the northern boundary fault of the rift system, over a distance of 2000 km. Quartz syenite and minor carbonatite are the principal rock types, but at Sept-Iles, a town located on the north shore of the Gulf of St. Lawrence, 800 km northeast of Montreal, an undeformed anorthosite complex was emplaced at the boundary between the Canadian shield and the St. Lawrence graben. Like the many other anorthosite bodies in the Grenville province this one has been considered Precambrian in age, but recent K-Ar and Rb-Sr dating by Higgins and Doig [15] indicate it to be only 540 m.y. old. The anorthosite within this body grades upward into a well layered gabbro which is overlain by granitic rocks containing alkali pyroxene and amphibole. This association is similar to that found in the Paleozoic ring complexes of Niger [3].

As Higgins and Doig [15] emphasize, this association of anorthosite with rifting and alkali granites may shed new light on the origin of anorthosites. Thus, the belt of Precambrian anorthosite bodies that extends through the Grenville province might owe its linear pattern to some earlier rift system that lay to the north of the St. Lawrence valley system.

The Ordovician-aged Chatham-Grenville and Rigaud stocks [23] consist of quartz syenite and granite, both of which contain aegerine and hastingsite. Although no carbonatites have been found with these bodies, igneous breccias containing abundant fragments of country rocks and syenite in a fine-grained syenitic matrix, and diatreme breccias are associated with the Chatham-Grenville stock.

The similarly aged intrusion at Bon Conseil is strikingly different from any of the previous igneous rocks associated with the St. Lawrence valley rift system. This poorly exposed body consists of two distinct parts, a mica-peridotite and a pyroxenite, with no contacts between the two being exposed. The pyroxenite, which is coarse-grained and consists of titanaugite and hastingsite with minor analcime, contains abundant irregular patches (up to 1 meter across) of analcime syenite which contain minor amounts of titanaugite and hastingsite of similar composition to that found in the pyroxenite (Fig. 2). Nowhere are dikes of either rock-type found cutting each other. Nor does there appear to be any age difference between the two, for crystals of pyroxene and amphibole cross the boundary between the pyroxenite and the syenite with no regard for the contact. The only difference, other than modal abundance, is that both the pyroxene and amphibole are noticeably more acicular in the syenite. While precisely the same relation between these rock types has not been found in any of the later coarse-grained plutonic bodies associated with this rift system, the close association of mafic and felsic rocks with a lack of intermediate ones typifies all of the later intrusions. And if fine-grained dike rocks are included in the comparison, many of the Monteregian ones have an ocellar structure which appears to be a finer scale version of the larger syenite patches (see below).

Following the emplacement of the Ordovician intrusions was a period of 300 m.y. during which no further intrusion occurred along the St. Lawrence valley rift system. In the meantime, the next rift-associated igneous activity in eastern North America came at the beginning of the Jurassic period with the voluminous extrusions of basalt which are now preserved in the down-faulted Triassic-Jurassic basins from the Bay of Fundy to northern Virginia.

These Jurassic basalts and associated dikes and sills are

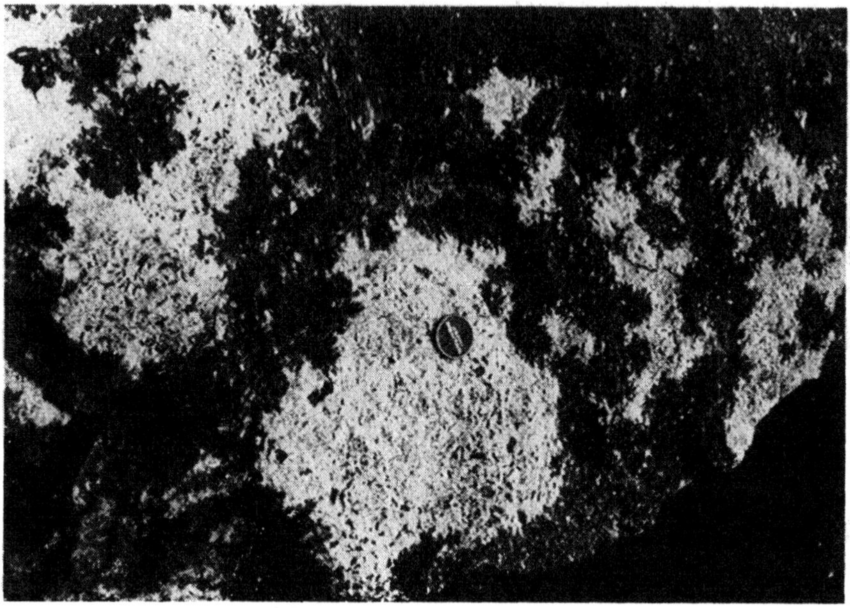

Figure 2. Analcime syenite patches in pyroxenite, Bon Conseil, Quebec.

well known from many local and regional studies. Although the basalts have been extensively analyzed, they tend to be partially altered, so that the most reliable indications of the compositions of magmas present during this episode are obtained from analyses of the much fresher diabase dikes, some of which occur within the Triassic-Jurassic basins, but most of which are found in the older surrounding rocks.

Weigand and Ragland [33] have made an extensive survey of these dikes and have shown that, while all are tholeiitic, four distinct types can be recognized, an olivine-normative type, a high TiO_2 quartz-normative type, a low TiO_2 quartz-normative type, and a high iron quartz-normative type. The quartz-normative types predominate north of Maryland, whereas the olivine-normative ones predominate to the south. Within each of the four groups there is limited chemical variation, suggesting that there was little opportunity for magmatic differentiation. This conclusion is supported by the almost complete lack of any igneous rocks other than basalt. The only exception is in northern Virginia where the basalts are inter-bedded with a trachytic crystal tuff [30]. A similar rock has also been found as fragments between basaltic pillows at the northern extremity of the Talcott basalt in Connecticut. Intrusion, in most cases must therefore have been rapid, with magma travelling from the source region to the surface of the Earth with no significant delays in intermediate magma chambers.

Although the complexes of the White Mountain magma series are not clearly related to a rift system, their initial activity coincides with the final subsidence of the down faulted Triassic-Jurassic basins, and their linear distribution appears to have been controlled by a major fracture system [5]. For this reason they are included here in the discussion of rift-associated igneous rocks.

The White Mountain complexes were intruded over a period of 70 m.y. [13] along a 300 km long belt extending from Mount Monadnock in northeastern Vermont to the Cape Neddick complex in southwestern Maine. Despite the rather long time over which intrusion occurred, most of the complexes are similar, being composed largely of granite which forms stocks and some ring dikes, and smaller amounts of syenite and quartz syenite, most of which occur in ring dikes [6]. Unlike the igneous rocks of the Triassic-Jurassic basins, gabbroic rocks are not common and are restricted to plug- or funnel-like bodies.

Because of the differences in age between complexes, no single large magma chamber could have supplied each of the complexes with the different rock types. Instead, each complex must have undergone its own differentiation process [13]. Chapman [6] shares this view and envisages a complex as being formed by the intrusion of a compound body of magma that is basaltic in its lower part and granitic in the upper part. The failure of the two magmas to mix does not, according to Chapman, infer they are immiscible. Instead, he believes a basaltic magma rising into the crust could selectively melt the intruded rocks and produce a capping of granitic magma which because of its low density and high viscosity would remain separate at the top of the magma chamber. With continued melting and upward intrusion the basaltic magma would fractionally crystallize and produce a residual syenitic magma which would underlie the granitic one. Foundering of roof rocks could then produce ring dikes of granite or syenite. According to Chapman [6], this model explains the rock types, their relative ages, and the main structures in each of the bodies.

The main igneous activity in the White Mountains began approximately 180 m.y. ago in northern New Hampshire in the vicinity of the White Mountain batholith. At approximately 150 m.y. ago intrusive centers appeared to the north (Gore Mountain) and south (Belknap complex) of this [13]. Then again between 120 and 110 m.y. ago intrusions were emplaced still farther north (Mount Megantic) and south (Mount Pawtuckaway). Still younger ages (110 to 95 m.y.) are encountered farther to the northwest if the Monteregian intrusions are included, and to the southeast the belt of intrusions could continue out under the continental shelf and appear on the ocean floor as the New England seamount chain. In fact, the linear distribution of the White Mountain complexes and their possible correlation with the seamount chain led Morgan [20] to suggest that they had formed as the North American plate travelled over a hot spot that today is marked by the Azores.

The Monteregian intrusions, while having the same age as the youngest White Mountain rocks and being geographically related to them, show significant differences. The distribution of Monteregian intrusions is clearly controlled by the Ottawa graben, and their ages generally decrease toward the west [14]. This is the opposite direction from that found for the White Mountain complexes and is difficult to reconcile with the model of intrusions having formed as the North American plate travelled westward over a hot spot. The biggest difference, however, is in their petrology. While the majority of the White Mountain rocks are mildly alkaline and oversaturated with silica, most of the Monteregian ones are strongly alkaline and critically undersaturated in silica. Also, mafic rocks are more abundant than felsic ones in the Monteregian province.

The Monteregian province consists of nine major stock-like intrusions spaced at rather regular intervals of approximately 15 km along a belt that extends eastward from the Ottawa graben across the flat floor of the St. Lawrence valley into the Appalachian fold belt [14,25]. In addition, there are many dikes, sills, and diatremes scattered throughout the province, although these are more abundant in the western part.

The rock types vary greatly, from kimberlite and carbonatite, through various melilite- and nepheline-bearing mafic rocks, to nepheline- and quartz-bearing felsic ones. Markedly different rock types occur in each intrusion, indicating that each magma chamber behaved independently, following its own peculiar path of differentiation. The variations between intrusions, however, are far from random, for there exists a very systematic variation in the silica content of rocks along the length of the belt [25, 26]. The most silica poor rocks, which occur at the western end, contain melilite and no calcic plagioclase. Farther east, in the vicinity of Montreal, plagioclase coexists with nepheline. Still farther east, nepheline- and quartz-bearing rocks occur together in the same intrusion, and at the extreme eastern end of the province, only oversaturated rocks are present. This silica variation does not take place gradually but occurs in steps which appear to coincide with major northeasterly striking normal faults that cross the east-west trend of the Monteregian province and, in part, control the location of the major intrusions. The movement on these faults indicates that the silica-rich rocks are associated with the areas of greatest subsidence.

Each of the main Monteregian intrusions consists of an early basic phase which is followed, in most cases, by a felsic phase with a complete lack of intermediate rock types. Many of the fine-grained regional dikes, that is ones not obviously associated with any main intrusion, contain ocelli, small spherical patches of felsic material. These have been interpreted, both from textural and experimental evidence, as resulting from liquid immiscibility between felsic and basic magmas. This immiscibility, then, is thought to have played a major role in producing the basic and felsic rocks within the main intrusions [26].

Although this summary of rift-associated igneous activity may at first appear to have introduced a large number of different rock types, this is not, in fact, the case. During the one billion year period dealt with here certain rock types have appeared repeatedly with very constant compositions. For instance, samples from the 568 m.y. old syenite from Mutton Bay (most northern locality in Fig. 1), from the 450 m.y. old Chatham-Grenville intrusion, and the 180 m.y. old White Mountain intrusions are all virtually indistinguishable. Similarly, the alkali granites from each of these intrusions are almost identical. In addition, a sample of one of the Precambrian diabase dikes from near the Ottawa graben could be slipped into a collection of Jurassic dikes from the eastern seaboard and go undetected (see Table 1).

Another striking feature is the different degrees to which differentiation has occurred in the various magmas. The Precambrian diabase dike swarm exhibits no compositional variation over distances of at least 600 km, and the Jurassic basalts show a similar constancy of composition. In contrast, each Monteregian intrusion is different from its neighboring one, and such contrasting rocks as peridotite and nepheline syenite can occur within a single small body. This could imply that differentiation within a tholeiitic magma is much more difficult than in an alkaline magma; however, this is not the case, for textural and experimental data on the Jurassic basalts indicate they were capable of undergoing strong differentiation at the time of their crystallization.

Before offering a model for the development of the various igneous rocks associated with these rift systems, a brief description will be given of a Jurassic basalt from Connecticut which has textures that provide evidence of an important process which may have played a dominant role in the generation of rift-associated igneous rocks.

4. JURASSIC THOLEIITIC BASALT FROM CONNECTICUT

Although most Jurassic basalts in the down-faulted basins along the eastern seaboard of North America are partially altered, one particular flow in the small Southbury basin of Connecticut is quite fresh. The apparently unusual feature of this rock is that, while having a similar composition to many of the analyzed diabase dikes, it exhibits indisputable evidence of silicate liquid immiscibility. The mesostasis, which constitutes 32% of this basalt, consists of two glasses, a brown high refractive index one (Fe-rich) and a clear lower refractive index one (Si-rich). These form droplets in each other, with larger droplets containing smaller droplets of the other glass (Figs. 3 and 4). Because such a large fraction of this rock consists of two glasses, differentiation into contrasting compositions could probably have occurred, had there been more time for the two liquids tc separate.

In view of the apparently unusual nature of this rock it is not unreasonable to suppose that minor chemical variations or peculiar conditions of crystallization may have been responsible for the abundance of immiscible liquids.

Comparison of an analysis of this basalt (Table 1) with those of Weigand and Ragland's high iron quartz-normative diabase, the average Deccan trap of India, the Precambrian diabase dike from north of the Ottawa graben, and the standard diabase W-1, reveals the rock to be a normal quartz tholeiite with its iron content in a rather reduced state.

	1	2	3	4	5
SiO_2	52.3	52.7	50.6	50.1	52.6
TiO_2	1.2	1.1	1.9	1.9	1.1
Al_2O_3	14.7	14.2	13.6	14.5	14.9
Fe_2O_3	1.9	13.9	3.2	3.7	1.4
FeO	10.6		9.9	10.4	8.7
MnO	0.2	0.2	0.2	-	0.2
MgO	5.3	5.6	5.5	6.2	6.5
CaO	9.9	9.9	9.4	9.8	10.9
Na_2O	2.6	2.5	2.6	2.6	2.1
K_2O	0.3	0.6	0.7	0.5	0.6
P_2O_5	0.2	-	0.4	0.3	0.1
H_2O+	0.6	-	2.1	-	0.5
H_2O-	0.5	-		-	0.1
Total	100.3	100.7	100.1	100.0	99.7

Table 1. Analyses of tholeiitic basalt and diabase. (1) Southbury, Connecticut; (2) Average high iron quartz-normative diabase [33]; (3) Average Deccan trap, India [32]; (4) Precambrian diabase north of Ottawa graben, N. Gray analyst; (5) Standard diabase W-1, Virginia [16].

An experimental investigation has been carried out (Philpotts, in press) to determine the factors controlling immiscibility for this particular composition. The experiments, which were performed at one atmosphere total pressure, reveal that two silicate liquids coexist over a wide temperature interval during crystallization, and that this interval is increased by raising the oxygen fugacity. Experiments were also done on the standard diabase W-1, and it too exhibits a period of liquid immiscibility during crystallization.

The conclusion seems inescapable that liquid immiscibility is to be expected during the crystallization of many tholeiitic magmas and at a sufficiently early stage of solidification to play a role in differentiation. Why, then, are more examples not known from rapidly cooled rocks? A few examples, of course, have been described [10,29], but far fewer than might be expected. The scarcity of truly fresh rocks containing glass is certainly a contributing factor, but the main cause is probably that immiscibility has not been looked for. A cursory examination of the collections at the University of Connecticut revealed several new occurrences, the most important of which is from the mid-Atlantic ridge in the vicinity of 26°N. The basalts from this locality form pillows, the cores of which are fine-grained and contain two

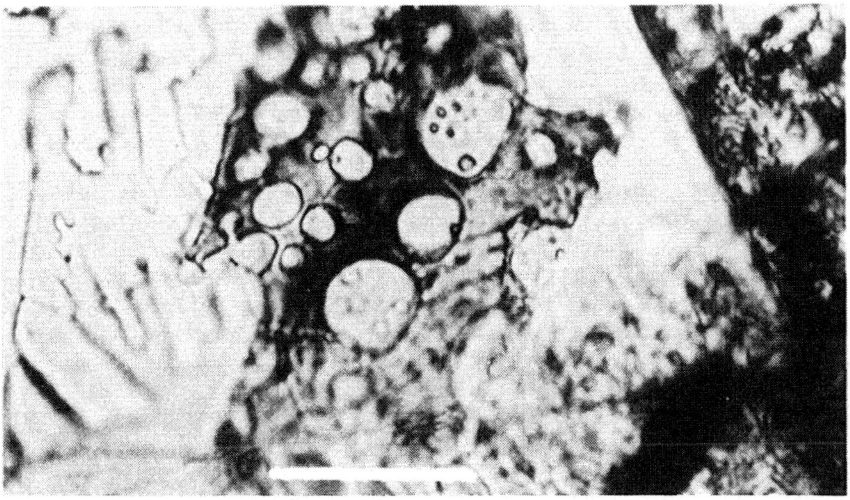

Figure 3. Patch of brown iron-rich glass containing spheres of clear silica-rich glass between plagioclase laths in basalt, Southbury, Connecticut. Note the smaller droplets of brown glass in the larger clear glassy spheres. Ordinary transmitted light. Bar is 10μ.

Figure 4. Euhedral plagioclase crystal surrounded by a mesostasis consisting of magnetite dendrites (white), spheres of iron-rich glass (light gray) in a silica-rich glass (dark). Grains at center top and lower right are augite. Note great variation in size of immiscible droplets. Reflected light. Bar is 10μ.

generations of plagioclase and olivine phenocrysts. A glassy mesostasis between laths of groundmass plagioclase consists of small magnetite dendrites and abundant brown glassy beads set in a clear glassy host (Fig. 5). The textures appear identical to those in the Connecticut basalt, suggesting that at least some of the lava related to the spreading of the Atlantic has not changed character significantly since the Jurassic.

5. FURTHER EVIDENCE FOR LIQUID IMMISCIBILITY

Evidence for liquid immiscibility is not limited to tholeiites. Alkaline magmas do not quench easily to glasses and, therefore, are not likely to preserve the least controversial textural evidence for immiscibility, that is, two glasses; however, they do commonly exhibit textures that have been interpreted as resulting from immiscibility. These have recently been discussed elsewhere

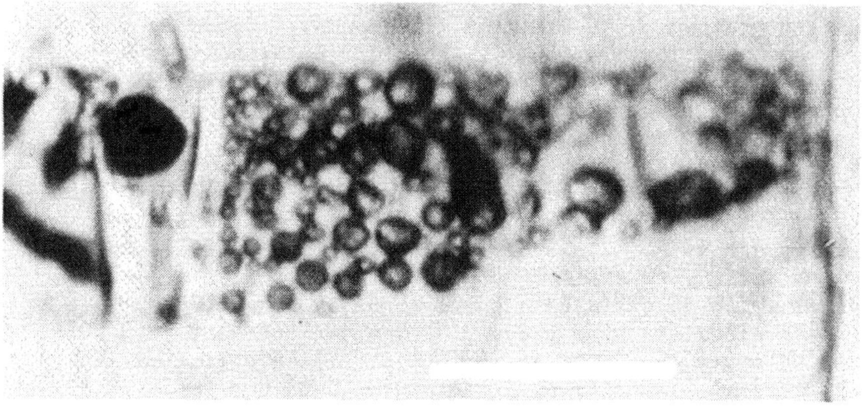

Figure 5. Mesostasis consisting of brown glassy beads in a clear glass between plagioclase laths in basalt from mid-Atlantic ridge at 26°N. Ordinary light. Bar is 10 μ.

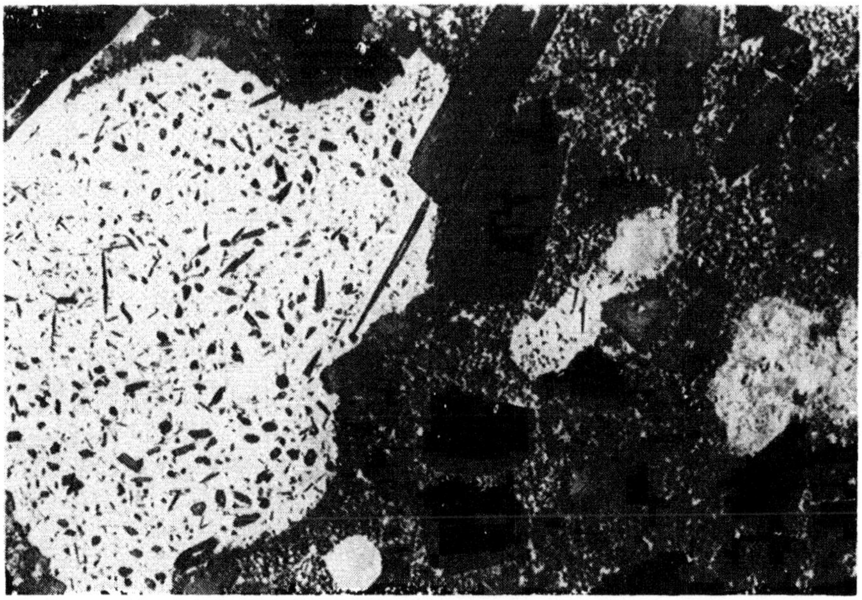

Figure 6. Analcime syenite ocellus in porphyritic monchiquite, Montreal, Quebec. Hastingsite and titanaugite crystals are acicular in the ocellus and stubby in the surrounding rock. Ordinary light. Height of ocellus is 1 cm.

[26] and so only brief mention will be made of them here.

Fine-grained mafic alkaline dike rocks, especially camptonites, commonly contain ocelli, spherical to globular bodies composed largely of alkali feldspar, but also containing minor amounts of the same ferromagnesian minerals as occur in the mafic host rock (Fig. 6). Analcime, nepheline, or quartz may also be present depending on the bulk composition of the rock. Most of the minerals in ocelli are typically igneous, in contrast with those in amygdales, and in some rocks, both ocelli and amygdales can be seen together as distinct structures. Ocelli have the ability to coalesce and form larger bodies of felsic material which in sills take the form of sheets which collect in the upper parts of these bodies. These textures indicate that ocelli probably originate as immiscible silicate liquids, an interpretation supported by experimental evidence for certain compositions.

Ocellar rocks are found associated with each of the Precambrian syenite-granite complexes along the St. Lawrence valley rift system. They are also found amongst the lamprophyric dikes associated with the White Mountain complexes, and they are extremely abundant in the Monteregian province. Thus, if the above interpretation is correct, immiscible liquids may have played an important role in the differentiation of all these intrusions.

Analyses of separated ocelli and their host rocks from the Monteregian province indicate that these two phases have appropriate compositions to produce, by accumulation, the contrasting mafic and felsic rock types of the main intrusions. Thus, with increasing silica content, mafic magmas will separate ocelli that range from nepheline monzonite, through quartz monzonite and syenite, to granite. For example, the ocelli and mafic host rock in Figure 6 have identical compositions to those of the pyroxenite and analcime syenite shown in Figure 2. This similarity extends even to the change in morphology of the pyroxene and amphibole crystals between the two phases, that is, stubby crystals in the mafic phase and acicular ones in the felsic phase.

6. POSSIBLE ROLE OF LIQUID IMMISCIBILITY IN RIFT-ASSOCIATED MAGMATISM

Since liquid immiscibility may have played an important role during the solidification of many rift-associated magmas, it may also have been important during their formation. Of course, differences in pressure between the source region and the final site of crystallization may alter phase relations and thus change the extent of immiscibility. Nakamura [22], for example, found that higher pressures diminish the size of the immiscibility field in the system fayalite-leucite-silica. The existence of an immiscibility field, however, depends on many factors and with existing data there are no reliable means of predicting the effect of changing pressures on the extent of immiscibility in a system as complex as a magma. Still, field observations and low pressure experiments demonstrate that these magmas can develop immiscible liquids. Thus, while acknowledging the uncertainties due to different pressures, it is proposed that liquid immiscibility plays a major role in controlling the initial compositions of some tholeiitic magmas and mildly alkaline syenites and granites. Furthermore, while immiscibility does not appear to have been involved with the generation of the strongly alkaline basic magmas of the Monteregian province, it did play a role in their differentiation at higher levels in the crust.

The Connecticut basalt, containing the abundant immiscible liquids, is almost identical to Weigand and Ragland's high iron quartz-normative type diabase; the Precambrian diabase from north of the Ottawa graben is also chemically similar (Table 1). Experiments on the basalt indicate that at low pressures this composition lies close to the intersection of the plagioclase-clinopyroxene cotectic with an immiscibility field (Philpotts, in press) and, thus, with fractional crystallization of plagioclase and pyroxene, or with the assimilation of siliceous rocks, the magma composition moves into the immiscibility field and a second liquid forms which is richer in silica and alkalies, and poorer in iron, magnesium, and calcium than the first liquid.

The proximity of this basaltic composition to the point of intersection of the plagioclase-pyroxene cotectic with the immiscibility field suggests the two are related and that perhaps under higher pressures the two points coincide. The relative constancy of composition of this particular magma would then be attributable ot the coexistence of a separate, more siliceous magma. Changes in the amount of melting in the source region would simply result in changes in the proportions of the two magmas and not in their compositions.

In order to obtain a magma of this basaltic composition, fractional melting must proceed through the immiscibility field. For example, laboratory melting of the high iron quartz-normative diabase results, first, in the melting of the patches of granophyre, followed by increased melting of sodic plagioclase, after which droplets of an immiscible iron-rich liquid appear as the main ferromagnesian minerals become involved with the melting. Continued heating increases the proportion of the iron-rich liquid until all of the silica rich liquid goes into solution in it. By analogy it is thought that the first fractional melting product in the source region would be granitic in composition and that with increased temperature the melt would become less siliceous, reaching a quartz syenitic to monzonitic composition at the boundary of the immiscibility field. On further heating, an iron-rich tholeiitic basalt would appear bearing an immiscible relation to the more siliceous magma. If, as Chapman [6] suggests happened in the White Mountain complexes, the rise of basaltic magma into the crust produced granitic magma by selective melting, this would be expected not to mix with the basaltic magma and thus give rise to the compound magma chambers.

With such a melting process it is envisaged that quartz syenitic and iron-rich tholeiitic magmas should be common and that the presence of one implies the existence of the other, although the proportions of each could vary greatly. For instance, mostly syenites were intruded in Cambrian time along the St. Lawrence valley rift system, except at Sept-Iles where gabbro and anorthosite were also emplaced. Similarly, in the White Mountain complexes mostly syenitic and granitic rocks were intruded with only minor amounts of gabbro. In the Triassic-Jurassic basins, however, the magmas were almost exclusively basaltic, but the one occurrence of trachyte in northern Virginia is evidence that other siliceous magmas may have been formed but not intruded.

The actual abundance of mafic and felsic rocks probably depends on the mechanism of intrusion and the tectonic setting at the time of emplacement as much as it does on the actual proportions of the two magmas produced. Throughout eastern North America dikes that are not closely related to main intrusions are either tholeiitic or lamprophyric. Syenitic and granitic dikes are limited to the vicinity of main intrusions from whence they presumably came. It is likely, therefore, that dikes, which are igneous bodies with the minimum thickness to volume ratio, would not permit viscous magmas, such as granitic or syenitic ones, to intrude fast enough to rise a significant distance before cooling. Pipe-like bodies, which have the maximum thickness to magma volume ratio, are probably the only ones that would do this.

The intrusion of any igneous body, of course, is dependent on there being a net upward force on the magma. Buoyant forces always exist on felsic magmas because their densities are lower than those of crustal rocks, but with mafic magmas this is not necessarily the case. For example, according to the partial molar volume data of Bottinga and Weill [2] a magma with the composition of the basalt from Southbury, Connecticut, would have a density of 2.70 gm cm^{-3} and, therefore, would not be forced to the surface by many crustal rocks. Additional forces would have been required which could have been provided by the magma source being located at some depth in the denser mantle, or by pressures generated by volume expansion on melting in the source region.

Regardless of the cause of the extra pressure, when the magma rose to a level in the crust where its density was more than that of the intruded rocks, lateral spreading of the magma and lifting of the overlying cover would have required less energy than continued intrusion to the surface, unless there were prominent fractures for the magma to follow [28]. The crustal extension in eastern North America associated with the initial spreading of the North Atlantic could certainly have provided such fractures. By Middle Jurassic, however, subsidence in the basins had ended and tensional fractures may have closed. Hence, although considerable basaltic magma may still have existed, it was largely the syenitic and granitic ones that were able to intrude buoyantly through the crust to form the White Mountain complexes.

Intrusion of dense magma along fractures through less dense rock must be a rapid process, otherwise lateral rather than upward migration of magma would occur. The rather close chemical groupings obtained by Weigand and Ragland [33] for Mesozoic diabases would also not be expected if intrusion had been slow. For instance, slow emplacement of the Southbury basalt could have brought about very strong differentiation, for in the experiments on this rock, two weeks was sufficient time to effect considerable coalescence of the droplets of immiscible liquid. In nature such coalescence would soon have resulted in the formation of bodies of felsic rock.

Rapid intrusion is also indicated by the results of a recent detailed study of a single 150 kilometer long, 50 meter wide diabase dike that extends through eastern Connecticut and Massachusetts (Fig. 1). By studying the chilled margin of this dike, Koza [17] found that almost all chemical variation along its length could be attributed to variations in the percentage of glomeroporphyritic aggregates of augite. The remainder of the rock, which consists of euhedral plagioclase and orthopyroxene phenocrysts in a fine-grained groundmass, is of remarkably constant composition. At one locality a second intrusion of diabase along the center of the dike must have occurred at least some tens of years later, because the second intrusion has a fine-grained chilled margin against the earlier diabase. Despite the time interval involved, this chilled margin is identical to the earlier one in terms of the composition of the groundmass and the percentage of plagioclase and orthopyroxene phenocrysts. These relations seem to imply the generation of a very constant composition starting material that contained varying proportions of refractory augite aggregates, and that this was injected rapidly before any segregation or separation of plagioclase or orthopyroxene phenocrysts could occur.

Intrusion of basic dikes in the Monteregian province also

appears to have been a rapid process, for many of them contain abundant xenoliths of underlying rocks which would have sunk, had there not been rapid upward flow. Some of these fragments are mantle derived, and contain hypersthene crystals which were certainly not in equilibrium with the critically undersaturated basic magmas at the depth at which solidification occurred. Again, had intrusion been slow, these xenoliths would have reacted completely with the surrounding magma. Some of the dikes contain so many xenoliths that injection appears to have been explosive.

These basic dikes differ from those associated with the Triassic-Jurassic basins in that they cover a wide range of compositions. They are all critically undersaturated in silica, but there appears to be no simple means of deriving them all from a common parental magma. Most of the different basic dikes more likely represent partial melting fractions from either a laterally zoned mantle or from different depths in a homogeneous or layered mantle.

The importance of the compositional range of these dikes is that a similar range of compositions is not found amongst the coarse-grained rocks of the main intrusions. In these, there is a significant lack of intermediate rocks with compositions between alkali gabbro and nepheline monzonite. Fine-grained dike rocks falling within this range are characterized by the ocellar texture which, as discussed previously, is interpreted as resulting from immiscibility. It was concluded, therefore, that at the depth of origin of these alkaline magmas there is no immiscibility field, but at lower pressures there is, so that dikes intruded rapidly toward the surface develop the ocellar texture before solidifying. In larger intrusions, however, slower cooling permits separation of the immiscible liquids and formation of contrasting rock types [26].

Although the Monteregian intrusions and White Mountain complexes are closely related geographically and some of the rock types are similar, the fact that each Monteregian intrusion is different from the next, whereas most of the White Mountain complexes are very similar to each other, indicated a profound difference between the two groups. The Monteregian intrusions are also clearly related to the extension of the Ottawa graben, whereas the White Mountain complexes are not directly related to a rift, but instead can be considered to cross the trend of the Triassic-Jurassic basins.

The rather constant, repetitive sequence of intrusion in the White Mountain complexes is consistent with these bodies having formed as the North American plate moved over a hot spot. Under these conditions melting may have occurred at a very high level in the mantle or even in the lower crust. The Monteregian intrusions, on the other hand, were probably formed by much smaller degrees of fractional melting at greater depths in the mantle along an ancient rift system that underwent readjustments at essentially the same time as the White Mountains were over the hot spot.

Immiscibility in rift-associated magmas may not be restricted to continental areas. The basalts from the mid-Atlantic ridge at 26°N appear identical to those from Southbury, Connecticut, and had they cooled in some plutonic body, segregations of felsic rock would have undoubtedly developed. Many of the composite dikes and flows of Iceland [1,34] could have resulted from such segregations. Development of these contrasting rock types through liquid immiscibility at spreading oceanic ridges may have played an important role during early Earth history in producing the felsic rocks necessary for the development of continents.

Although seriously doubted for over fifty years [4], the existence of liquid immiscibility at low pressures in certain important magmas associated with rift systems can no longer be questioned. However, determination of the validity of the speculations presented here concerning the part played by immiscibility in the initial generation of these magmas must await the results of high pressure experimental studies.

REFERENCES

1. D.H. Blake, R.W.D. Elwell, I.L. Gibson, R.R. Skelhorn and G.P.L. Walker, Quart. Jour. Geol. Soc. London, **121**, 31, 1965.
2. Y. Bottinga and D.F. Weill, Am. Jour. Sci., **269**, 169, 1970.
3. P. Bowden and D.C. Turner, Peralkaline and associated ring complexes in the Nigeria-Niger province, West Africa; in: The Alkaline Rocks, ed. by H. Sørensen, New York, John Wiley & Sons, 1974.
4. N.L. Bowen, The evolution of the igneous rocks, Princeton University Press, Princeton, New Jersey, 1928.
5. C.A. Chapman, A comparison of the Maine coastal plutons and the magmatic central complexes of New Hampshire, in: Studies in Appalachian geology: Northern and Maritime, ed. by E-an Zen, W.S. White, J.B. Hadley and J.B. Thompson Jr., New York, Interscience Pubs., Inc., 1968.
6. C.A. Chapman, Structural evolution of the White Mountain magma series, in: Studies in New England Geology, ed. by P.C. Lyons and A.H. Brownlow, Geol. Soc. America Mem., **146**, 1976.
7. T.H. Clark, Montreal area., Quebec Dept. Nat. Resources Geol. Rept., **152**, 1972.
8. B. Cornet, Geological history of the Newark Supergroup in capsule form. Unpub. ms., 1975.
9. B. Cornet and A. Traverse, Geoscience and Man., **11**, 1975.

10. A. De, Geol. Soc. Amer. Bull., 85, 471, 1974.
11. R. Doig and J.M. Barton Jr., Canadian Jour. Earth Sci., 5, 1401, 1968.
12. W.F. Fahrig and R.K. Wanless, Nature, 200, 934, 1963.
13. K.A. Foland, A.W. Quinn and B.J. Giletti, Am. Jour. Sci., 270, 321, 1971.
14. D.P. Gold, Alkaline ultrabasic rocks in the Montreal area, Quebec; in: Utramafic and Related Rocks, ed. by P.J. Wyllie, John Wiley and Sons, New York, 1967.
15. M.D. Higgins and R. Doig, Nature, 267, 40, 1977.
16. C.O. Ingamells and N.H. Suhr, Geochim. et Cosmochim. Acta., 27, 897, 1963.
17. D.M. Koza, Petrology of the Higganum diabase dike in Connecticut and Massachusetts. M.Sc. thesis, University of Connecticut, 1976.
18. P.S. Kumarapeli and V.A. Saull, Canadian Jour. Earth Sci., 3, 639, 1966.
19. P.R. May, Geol. Soc. America Bull., 82, 1285, 1971.
20. W.J. Morgan, Nature, 230, 42, 1971.
21. G.S. Murthy, Canadian Jour. Earth Sci., 8, 802, 1971.
22. Y. Nakamura, Carnegie Inst. Washington Year Book 73, 352, 1974.
23. E.F. Osborne, Quebec. Trans. Roy. Soc. Canada, 28, 49, 1934.
24. V.S. Papezik, J.P. Hodych and A.K. Goodacre, Canadian Jour. Earth Sci., 12, 332, 1975.
25. A.R. Philpotts, The Monteregian province; in: The Alkaline Rocks, ed. by H. Sørensen, New York, John Wiley & Sons, 1974.
26. A.R. Philpotts, Am. Jour. Sci., 276, 1147, 1976.
27. A.R. Philpotts, Quebec Dept. Nat. Resources Geol. Rept. 156, 1976.
28. J.L. Roberts, The intrusion of magma into brittle rocks, in: Mechanism of Igneous Intrusion, ed. by G. Newall and N. Rast, Gallery Press, Liverpool, 1970.
29. E. Roedder and P.W. Weiblen, Petrology of silicate melt inclusions, Apollo 11 and 12, and terrestrial equivalents, in: Second Lunar Sci. Conf. Proc. Geochim. et Cosmochim. Acta Suppl. 2, 1, 507, 1971.
30. E.C. Toewe, Virginia Div. Mineral Resources Rept. Inv., 11, 1966.
31. F.B. van Houten, Am. Assoc. Petroleum Geologists Bull., 61, 79, 1977.
32. H.S. Washington, Geol. Soc. America Bull., 33, 765, 1922.
33. P.W. Weigand and P.C. Ragland, Contr. Mineral Petrol., 29, 195, 1970.
34. H.S. Yoder Jr., Am. Mineralogist, 58, 153, 1973.

BASALTIC AND ALKALIC ROCKS OF SOUTHERN BRAZIL: A REVIEW

Norman Herz

Department of Geology, University of Georgia
Athens, Georgia 30602, USA

ABSTRACT. General domal uplift may have begun in early Mesozoic in the South Atlantic preceding volcanism by at least 50 m.y. Rifting and the start of basaltic activity began 147 m.y. ago: diabase dikes filled in rifts parallel to the coast and alkalic volcanism took place at Jacupiranga and Anitápolis 122-138 m.y. ago. The Paraná basalts include a complete range of types, but chemically, they are largely alkalic or high alumina. Rüegg [1] divides them into four general geochemical areas and concludes they originated by fractional crystallization from an olivine tholeiite magma, underwent differentiation and crystal fractionation at some depth, show some degree of contamination, and ascended slowly or in stages to the surface.

Alkalic rocks are divided into five sets: (1) Jacupiranga and Anitápolis nodes, 119-138 m.y. with high nepheline types of the urtite-ijolite series as well as carbonatites and pyroxenites; (2) Tunas and Lajes groups, 65-110 m.y., with rock types similar to (1); (3) Minas Gerais-Goiás belt, 64-91 m.y., with the most K- and rare earth-enriched compositions of all; (4) São Paulo-Rio de Janeiro littoral belt, 51-80 m.y., salic nepheline syenites with an Na to alkali ratio of about 65: (5) Atlantic Islands, 0.2-3.6 m.y. with high Na to alkali ratios and rock types including tannbuschites, nepheline and phonolite pyroclastics and flows, and ankaratrites. (1) coincides with triple junction development; (2) with actual plate separation and deposition of marine sediments.

E.-R. Neumann and I.B. Ramberg (eds.), Petrology and Geochemistry of Continental Rifts, 155–162.

1. INTRODUCTION

Southern Brazil is a classical area for igneous petrologists; the great Paraná flows and the alkalic rocks have long been known and described in the world's geological literature. In recent years, more detailed information on the geochemistry and geochronology of the basalts and some alkalic rocks as well as on the stratigraphy and geophysics of the continental margins has become available [1,2]. Plate tectonic modelling of these data have led to a greater understanding of early magmatic and tectonic processes associated with early continental rifting in the South Atlantic [3,4]. These models agree generally with earlier ones based on evidence from deep sea drilling and extrapolation from magnetic anomalies present near the central part of the ocean [5].

Martin [6] has pointed out evidence for general domal uplift previous to 147 m.y. ago, the time of the first eruption of the Paraná basalts. Upper Paleozoic drainage across the Precambrian Damara fold belt basement of Southwest Africa was impeded starting in early Mesozoic. Triassic to lower Cretaceous sediments and lava flows overstep the highest elevations, suggesting that uplift and initial rifting, but not spreading, had already started in Triassic times. Uplift of the rift shoulder reversed the ancient gradient, causing aggradation of sediments and lavas. Studies of sedimentary basins in Brazil [7] also suggest that a huge domal structure was developed in the south about this same time. Thus domal uplifting may have preceded rifting and volcanism by at least 50 m.y.

Domal uplift was then followed by eruption of the Paraná basalts in southern Brazil and neighboring South America, starting at 147 m.y. and climaxing at 122-133 m.y. ago [8], or about the same time as dolerites in Southwest Africa. The first nodes of alkalic activity, at Jacupiranga and Anitápolis, dated 122-138 m.y., probably mark the time of development of the mantle plume-triple junctions in the area (fig. 1).

Models proposed for other continental rifts can also explain these data [6,9]. Initial uplift is a result of thermal expansion in the upper mantle with concomitant emplacement of an aestheno-lith cushion into the base of the sialic crust. The crust thins by a combination of poorly understood subsurface processes, including subcrustal erosion by currents, loading by dense intrusions, oceanization, and gravitational spreading of less dense continental crust over oceanic lithosphere. Subaerial erosion, depending on the climate during uplift, can remove several times by the initial uplift; since vertical uplifts are also controlled by isostatic compensation, about 3.0 km of the crust may be removed for an uplift of only 1.5 km [9].

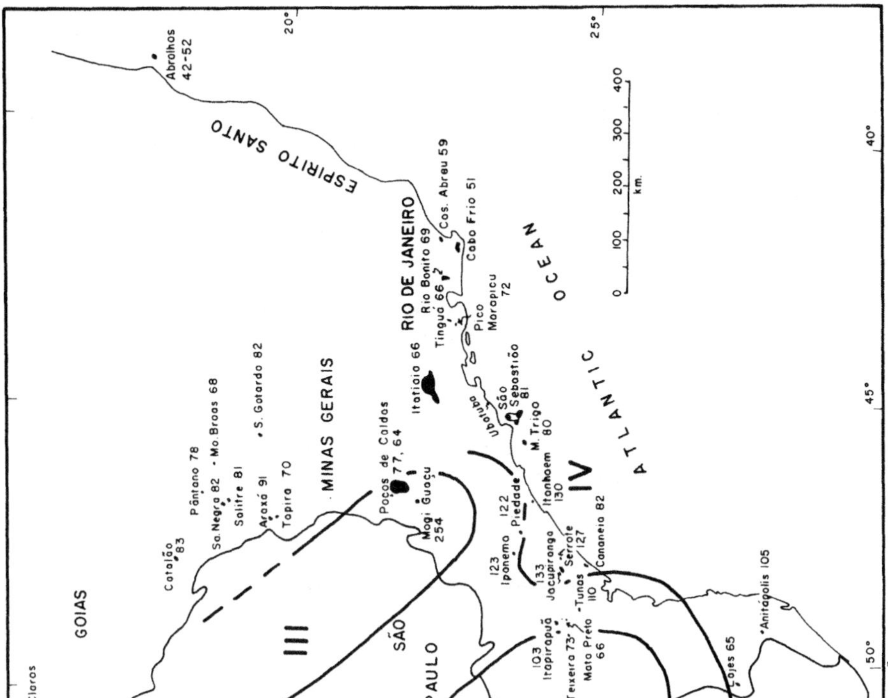

Figure 1. LOCATION MAP, AGES in m.y, BASALT areas [1,4]

2. PARANÁ BASALTS

By uppermost Jurassic, subsidence, extensive normal faulting and rift formation was well under way. The first appearance of the Paraná flood basalts was during Tithonian; they eventually covered an area of 1,200,000 km^2 in Brazil and about 4 million km^2 in all of South America [1] or about 2.5 per cent of the total land area of the earth. Although previous writers attempted to do so, the Paraná basalts could not be classified into any one accepted classical type. Mineralogically, they are made up of pyroxenes (augite and pigeonite), and plagioclase An_{50-70} which includes both high and low temperature structural types; normative plagioclase is generally andesine. Orthopyroxene and olivine as ragged inclusions within clinopyroxene have been described only rarely. Principal oxide minerals are titanomagnetite and ilmenomagnetite.

Rüegg [1] has shown that a great variation in chemical compositions does exist and all types, including normal tholeiite, olivine tholeiite, and alkali basalt are present. He delineated four geochemical areas (fig. 1), based on trend surface maps for the major elements [1, figs. 3.2-3.9]. On a Yoder and Tilley

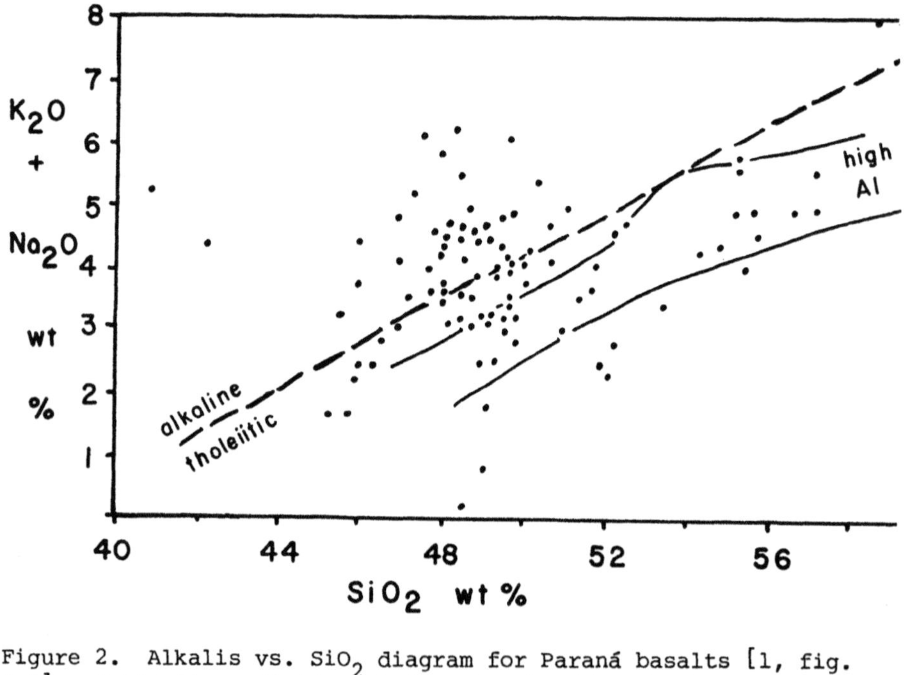

Figure 2. Alkalis vs. SiO_2 diagram for Paraná basalts [1, fig. 4.1]. Solid lines divide alkaline basalt (above), high alumina, and tholeiitic (below) fields (Kuno, 1966); broken line alkaline and tholeiitic (MacDonald and Katsura, 1964).

(1962) diagram, most of the analyses are supersaturated tholeiites [1]; on a Kuno (1966) diagram, they are mostly alkali or high alumina (fig. 2); on a K vs. K/Rb diagram [1] they fall into a general alkaline basalt field. It is apparent that the Paraná basin does not represent a uniform basaltic province. In area I, Rüegg [1] concludes that the rocks differentiated from tholeiitic to almost rhyolitic in composition. They are largely extrusive and underwent differentiation before emerging. Areas II and III show a differentiation trend towards andesites. Fractional crystallization, with a systematic increase in alkalis, SiO_2, as well as Ti and Fe, and with Na_2O/K_2O much higher than I characterizes these areas. Area IV has examples of the most advanced stages of differentiation; trace elements such as Ba, Rb, Sr, Zr, are high while Co, Cr, Ni, and V are low [1, 3.1-3.2].

Rüegg [1] concludes that the basaltic rocks of the Paraná basin: (1) may have originated by fractional crystallization following segregation of olivine at low pressures, from a primary magma of olivine tholeiite composition, generated by partial fusion of mantle material [10], (2) underwent differentiation and fractional crystallization at some depth: lower pressure for areas I and III and intermediate for II and IV, (3) show evidence of contamination by crustal materials: extensive in the case of area I basalts, shown by low K/Rb and Rb/Sr ratios, and moderate elsewhere, shown by abnormally high (for tholeiites) K, TiO_2 (in II and III), Ba, Rb, Zr, and Rb/Sr ratio. Contamination and crystal fraction imply that the basaltic magma either ascended slowly, or ascended in stages during which fractional crystallization and some contamination took place.

3. ALKALIC ROCKS

Alkalic rocks of southern Brazil associated with early continental rifting and presumably related to the Paraná basalts can be divided into five distinct sets, based on absolute ages [11] and geographic location [4; fig. 1]:

(1) synchronous with Paraná basalt volcanism, 119-138 m.y. ago: the Jacupiranga and Anitápolis nodes,
(2) younger alkalic rocks, associated geographically with (1), 65-110 m.y. old: the Tunas group and Lajes,
(3) Minas Gerais-Goiás belt, 64-91 m.y. old,
(4) São Paulo-Rio de Janeiro littoral belt, 51-80 m.y. old,
(5) Atlantic Islands, 0.2-3.6 m.y. and Abrolhos, 42-52 m.y. old.

(1) Emplacement of alkalic rocks began about the same time as the start of mafic dike emplacement in rifts along the São Paulo coast. From about 133 to 122 m.y., alkalic rocks of the Jacupiranga node were emplaced in an area roughly 200 km ENE by 125 NNW. Rock types

include ultramafic alkalics such as jacupirangite, pyroxenite, as well as carbonatite, shonkinite, nepheline syenite, and tinguaite. About 129 m.y. ago, the Anitápolis node was also active with alkalic mafic to ultramafic rocks. Country rocks were fenitized in both areas.

(2) After a quiescent period of more than 10 m.y., alkalic activity was renewed 110 m.y. ago with the emplacement of the Tunas group of chimneys consisting of quartz syenites, syenodiorites with minor gabbro, breccia, and trachytic dikes, and Lajes where tinguaite as well as sheetlike bodies of nepheline syenite and phonolite were emplaced.

The first true marine environment developed in the area about Aptian time with the appearance of thin-bedded limestone and evaporites overlying basal conglomerates and sandstones of a transitional terrestrial deposition. This marks the first significant incursion of the embryonic Atlantic Ocean and suggests that spreading had begun by about 112 m.y. ago. The Tunas Group of alkalic rocks is aligned on the Ponta Grossa Arch and may signify volcanism associated with the triple junction that was generated by a hot spot along which actual plate separation had begun. Jacupiranga and Anitápolis may represent surface expressions of mantle plumes.

(3) Twenty m.y. after the start of Tunas Group volcanism, alkalic volcanism began along an ancient Brazilian Cycle foldbelt suture that divides the Archean Guaporé Platform to the west from the Proterozoic São Francisco Platform to the east. These rocks lie in a rough belt that is perpendicular to the coast and skirts around the eastern and northern sides of the Paraná basalts. The rocks of the belt include a great variety of alkalic types from ultramafic to nepheline syenites. The belt has the highest K_2O/Na_2O ratios of the Brazilian alkalic rocks and is well known for its economic mineral concentrations, including deposits of rare earths, niobium, zirconium, uranium, thorium, titanium, and phosphorus. The largest diamonds of Brazil have been found in the western part of Minas Gerais, an area which has kimberlitic plugs, although all mining to date has been alluvial.

(4) From about 83 to 51 m.y. ago, a belt of alkalic rocks was emplaced along the present-day São Paulo-Rio de Janeiro littoral. About 80 m.y. ago, the spreading center pole of rotation shifted from 19°N lat, 40°W long to its present 67°N lat, 40°W long [5]. Presumably, at this time the Atlantic Ocean had become wide enough so that the constraints imposed by overlapping continental parts of South America and Africa were no longer operative. The spread of ages in the littoral belt, from 83 to 51 m.y. is fairly uniform and becomes younger west to east, which suggests the passage of the plate over a hot spot. Other evidence for a single hot spot origin is seen in the remarkable uniform compositions in the belt: the rocks are largely salic nepheline syenites with an average $Na_2O/(Na_2O + K_2O)$ ratio of about .65 and most of their

chemical analyses plot in a small area of the $NaAlSiO_4$-$KAlSiO_4$-SiO_2 diagram near a thermal minimum [4].
(5) The islands of Trindade and Martin Vaz are about 1.300 km east of the coast and lie on the eastern end of the Columbia Seamounts. They include, from oldest to youngest, tannbuschite, nepheline and phonolitic pyroclastics, nephelinite flows, and ankaratrites.

4. THE MODEL

Precursor - doming: at least 200 m.y. ago

Stage I, Tithonian-Hauterivian: first basalts, 147 m.y. ago; start of rift and graben tectonics; mafic dike emplacement parallel the coast and Jacupiranga and Anitápolis alkalic nodes 122-138 m.y. Triple junction outlined: two arms parallel coast and failed arm trending northwest into thickest section of Paraná basalts. Sedimentation rift-filling sandstones and conglomerates, deltaic-lacustrine types. Normal faults have throws of 3,000-5,000 m.y. Doming renewed in Barremian with gap in volcanic and sedimentary record.

Stage II, Aptian-Santonian: Tunas and Lajes alkalic rocks and basalts erupt 110 m.y. ago; Minas Gerais-Goiás belt emplaced starting 91 m.y. ago. Continental margin sandstones and shales with evaporites show plate separation started.

Stage III, Santonian-Early Paleocene: littoral alkalic belts develop 82 to 51 m.y. ago and suggest traverse over a hot spot. Alkalic rocks also emplaced in Minas Gerais-Goiás belt and Tunas. Sedimentation continental margin and marine; unconformity in Upper Cretaceous marking northerly shift in pole of rotation and accelerated plate separation velocity.

Stage IV, Early Paleocene-Holocene: mafic alkalic volcanism continues at Trindade which may mark a present-day hot spot.

ACKNOWLEDGMENTS

I am indebted to H.E. Asmus for many helpful comments and to J.A. Whitney for critically reviewing the manuscript.

REFERENCES

1. N.R. Rüegg, Modelos de variaçao química na província basáltica do Brasil Meridional, Univ. São Paulo, Inst. Geoc., v. I, II, 1975.

2. F.F.M. de Almeida, ed., Continental margins of Atlantic type, An. Acad. Bras. Cien., 48, supl., 1976.
3. F.C. Ponte and H.E. Asmus, The Brazilian Marginal Basins: current state of knowledge, in 2, 215.
4. N. Herz, Geol. Soc. Am.B. 88, 111, 1977.
5. X. Le Pichon and D.E. Hayes, J. Geophys. Res. 76, 6283, 1971.
6. H. Martin, A geodynamic model for the evolution of the continental margin of Southwestern Africa, in 2, 169.
7. H.E. Asmus, Rev. Bras. Geoc. 5, 160, 1973.
8. G. Amaral, U.G. Cordani, K. Kawashita and J.H. Reynolds, Geochim. et Cosmochim. Acta 30, 1959, 1966.
9. N.H. Sleep, Geophys. J. Roy. Astr. Soc. 24, 325, 1971.
10. D.H. Green and A.E. Ringwood, Cont.Min.Pet. 15, 103, 1967.
11. G. Amaral, J. Bushee, U.G. Cordani, K. Kawashita and J.H. Reynolds, Geochim. et Cosmochim. Acta 31, 117, 1967.

THE GARDAR IGNEOUS PROVINCE: EVIDENCE FOR PROTEROZOIC CONTINENTAL RIFTING

B.G.J. Upton and D.J. Blundell

Department of Geology, Edinburgh University, Edinburgh, U.K. and Department of Geology, Chelsea College, London, U.K.

ABSTRACT. Three principal episodes (at c. 1300, 1250 and 1170 Ma.) of faulting and associated alkaline magmatism are recognised in the evolution of the Gardar province in south Greenland.

Terrestrial sediments and lavas, accumulated in fault-bounded troughs and cut by Early Gardar alkaline complexes, constitute evidence for continental rifting and concomitant volcanism in the first episode. Subsequent (c. 1250 Ma.) basic dykes, followed by faulting and intrusion of alkaline central complexes are related to a Mid-Gardar episode of continental rifting.

The Late Gardar episode involved intrusion of ENE and NE trending dyke swarms with further faulting and emplacement of alkaline complexes, within a zone c. 70 km broad. One major sub-zone (Tugtutôq - Narssaq) characterised by massive 'giant dykes', followed by smaller dykes and cross-cutting central complexes, is centred upon a closed gravity 'high', c. 80 km long by 25 km broad. This anomaly is attributed to an underlying mass of high density rocks at shallow depth.

Evidence from the Gardar province suggests that continental rifting was recurrent over more than 100 Ma. and that the magmatic history was controlled by the structure and character of the underlying lithosphere.

1. INTRODUCTION

The Precambrian Gardar igneous province [1] represents a 200 Ma. period of persistent fracturing and alkaline magmatism starting

E.-R. Neumann and I.B. Ramberg (eds.), Petrology and Geochemistry of Continental Rifts, 163–172.

about 1350 Ma. before present. Although it lies mainly within an area of plutonic rocks generated during the Ketilidian orogeny (1850-1600 Ma; [2]), in the north-west of the province Gardar magmas intruded the marginal parts of the Archaean cratonic region which composes much of southern Greenland north of the Ketilidian belt. Rb-Sr isochron dating of the intrusive rocks [3] has allowed recognition within the province of two major and one minor intra-plate tectono-magmatic episodes, each of which is likely to have been associated with features which, in a Neogene setting, would be regarded as continental rift structures. The cratogenic association of alkaline igneous rocks, abundant faulting and the presence of sediment-lava piles clearly accumulated in a block-faulted environment has long suggested the idea of continental rift structures, comparable to those of East Africa, being associated with the evolution of the Gardar province [4]. These ideas may be assessed and more critically examined now that (a) the 1:20 000 scale mapping of the region is complete, (b) most of the major intrusions have been dated radiometrically and (c) gravity surveys have been conducted across some of the critical areas.

2. EARLY GARDAR EVENT (FIG. 1)

The three major nepheline syenite intrusive complexes comprising Grønnedal-Ika [5] and the two oldest centres of Igaliko [6] dated at 1327 $\pm$ 17, 1295 $\pm$ 61 and 1310 $\pm$ 31 Ma. respectively [3] define what may be loosely referred to as the 1300 Ma. event. A sequence of subaerial alkaline to sub-alkaline (predominantly basic) lavas and continental sediments (the Eriksfjord Formation; [7,8]) rests unconformably on the Ketilidian basement and is itself cut by the Igaliko intrusions. Geological evidence suggests that the Eriksfjord Formation accumulated rapidly in subsiding faultbounded graben, the sediments being derived from the surrounding highlands and the lavas erupted within the subsiding blocks. Although isotopic dating of the lavas (which are always more or less altered) has not been feasible, it is likely that they were erupted shortly before the emplacement of the nepheline syenite central complexes and that the petrogenesis of the intrusions and lavas is closely related. Further, it is inferred that the continental rifting that permitted accumulation of over 3 km thickness of Eriksfjord supracrustal materials followed the regional uplift of a peneplaned Ketilidian land surface. The faults bounding the rifts appear to have had predominantly E-W orientations.

3. MID GARDAR EVENT (FIG. 1)

In the western part of the province (Ivigtut region), NE and ENE-trending alkali olivine dolerite dykes are abundant. Some of these transect the Grønnedal-Ika complex, whereas others are cut

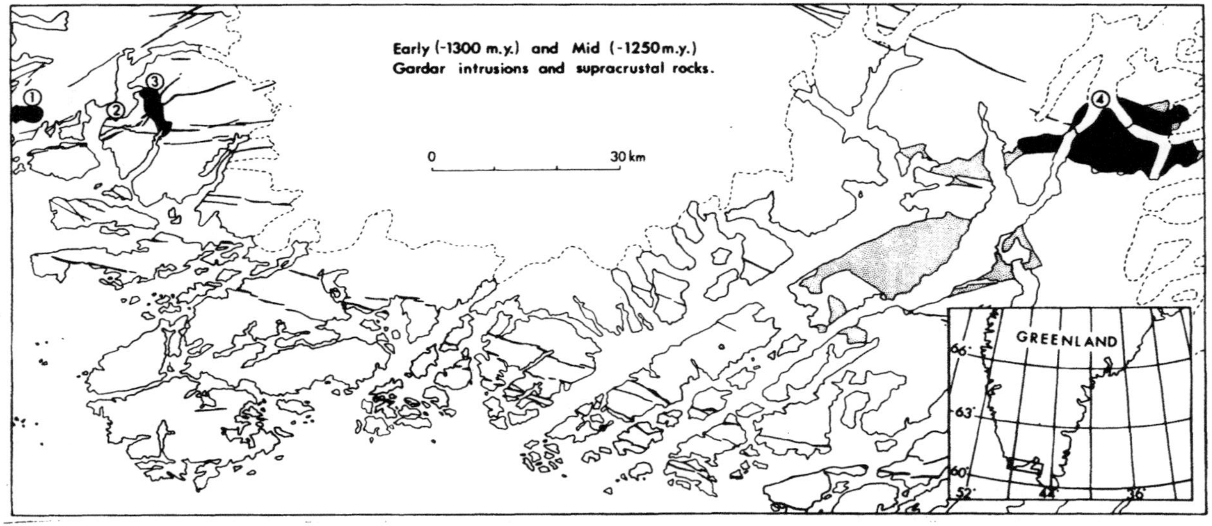

Fig. 1 Sketch map indicating relationships of Early and Mid Gardar rocks.

1. Kûngnât: 2. Ivigtut: 3. Grønnedal-Ika: 4. (Early) Igaliko: Stippled ornament - Eriksfjord Formation lavas and sediments.

by the alkaline complex at Kûngnât (1245 ± 17 Ma. [9,3]), so that the dyke swarm is largely or wholly bracketed within the interval 1327 ± 17 Ma. and 1245 ± 17 Ma. Recent Rb-Sr dating on five of these dykes [10] has provided confirmation of this, showing them to have ages between 1280 and 1250 Ma. Furthermore, other extensive basic dykes trending ESE-WNW, to the south and southeast of the Ivigtut region, have also yielded emplacement ages of approximately 1250 Ma [10].

Roughly mid-way between the Kûngnât and Grønnedal-Ika complexes lies the small Ivigtut complex (1248 ± 25 Ma. [11,12,3]). The episode of central-type activity that gave rise to the Kûngnât and Ivigtut instrusives with their dominantly silica-oversaturated rock suites is petrologically and geochronologically quite distinct from the earlier episode of central-complex formation which yielded the larger, strongly undersaturated complexes of Grønnedal-Ika and the (early) Igaliko centres. The three early Gardar complexes are severely affected by transcurrent faulting comprising left-lateral motions along a set of fractures trending approximately E-W, and right-lateral motions along another set of faults trending approximately N-S. These faults also affect the dykes but, as they are not observed in the immediate vicinity of the complexes of Kûngnât or Ivigtut, it is possible that the faulting occurred after the tensional (dyke) phase but before the ascent (by stoping and ring-faulting) of the Kungnat and Ivigtut centres. The evidence is compatible with the postulate of a mid-Gardar episode of crustal extension with dyke-emplacement terminated by (a) block faulting and (b) ascent of syenitic and granitic magmas.

4. LATE GARDAR EVENT (FIG. 2)

Some 100 km to the ESE of the Ivigtut area lies the island of Tugtutôq [13] whose coastlines, in common with those of many of the other islands and peninsulas in the region, are principally controlled by major ENE-trending fjords. Severe crushing of the Ketilidian basement close to these shores suggests that the fjords have been eroded along major fracture zones. Approximately midway between the postulated bounding faults (See Fig. 2), the island is traversed by a group of 'giant' bifurcating dykes with widths of up to 800 m, consisting of alkali gabbros and a wide spectrum of alkaline salic differentiates. These dykes, which in part exhibit internal layered structures comparable to those of the Rhodesian Great Dyke, resulted from intrusion, and subsequent slow cooling, of mildly alkaline to transitional olivine basalt magma. These dykes, dating from 1175 ± 9 Ma. [3] marked the commencement of a period of crustal dilation associated with the injection of a major swarm of ENE-trending dykes which tend to diminish in width while increasing in degree of fractionation with time. Further intrusion of major alkaline central complexes

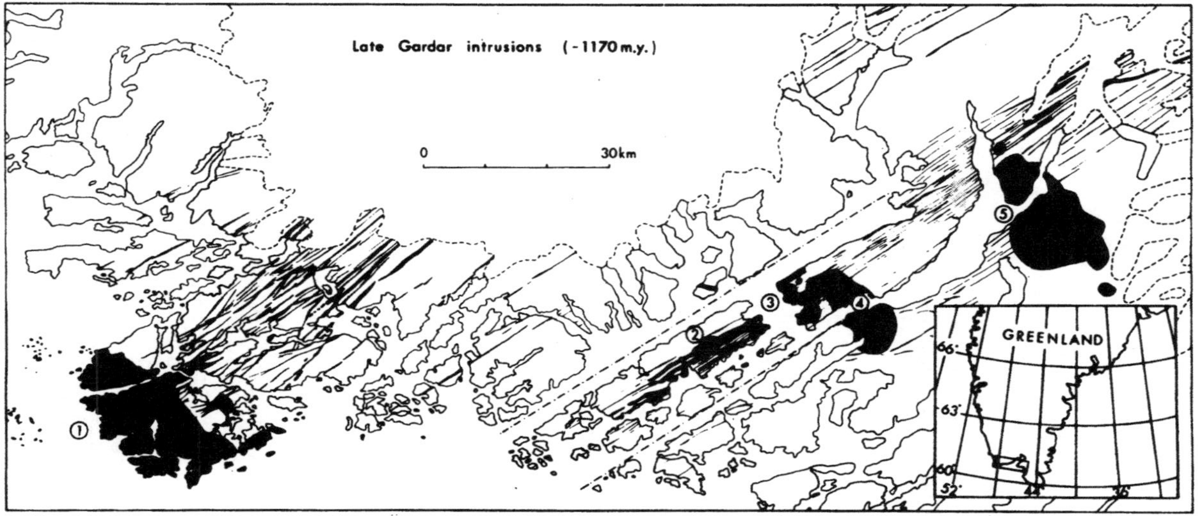

Fig. 2 Sketch map indicating relationships of Late Gardar intrusions.

1. Nunarssuit: 2. Central Tugtutôq: 3. Narssaq: 4. Ilímaussaq: 5. (Late) Igaliko. Approximate positions of the faults bounding the inferred Tugtutôq-Ilímaussaq graben indicated by dash-dot lines.

also took place during this late period of crustal extension, including the Narssaq Complex [1] and most of the remaining under-saturated syenites of the Igaliko complex. Vigorous E-W left-lateral faulting then affected much of the area and caused dislocations in the Narssaq and later Igaliko centres. This faulting, however, appears to have terminated before a final suite of alkaline central complexes was emplaced (the Central Tugtutôq Complex, 1168 ± 37 Ma. [14], Ilímaussaq, 1168 ± 21 Ma. [15], late units of the Igaliko complex, 1167 ± 15 Ma. [3], Klokken, 1159 ± 11 Ma. [16]. The main inferences to be drawn from these late Gardar events within this eastern zone are that: (i) the Late Gardar tectonic and magmatic activity occurred within the period 1184-1131 Ma., (and possibly within a much narrower timespan around 1170 Ma.). (ii) crustal dilation within a zone about 15 km broad amounted to approximately 1.5 km. (iii) approximately 75 per cent of this dilation occurred at a very early stage with the formation of the giant dykes, the subsequent waning of dilation being attended by intermittent ascent of large volumes of mantle-derived alkaline felsic magma. (iv) genesis of the felsic magmas to form central complexes was intimately associated with the introduction of large bodies of basic magma along a linear ENE-trending zone of lithospheric failure.

Assuming that surface volcanism accompanied the intrusive activity it may be inferred that early fissure-erupted alkali and transitional olivine basalt lavas were succeeded by a sequence of trachybasaltic, trachytic, phonolitic and alkali rhyolitic volcanics emitted from both fissure and central vents. Tugtutôq and the Ilímaussaq peninsula may have constituted part of a narrow graben system within which volcanics accumulated.

Approximately midway between Ivigtut and Tugtutôq lies the Nunarssuit and Isortoq region, (Fig. 2) which, like the Tugtutôq region is geographically defined by broad, straight ENE-trending fjords and geologically defined by abundant basic Gardar dykes with trends between NE and ENE. The bounding fjords are again believed to reflect major fracture zones on either side of another Gardar fault block, approximately twice the width of the Tugtutôq block. The dyke swarm, predominantly composed of alkali dolerites and gabbros [17] includes some giant dykes (>0.5 km wide) similar to those of Tugtutôq. It has not, however, been demonstrated that these are the earliest components of the swarm (as on Tugtutôq). This broad dyke swarm is cross-cut by the Nunarssuit alkaline complex. Rb-Sr whole-rock isochron ages [3] from four units in this complex suggest that the whole complex was emplaced within the space of a few million years and is chronologically indistinguishable from the Late Gardar complexes to the east.

Whereas the ENE dyke-emplacement in the east is known from the Tugtutôq region to have begun 1175 ± 9 Ma., no age for the commencement of NE and ENE dyke intrustion in the Nunarssuit-Isortoq zone is available. However, in both places, dykes showing these trends are younger than WNW-trending dolerite dykes, (referred to as BD_0 dykes in the literature) now known to date from the 1250 m.y. activity [10]. Thus, despite some uncertainty concerning the start of NE and ENE dyke-injection in the Nunarssuit-Isortoq block, the existing data suggest that these dykes are broadly contemporaneous with those of the eastern part of the province. It is suggested that the Nunarssuit-Isotoq block also behaved as a subsiding unit that experienced intense fissuring parallel to its boundary faults and in which the massive uprise of mildly alkaline basaltic magma during an initial tensional phase was succeeded by large volumes of felsic alkaline magmas emplaced by stoping.

As on Tugtutôq, the crustal extension in the Nunarssuit-Isortoq block amounts to approximately 1.5 km so that, treating the two blocks as components of a larger composite and asymmetric rift structure some 70 km broad, the total dilation due to dyke emplacement is about 3 km.

5. GRAVITY SURVEY RESULTS

Gravity surveys across the Tugtutôq-Narssaq region [18] demonstrate the presence of a pronounced gravity 'high' of up to 300 g.u., elongate parallel to and axially near-coincident with, the principal zone of dyking. The anomaly is inexplicable in terms of the individual central complexes and observed dykes and is believed to be due to the sub-surface presence of a major basic intrusive some 80 km long by 25 km broad. This is deduced to rise close to the present erosion level in the vicinity of Narssaq (locality 3, Fig. 2) but lies at greater depth (3-5 km) further to the WSW.

In the East African Rift gravity studies [19] have shown that the active rift has a broad dominant 'low' upon which is superimposed a central (axial) 'high' c. 55 km wide and with an amplitude of some 200 g.u. Fairhead regards the 'low' as due to the underlying presence of relatively low density anomalous mantle (which may be partially melted), while a 15 km broad dyke-like (gabbroic?) intrusion within the crust is responsible for the axial 'high'. It might be anticipated, that, with solidification of the anomalous partial-melt zone in the mantle and maintenance of isostatic equilibrium, the gravity 'low' would be eliminated while the axial 'high' would remain as a residual feature. We suggest that the Tugtutôq-Narssaq 'high' is just such a feature, comparable to the axial 'highs' of other continental rift structures, e.g. that of the Oslo graben [20]. The evidence can thus be construed as lending support to the suggestion [21] that the Tugtutôq region represents a deeply dissected structure analogous to the southern Kenya rift. The zone may have developed through failure of the crust along a zone of weakness controlled by earlier (Ketilidian) structures allowing crustal thinning and admittance to high levels of mantle-derived basic rocks.

However, further gravity surveys in the Kûngnât and Nunarssuit-Isortoq regions [22] have failed to identify any major anomaly such as that of the Tugtutôq-Narssaq 'rift'. It may be that these represent more diffuse zones of crustal thinning (in the Mid- and Late-Gardar respectively) in which large bodies of mantle-derived basic rocks were not able to penetrate to such shallow levels. It should be noted, however, that studies of anorthosite xenoliths [23] suggested the sub-surface presence of a large body (or bodies) of anorthosite beneath virtually the whole Gardar province.

6. SUMMARY AND CONCLUSIONS

The province provides evidence for (1) an early rift system at about 1300 m.y. before present, from which fragmentary and much-faulted sequences of lavas and sediments remain together with large undersaturated syenite volcanic centres; (2) a period of rifting at about 1250 m.y. before present, marked by the intrusion of alkaline basic dykes, transcurrent faulting and finally, injection of silica over-saturated central complexes; (3) a relatively well-defined Late Gardar (c. 1170 m.y.) asymmetric rift zone about 70 km across with a 3 km extension associated with large scale emplacement of alkaline basic dykes, rejuvenation of the older transcurrent fault systems and late-stage emplacement of a number of alkaline central complexes. On the south eastern side of this zone a marked gravity 'high', more or less coincident with the principal axis of dyking around Tugtutôq-Narssaq is believed to result from a large basic complex at shallow depth. The Late Gardar rift zone lies some tens of kilometres south of, and approximately parallel to the margin of the south Greenland Archaean craton, wholly within the Ketilidian mobile belt.

By analogy with modern continental rifts, it is likely that the Gardar rifts were associated with crustal upwarps. The Late Gardar rift zone traverses the Ketilidian granite ('Julianehåb Granite') outcrop which is bounded north and south by Ketilidian metasedimentary and metavolcanic outcrops, suggesting a broad, domelike structure in the pre-Gardar basement. Although it has not been shown that this feature is of Gardar age, the axial disposition of the rift system with respect to it raises the possibility that the structure represents the deeply eroded continental upwarp which accompanied the 1170 m.y. event.

The geochemical/petrological similarities of the magma suites attending each of the identifiable episodes of Gardar rifting [1] are such as to suggest close identity of the mantle source rocks involved in each as well as of the conditions under which the primary melts were generated.

It is concluded that similar primary Gardar magmas were produced at intervals of 50-100 m.y. from a large and quasihomogeneous upper mantle source region during tensional phases incidental to large-scale Late Proterozoic plate motions [24,25].

ACKNOWLEDGEMENTS

The authors are indebted to the colleagues with whom the ideas presented in this paper have been discussed. In particular, they wish to thank Dr. A.B. Blaxland for his assistance in the preparation of the manuscript, and the Director of the Geological Survey of Greenland for permission to publish.

REFERENCES

1. C.H. Emeleus and B.G.J. Upton, The Gardar period in southern Greenland, in: The Geology of Greenland, ed. by A. Escher and W.S. Watt, 1976.
2. O. van Breemen, M. Aftalion and J.H. Allaart, Bull. Geol. Soc. Am., 83, 3381, 1974.
3. A.B. Blaxland, O. van Breemen, C.H. Emeleus and J.G. Anderson, Age and origin of the major syenite centres in the Gardar province of South Greenland: Rb-Sr studies. Bull. Geol. Soc. Am., in press.
4. H. Sørensen, Can. Min., 10, 229, 1970.
5. C.H. Emeleus, Bull. Gønlands Geol. Unders., 45 (also Meddr Grønland 172, 3), 1964.
6. C.H.Emeleus and W.T. Harry, Bull. Grønlands Geol. Unders., 85 (also Meddr Grønland 186, 3), 1970.
7. J.W. Stewart, The earlier Gardar igneous rocks of the Ilímaussaq area, South Greenland. Unpublished Ph.D. Thesis, Univ. Durham, England, 1964.
8. V. Poulsen, Rapp. Grønlands Geol. Unders., 2, 1964.
9. B.G.J. Upton, Bull. Grønlands Geol. Unders., 27 (also Meddr Grønland 123,4), 1960.
10. P.J. Patchett, Rb-Sr geochronology and geochemistry of Proterozoic basic intrusions in Sweden and South Greenland. Unpublished Ph.D. Thesis. Univ. Edinburgh, 1976.
11. H. Pauly, Neues Jb. Min. Abh., 94, 121, 1960.
12. A. Berthelsen, Geol. Rundsch., 52, 269, 1962.
13. B.G.J. Upton, Bull Grønlands Geol. Unders., 34 (also Meddr Grønland 169,8), 1962.
14. O. van Breemen and B.G.J. Upton, Bull. Geol. Soc. Am., 83, 3381, 1972.

15. A.B. Blaxland, O. van Breemen and A. Steenfelt, Lithos, 9, 31, 1976.
16. A.B. Blaxland and I. Parsons, Bull. Geol. Soc. Denmark, 24, 27, 1975.
17. I. Winstanley, Petrology of the Gardar basic dykes, South Greenland. Unpublished Ph.D. Thesis, Univ. Edinburgh, 1975.
18. D.J. Blundell, A gravity survey across the Gardar igneous province, S.W. Greenland, (in prep.).
19. J.D. Fairhead, Tectonophysics, 30, 269, 1976.
20. I.B. Ramberg, Norges Geol. Unders., 325, 1976.
21. B.G.J. Upton, The alkaline province of South West Greenland, in: The Alkaline Rocks, ed. by H. Sørensen, New York: Wiley, 1974.
22. D.J. Blundell, A gravity survey of the region around Ivigtut and Nunarssuit. Geological Survey of Greenland, internal report, 1976.
23. D. Bridgwater and W.T. Harry, Bull. Grønlands Geol. Unders., 77 (also Meddr Grønland 185,2), 243, 1968.
24. K. Burke and D.F. Dewey, J. Geol., 81, 406, 1973.
25. J.D.A. Piper, Nature, 251, 381, 1974.

EVIDENCE FOR AN EARLY PERMIAN OCEANIC RIFT IN THE NORTHERN NORTH ATLANTIC

M.J. Russell

Department of Applied Geology,
University of Strathclyde,
Glasgow, Scotland.

D.K. Smythe

Institute of
Geological Sciences,
Edinburgh, Scotland.

ABSTRACT. The Greenland-Svalbard and Proto-Bay of Biscay fault zones are small circles to a common pole, and it can be shown that transform motion on these faults opens the Rockall Trough and the eastern Norwegian Sea. Arthaud and Matte calculate a dextral movement along the Proto-Bay of Biscay fault of more than 100 km. This movement took place between the emplacement of the Variscan granites and the beginning of the Triassic, and we argue here that this was one of the results of ocean floor spreading in Rockall Trough and the eastern Norwegian Sea. Widespread intrusions of dolerite in Sweden, Norway, England and Scotland c. 290 Ma imply that the lithosphere had reached its limit of strength in relative tension and it is surmised that this is the time that the lithosphere began to separate between Greenland and Northwest Europe. Magmatism continued in the Early Permian as evidenced by the Oslo extrusives and fluorite deposits of magmatic derivation in England. Rosemary Bank is an extinct volcano in the northern Rockall Trough, and judging from the remanent magnetic vector, this seamount was probably active during the Permian.

In mid Permian times the Zechstein Sea, which presumably occupied Rockall Trough and the Faeroe-Shetland Channel, as well as the eastern Norwegian Sea, flooded parts of East Greenland, Ireland, England, the North Sea and north-central Europe.

1. INTRODUCTION

The Rockall Trough, the Faeroe-Shetland Channel and the eastern

E.-R. Neumann and I.B. Ramberg (eds.), Petrology and Geochemistry of Continental Rifts, 173–179.

Norwegian Sea comprise a contiguous pre-anomaly 24 rift (the Proto northern North Atlantic) probably floored by oceanic basalts [1,2]. The Moho is about 14 km deep in the Rockall Trough [3] and the Vøring Plateau [4]. Here we provide evidence to support the proposal that the Proto northern North Atlantic Rift began to form at the beginning of the Permian [5,1].

2. GEOMETRY

The margins of the pre-anomaly 24 rift are remarkably parallel and a little over 200 km apart [6]. There is a generally held assumption [7] that the Rockall Trough opened as a northerly propagation of the central North Atlantic. The northern limit of this rift must have been a transform fault between Svalbard and Greenland. Roberts [8] has argued that the Rockall Trough opened at the same time as the Bay of Biscay, but since the proto-Bay of Biscay Fault trace is a small circle with a common pole to the Greenland-Svalbard Fault (which is presumably more than coincidental), we suggest instead that the Proto northern North Atlantic opened before the Bay of Biscay, and that the northern and southern limits of the rift were these transform faults. Furthermore, a rotation of Greenland-North America, relative to Europe of 2.4°, about this pole, opens the Rockall Trough to its present configuration [2].

A test of this geometrical hypothesis would be demonstrable dextral movement along these faults.

As the Svalbard-Greenland Fault parallels the orogenic grain the theory is as yet untestable in the north. However, Hercynian structures do cross the proto-Bay of Biscay Fault, and Arthaud and Matte [9] have shown that the Variscan leucogranites provide a good marker. When Iberia and Brittany are reassembled [10] it is clear that there has been a dextral offset along the proto-Bay of Biscay Fault of at least 100 km and possibly about 200 km (Figure 1).

Arthaud and Matte [9] argue that this movement took place after the intrusion of the leucogranites (c. 300 Ma) but before the Triassic, since associated faults do not affect overlying Triassic basin sediments. Their preferred age for this lateral movement is Early Permian. The proto-Bay of Biscay Fault cannot continue between Greenland and Labrador as demonstrated by the contiguous margin to the Superior chelozone [11], (Figure 1).

From the refit of Figure 1 (amended from Bullard _et al_ [12]) we can see that this significant dextral motion along the proto-Bay of Biscay Fault must therefore have opened the Rockall Trough.

3. TECTONICS

Immediately preceding a period of rifting the crust normally upwarps to form a dome or series of domes [13]. It is known that domes did form around the time of the Carboniferous-Permian boundary in the Oslo region [14] and in the Northern Pennines [15]. We know also that there was uplift at this time in Scotland where there is a non-sequence between probable upper Stephanian red beds and the underlying Westphalian containing marine bands [16]. The Westphalian sediments have been oxidised down to a maximum depth of half a kilometre [17]. There was also a period of non deposition in East Greenland at about the same time [18].

The dome in the Oslo region finally collapsed to form a lower Permian rift, and we argue that this was a microcosm of what was happening on the site of Proto northern North Atlantic at the beginning of the Permian.

McLean [19], elaborating on the earlier views of Bott [20] argues that the belt of Permo-Triassic basins parallelling and within 150 km of the continental margin northwest of Britain is best explained in terms of flow of ductile lower lithosphere towards an Early Permian ocean.

4. MAGMATISM

About 290 Ma there was sudden intrusion and extrusion of basic magma in northern Britain, Norway and southern Sweden (Figure 1) [21-25]. In Britain and southern Sweden the magmatism comprised a short lived episode of dolerite and quartz dolerite intrusion. In Norway volcanic activity was of an alkaline type and significant magmatism continued for about 20 Ma, accompanying rifting.

Dolerite dyke swarms are often associated with the onset of rifting [26] and we suggest that 290 Ma is the date when the lithosphere reached its limits of strength in relative tension and began to separate.

It may be that towards the end of this phase of ocean floor spreading, Rosemary Bank formed as a large volcano. This conjecture is based on Scrutton's [27] calculation of the remanent magnetic vector, which suggests that the latitude of formation was about 16° ($\pm$ 20°)N. A 16°N position would place the formation of the Bank in the Permian [28].

5. MINERALISATION

There was a significant period of mineralisation c. 280 Ma in Britain, Norway and southern Sweden [29]. Most of the ores

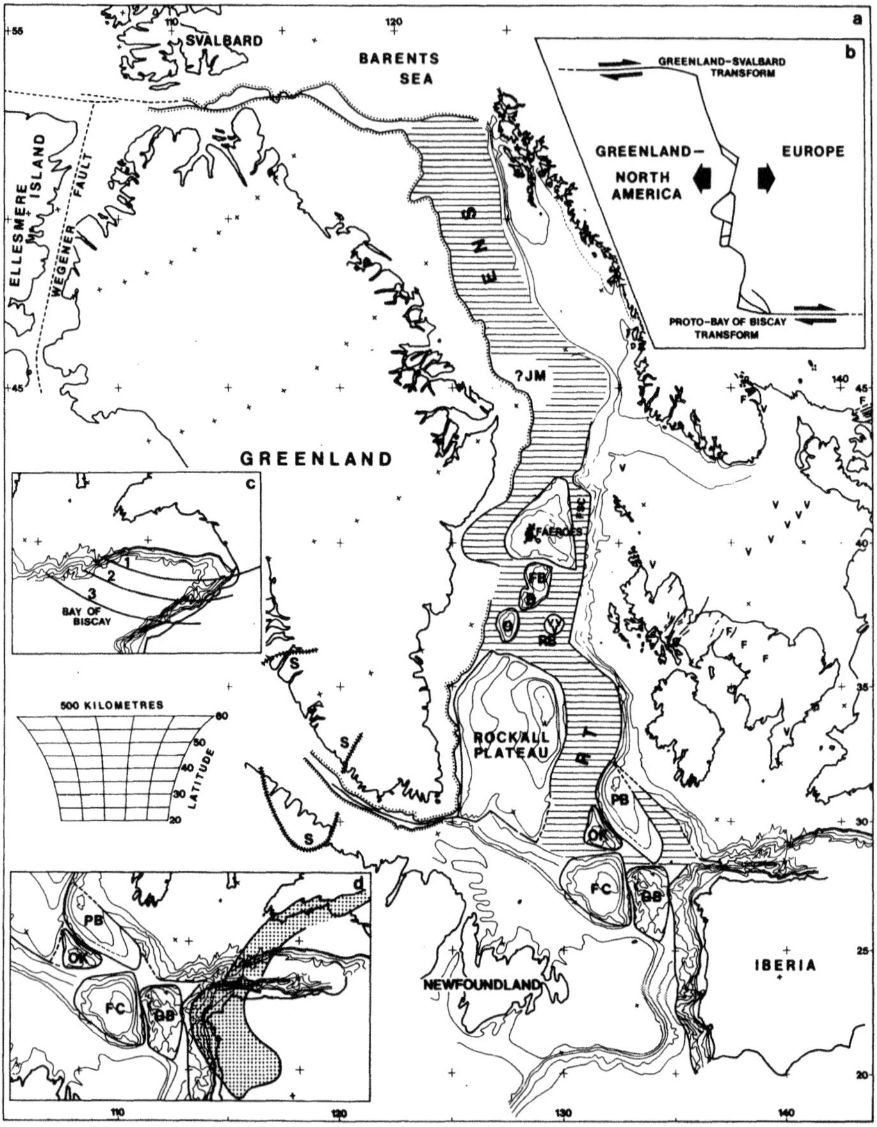

Fig. 1. a, Pre-anomaly 32 northern North Atlantic reconstruction on an oblique Mercator projection. The north pole of the oblique aspect is at 55°N, 100°E, with Europe in its modern position (basic intrusive and extrusive ∿ 290 Ma in black). b, The projection pole is the pole of rotation of Europe away from Greenland - North America; transform faults should then approximate to parallels of latitude. c, transform fault traces in the Bay of Biscay. d, Variscan leucogranite belt (crosses) on a pre-drift reconstruction.

include fluorite. For example fluorite occurs as a primary mineral in the border zones of the alkali granite in the Ramnes cauldron of the Oslo Graben [30]. Fluorite deposits also occur to the west and southwest of the Graben at Lassedalen and Gjerpenfeltet respectively [31,32].

Smith [33] has shown that the yttrium content of fluorite in the North Pennines, England, is in the range of 120 to 815 ppm, and so supports Sawkins [34] in considering that the ores have an igneous source. Fluorite in the Central Pennines probably has a similar origin, and we may assume that the fluorite distribution represents the extent of partial melting of the upper mantle in the Early Permian [29].

Vokes [35] and Sawkins [36] have previously pointed out the apparent association between this mineralisation and rifting.

6. STRATIGRAPHY

The Zechstein and Bakevellian Seas of mid Permian age supported a Boreal fauna, and are best explained as incursions from a Permian ocean in the Proto northern North Atlantic [1]. Talwani and Eldholm [37] have suggested that Zechstein strata may occur near the base of the sedimentary pile in the eastern Vøring Plateau, although they assume the basement to consist of subsided continental crust rather than the ocean floor postulated by us [1,2].

7. PREDICTIONS

Drilling in the Rockall Trough, the Faeroe-Shetland Channel and the

---

←
Closure of the Bay of Biscay along the best fit transforms 1-3 (1c) corresponds closely to the position of Iberia relative to Europe (1a) after the Rockall Trough has opened, whereas continuity of the Variscan leucogranite belt is achieved on the pre-drift reconstruction (1d). Geological arguments (see text) place the opening of the Proto northern North Atlanitic (viz. from the configuration shown in 1b,d to that of 1a) in the Early Permian.

ENS; eastern Norwegian Sea. FSC; Faeroe-Shetland Channel.
RT; Rockall Trough. JM; Jan Mayen continental fragment.
FB; Faeroe Bank. B; Bill Bailey Bank. O; Outer Bailey Bank.
RB; Rosemary Bank. PB; Porcupine Bank. OK; Orphan Knoll.
FC; Flemish Cap. GB; Galicia Bank. S; Superior chelozone.
V; Permian volcanics. F; Permian fluorite deposits.

Sources: 1,2,9-12,15,21,25,29-33,38-41.

Eastern Norwegian Sea will intersect sediments down to Permian age and then pass into Lower Permian ocean floor basalts. Jurassic sediments will be present and may be oil-bearing [2].

Fluorite and metalliferous deposits may be found spatially associated with 290 Ma intrusives.

8. ACKNOWLEDGMENTS

We thank Mike Leeder for drawing our attention to Arthaud and Matte's work, and Roger Scrutton for discussions on Rosemary Bank.

REFERENCES

1. M.J. Russell, Scott. J. Geol., 12, 315, 1976.
2. D.K. Smythe, N. Kenolty, M.J. Russell and R.A. Scrutton, in preparation.
3. E.J. Jones, M. Ewing, J.I. Ewing and S.L. Eittreim, J. Geophys. Res., 75, 1655, 1970.
4. K. Hinz, Meteor Forsch-Ergebnisse, 10C, 1, 1972.
5. M.J. Russell, Scott. J. Geol., 8, 75, 1972.
6. I.R. Vann, Nature, 251, 209, 1974.
7. A. Hallam, J. Geol., 79, 129, 1971.
8. D.G. Roberts, Structural development of the British Isles, the Continental Margin, and the Rockall Plateau, in: The Geology of Continental Margins, ed. by C.A. Burk and C.L. Drake, Springer-Verlag, Berlin, 1974.
9. F. Arthaud and P. Matte, Tectonophysics, 25, 139, 1975.
10. X. Le Pichon, J. Bonnin, J. Francheteau and J-C. Sibuet. Une hypothése d'évolution tectonique du golfe de Gascogne, in: Histoire structurale du golfe de Gascogne, tome 2, Publications de l'Institut Francais du Pétrole, Paris, VI, 11-1, 1971.
11. F.J. Fitch, Phil. Trans. R. Soc. Lond., 258A, 191, 1965.
12. E.C. Bullard, J.E. Everett and A.G. Smith, Phil. Trans. R. Soc. Lond., 258A, 41, 1965.
13. J.G. Gass, Phil. Trans. R. Soc., 267A, 369, 1970.
14. I. Ramberg, Norges Geol. Unders., 325, 1, 1976.
15. K.C. Dunham, Geology of the Northern Pennine orefield, Geol. Surv. Gt. Br., 1949.
16. W. Mykura, Bull. Geol. Surv. Gt. Br., 26, 23, 1967.
17. E.H. Francis and C.J. Ewing, Geol. Mag., 99, 145, 1962.
18. J. Haller, Geology of the East Greenland Caledonides, Interscience, London, 1971.
19. A.C. McLean, Geol. J. Special Issue, in press.
20. M.H. Bott, Tectonophysics, 11, 319, 1971.
21. H.N. Priem et al, Phys. Earth Planet. Interiors, 1, 373, 1968.
22. F.J. Fitch, J.A. Miller and S.C. Williams, Isotopic ages of British Carboniferous Rocks. C.R. 6th Int. Congr. Carb. Strat. Geol., 2, 771, 1970.

23. B. Sundvoll, Rb/Sr relationship in the igneous rock of the Oslo palaeorift. European Colloquium on Geochronology, Abstract, Amsterdam, 89, 1976.
24. R.B. Faerseth, R.M. Macintyre and J. Naterstad, Lithos, 9, 331, 1976.
25. I. Klingspor, Geol. Foren. Stockh. Forh., 98, 195, 1976.
26. P.R. May, Bull. Geol. Soc. Am., 82, 1285, 1971.
27. R.A. Scrutton, Geophys. J.R. Astr. Soc., 24, 51, 1971.
28. A.G. Smith, J.C. Briden and G.E. Drewry, Phanerozoic world maps, in: Organisms and Continents through time, ed. by N.F. Hughes, Palaeontological Association, London, 1973.
29. M.J. Russell, Geol. J. Special Issue, in press.
30. R. Sørensen, Norges Geol. Unders. 321, 67, 1975.
31. J. Willms, Fluorite- Vorkommen in Telemark Südnorwegen, Diplomarbeit, Hamburg, 1975.
32. P.O. Kaspersen, En malmgeologisk undersøkelse av flusspat-mineraliseringen i Gjerpenfeltet nord for Skien, Unpublished thesis, University of Trondheim, 1976.
33. F.W. Smith, Factors governing the development of fluorspar orebodies in the North Pennine orefield, Ph.D. thesis, University of Durham, 1974.
34. F.J. Sawkins, Econ. Geol., 61, 385, 1966.
35. F.M. Vokes, Metallogeny possibly related to continental break-up in southwest Scandinavia, in: Implications of Continental Drift to the Earth Sciences, Vol. 1, ed. by D.H. Tarling and S.K. Runcorn, Academic Press, London, 1973.
36. F.J. Sawkins, J. Geol., 84, 653, 1976.
37. M. Talwani and O. Eldholm, Bull. Geol. Soc. Am., 83, 3575, 1972.
38. F.E. Wickman et al, Ark. Min. Geol., 11, 193, 1963.
39. F.W. Dunning, Tectonic map of Great Britain and Northern Ireland, Ordnance Survey, 1966.
40. K.S. Heier and W. Compston, Lithos, 2, 133, 1969.
41. W.H. Ziegler, Outline of the geological history of the North Sea, in: Petroleum and the Continental Shelf of North-west Europe, Vol. 1, Geology, ed. by A.W. Woodland, Applied Science Publishers, 1975.

RB/SR - RELATIONSHIP IN THE OSLO IGNEOUS ROCKS

B. Sundvoll

Norges Geologiske Undersøkelse, p.t. Mineralogisk-Geologisk Museum, Sarsgt. 1, Oslo 5, Norway.

ABSTRACT. The preliminary conclusions from an extensive study of the Rb/Sr-relationships of the Oslo igneous rocks are reported with emphasis on the time development of the magmatic activity and the petrogenetical implications of the initial Sr-isotopic ratios.

1. INTRODUCTION

Due to the exposure of huge quantities of subsurface intrusives, the igneous rock complex of the Oslo Region represents one of the most outstanding examples of alkaline magmatic activity connected with continental rifting. A survey study of the Rb/Sr isotopic relationships in this complex has therefore been carried out in order to investigate petrogenetic aspects of rift magmatism as well as development in time of this particular geological feature.

Approximately 350 samples from all over the Oslo Region, representing all major rock types and units have been collected and analyzed. Details of the analytical procedures and a full discussion of the individual results will be reported elsewhere. This paper only summarizes the main conclusions which can be deduced from the material at this preliminary stage.

2. AGE RELATIONS

It has previously been assumed that the effusive and intrusive events in the Oslo rift only lasted a few million years [1,2]. The first objective, therefore, was to obtain data to establish the actual time span of the magmatic activity in the rift. The lava

E.-R. Neumann and I.B. Ramberg (eds.), Petrology and Geochemistry of Continental Rifts, 181-184.

remnants in the area are with few exceptions older than the intrusives and the lowermost flows therefore give an indication of the earliest manifestations of igneous activity. Rb/Sr whole rock ages on the B_1 basalt flow and the overlying rhombporphyry flows $RP_{1\&2}$ at Krokskogen are 292 ± 8 and 294 ± 5 m.y. respectively. These results probably represent minimum ages as we have good reason to suspect that earlier lava flows are erupted further south, i.e. in the Holmestrand-Jeløya and Skien areas [3]. The youngest rock so far encountered in the Oslo Region is the Tryvann granite, Rb/Sr whole rock age: 247 ± 4 m.y. Thus a minimum time span of at least 40 million years is now established for the development of the Oslo igneous complex.

In terms of the most recent phanerozoic timescales [4,5] the volcanic activity was initiated close to the boundary between the Carboniferous and the Permian, and the magmatism continued throughout most of the Permian.

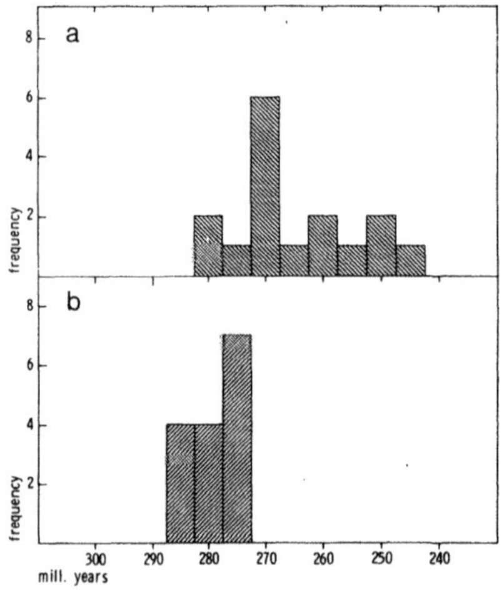

Figure 1. Frequency distribution of Rb/Sr ages from a) the Nordmarka-Hurdal batholith, and b) the Vestfold batholith.

Looking at the frequency distribution of the whole rock ages resulting from this study (Fig. 1), we find that the southern (Vestfold) batholith complex of intrusives is generally older than the northern one (Nordmarka-Hurdal), the spread in ages being 286-275 m.y. (Vestfold) and 280-247 m.y. (Nordmarka-Hurdal). This result confirms earlier conclusions based on geophysical and magmatectonic studies [6,7], that the rift structure has a trend of development from the southwest towards the northeast. The main sequence of intrusions in both of the composite batholiths is generally speaking as follows: monzonites and biotite granites, syenites, alkalisyenites and/or nepheline-syenites, peralkaline and alkaline granites. This sequence may be exemplified by the following Rb/Sr whole rock ages from the Vestfold batholith complex: Skrim larvikite (monzonite) 285 ± 5 m.y., Nordagutu granite (biotite-granite) 284 ± 7 m.y. [8], Siljan nordmarkite (syenite-alkalisyenite) 277 ± 4 m.y. and the Eikern ekerite (peralkaline granite) 275 ± 3 m.y.

Cauldron subsidences which can be dated from the ages of ring-dyke and central dome intrusives also seem to have occurred later in the northeastern rift seqment (Akershus-graben [7]) than in the southwestern segment (Ringerike-Vestfold graben). The age intervals are 280-275 m.y. for cauldrons in the southwestern segment, and 274-250 m.y. for those in the northeastern segment. It is suspected, however, that some dykes and central domes are not directly related to the subsidence itself, but are later intrusives that have followed previously existing structures and replaced the original cauldron intrusives.

3. INITIAL SR RATIOS

A summary of the initial $^{87}Sr/^{86}Sr$ ratios of the main rock types in the Oslo Province is presented in Table 1. The relatively small spread in ratios (from 0,7034 to 0,7074) is clearly compatible with a simple petrogenetic model in which all the igneous rocks, including the biotite granites, have been derived from a mantle source magma. This mantle source most probably had an initial Sr ratio close to those of the monzonites and rhombporphyry-ries, i.e. 0.7038-0.7041, because these rocks are volumetrically the most abundant at the present surface and have also formed early enough in the development scheme that they have had the best chances of preserving the parental ratio.

The later syenitic and granitic derivatives have, without exceptions, higher ratios than the monzonites. The very moderate increase in the ratios of the syenites (0,0007-0,0009) can readily be explained by the time differences and by the increased Rb/Sr ratios in these rocks compared to the monzonites without postulating any interaction with the surroundings rocks. Similar arguments also apply to the alkaline and peralkaline rocks, but in these cases the Sr content has decreased so dramatically (approx. averages: 700 ppm in monzonites, 150 ppm in syenites and 10 ppm in peralkaline granites) that very small amounts of any contaminant and/or isotopic equilibrization with the country rock would influence the Sr-isotopic ratios.

Significant contamination by crustal material has only been observed in intrusive bodies with macroscopical amounts of xenoliths of paleozoic sediments. Large scale assimilation of country rock does therefore not seem to have played a substantial role in

Table 1. Summary of $^{87}Sr/^{86}Sr$ initial ratios.

Extrusives:	
Basalts (Skien-Porsgrunn)	0.7034-0.7046
" (Holmestrand-Jeløya)	0.7038-0.7054
" (Krokskogen)	0.7053
Rhombporphyry (RP_1K & RP_2K)	0.7040
Trachyte (T_1V)	0.7066
Intrusives:	
Gabbros (Oslo-essexites)	0.7038-0.7051
Monzonites (Larvikites etc.)	0.7039-0.7041
Plagifoyaites (Lardatites)	0.7038
Syenites (Pulaskites, etc.)	0.7040-0.7047
Alkali-syenites (Nordmarkites)	0.7042-0.7059
Nephelinesyenites (Foyaites)	0.7046
Alkali-granites	0.7048-0.7057
Peralkaline granites (Ekerites)	0.7068-0.7073
Biotite granites	0.7044-0.7067

the petrogenetic history of the Oslo igneous rocks. The pre-existing material of the upper crust in the grabens most probably dropped down into the rising magmas and are either present as partly modified xenoliths at some depth, or if anatectic melts were indeed formed, they never reached the surface in any quantity. On the other hand, remelting and assimilation of pre-emplaced Oslo intrusives seem to have taken place quite extensively and most of the syenitic rocks termed akerite in older works are isotopically of a hybrid character.

REFERENCES

1. C. Oftedahl, Geol. Rundschau, 57, 203, 1967.
2. I.S.E. Carmichael, F.J. Turner and J. Verkoogen, Igneous Petrology, Mc Graw-Hill Book Company, 504, 1974.
3. C. Oftedahl, Studies on the igneous rock complex of the Oslo Region XII. Skr. Norske Vidensk.-Akad. i Oslo. I. Mat.-naturv. KL. 1952, 1952.
4. W.B. Harland, A. Gilbert Smith and B. Wilcock (editors), Geol. Soc. London, Special publ. no. 1, 1964, 1964.
5. F.J. Fitch, J.A. Miller and S.C. Williams, Compte Rendu 6e Congrés Intern. Strat. Géol. Carbonif. 1967 Vol II, 771, 1970.
6. I.B. Ramberg, Norges Geol. Unders., 325, 1976.
7. I.B. Ramberg and B.T. Larsen, Tectmomagmatic evolution, in: The Oslo Paleorift, Part 1, A Review, ed. by J.A. Dons and B.T. Larsen, (Preliminary edition), 1977.
8. S.B. Jacobsen and G. Raade, Norsk Geol. Tidsskr., 55, 171, 1975.

DISTRIBUTION OF Th, U, K IN THE PLUTONIC ROCKS OF THE OSLO REGION, NORWAY

Gunnar Raade

Institutt for geologi, Universitetet i Oslo,
Blindern, Oslo 3, Norway.

ABSTRACT. The plutonic rocks of the Oslo Region can be divided into two main groups on the basis of their Th/U ratios: (1) an older group with uniform ratios and mean values in the range 3.5-4.5, (2) a younger group with variable ratios and mean values in the range 4.5-6.5. The petrogenetic implications of this two-fold division are discussed.

1. INTRODUCTION

Concentrations of Th, U and K have been determined by γ-ray spectrometry in 966 samples of plutonic rocks from the Oslo Region. The sample sites were chosen to achieve a uniform regional distribution (Fig. 1). Care was taken to collect only fresh and unaltered material, and most of the samples were taken from road-cuts and quarries.

This paper is a short summary of a part of the author's unpublished cand.real. thesis [12]. In addition, the thesis contains data from the areas of cauldron subsidence, some additional less common rock types, U/K and Th/K ratios and heat production data plus trend surface analyses of individual plutonic bodies.

2. PETROGRAPHY OF PLUTONIC ROCKS

The main plutonic rocks of the Oslo Region are listed in Table 1. The local rock names are used in this paper for convenience. It should be noted that the Oslo-essexite series of volcanic necks [2] has not been studied in this work.

E.-R. Neumann and I.B. Ramberg (eds.), Petrology and Geochemistry of Continental Rifts, 185-192.

Thin sections of all the samples were examined in order to classify the rocks petrographically and to trace petrographic variations within single plutonic bodies. This information plotted on ten detailed maps can be supplied by the author on request.

3. ANALYTICAL METHOD

The theory and practice of γ-ray spectrometry have been discussed by Adams and Gasparini [1]. The equipment at the Mineralogical-Geological Museum in Oslo consists of a 5"x5" NaI(Tl) detector crystal and a 400 channel pulse-height analyzer. Coarse-crushed samples weighing 600-800 g were used for the analysis. The K values are corrected for mass absorption of low-energy γ-radiation, based on a linear dependence on the weight of the sample. Similar corrections for Th and U were not deemed necessary. The study began with a detailed evaluation of the laboratory equipment, its calibration, background fluctuations, sample preparation and the effect of sample packing density [12].

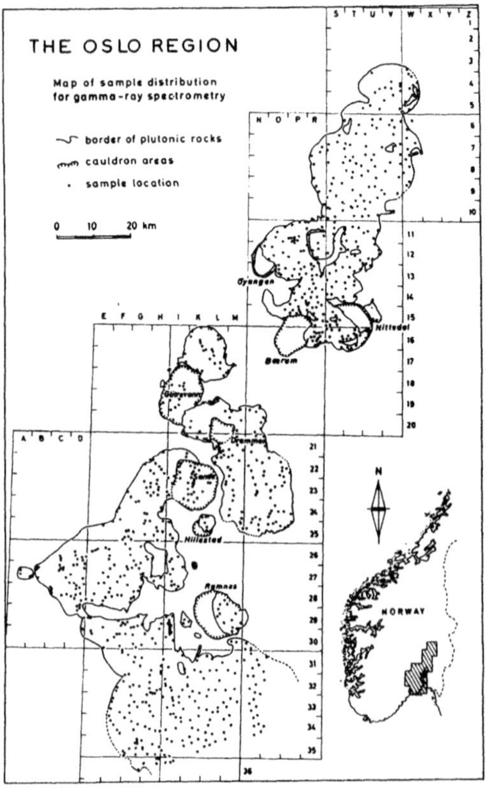

Fig. 1. Sample locations for plutonic rocks analyzed by γ-ray spectrometry. Three subregions of plutonic rocks can be recognized: (1) a northern region of monzonitic to granitic rocks (Hurdalen-Nordmarka), (2) a central region of biotite granites (Finnemarka-Drammen), (3) a southern region dominated by larvikites (Skrim-Larvik). The grid system is based on the map series M 711 in scale 1:50,000.

TABLE 1. Survey of the main plutonic rocks of the Oslo Region.

Rock group	Local name	Characteristic mafic minerals	Streckeisen classification
Diorites and syenodiorites	Sørkedalite	Olivine (diopside, biotite)	Diorite
	Akerite[1]	Augite, hornblende, biotite	Syenodiorite (monzodiorite and monzonite)
	Kjelsåsite	Augite, biotite (amphibole)	Monzonite to monzodiorite
	Larvikite[2]	Augite, biotite (amphibole)	Monzonite to syenite
Nepheline-bearing rocks	Ditroite	Augite, barkevikite, biotite	Foyaite
	Lardalite	Pyroxene, biotite	Plagifoyaite ?
	Foyaite - Hedrumite	Aegirine to diopside	Foyaite to alkali syenite
Syenites	Hedrumite ring-dike	Augite, biotite, amphibole	Alkali syenite
	Pulaskite	Diopside to augite, biotite, amphibole	Alkali syenite
	Grefsen syenite	Biotite, hornblende	Syenite
	Nordmarkite	Alkali amphibole, aegirine (augite, biotite)	Alkali syenite
Granites	Ekerite	Alkali amphibole, aegirine	Alkali granite
	Drammen granite	Biotite	Granite
	Finnemarka granite	Biotite	Alkali granite
	Other biotite granites	Biotite	Granite or alkali granite

1 Including monzodiorite from Finnemarka and akerite-kjelsåsite transitional rock from Nordmarka.

2 Including the red-coloured variety tønsbergite.

The precision is assumed to be better than 5% for U, 3% for Th and 1% for K, and the accuracy is on the same level. Further discussion on the precision and accuracy of the equipment is furnished by Killeen and Heier [9].

4. STATISTICAL TREATMENT OF DATA

Th, U, K and Th/U data for the most abundant plutonic rocks of the Oslo Region are given in Tables 2 and 3. Correlation coefficients are listed in Table 4.

For the purpose of comparing different populations, use of the arithmetic mean and other statistical parameters based on a normal distribution is thought to be warranted, although the frequency distributions in most cases are slightly skewed to the right (near lognormal) and occasionally even more irregularly shaped (Fig. 2). This would also seem to be justified by considering the generally very close proximity between the arithmetic and geometric means, the latter being equal to the median value for a lognormal distribution (Tables 2 and 3). Further, the average values are in most cases within, or very close to, the mode intervals on the frequency distributions. It should also be noted that we are here not comparing positively skewed distributions with negatively skewed ones, as this might conceivably lead to erroneous conclusions. At any rate one should be reluctant to use the mode, as this is unrelated to the shape of the frequency curve. The risk of introducing biased estimates by applying lognormal theory to pseudolognormal distributions was discussed by Link and Koch [10].

5. RESULTS

Large variations in Th and U contents are found within the same rock type (Table 2). The Oslo plutonic rocks are, however, characterized by high average values; the range in the averages of all rock types under consideration being 7.9-34.8 ppm for Th and 2.2-9.1 ppm for U. Petrologically more differentiated rocks like nepheline syenites and granites tend to show the highest average Th and U contents.

A large variability in concentrations of K (and other major elements) is shown by the kjelsåsite-larvikite series (1.73-5.56% K) and the nordmarkites (2.88-5.49% K). The biotite granites are peculiar in this respect, having remarkably uniform K contents, cf. Table 3 (standard deviations) and Fig. 2.

Two main groups of plutonic rocks can be distinguished from their distribution of Th/U ratios. The first group, having uniform Th/U ratios and low mean values in the range 3.5-4.5, comprises the kjelsåsite-larvikite series and related rocks, including nepheline syenites, and the Grefsen syenite. The second group, with variable Th/U ratios and high mean values in

TABLE 2. Th and U data for plutonic rocks of the Oslo Region.

		Th (ppm)						U (ppm)					
Rock type	No. of samples	Range	Geom. mean	Arithm. mean	Std. dev.	Coeff. of var.	Std. error	Range	Geom. mean	Arithm. mean	Std. dev.	Coeff. of var.	Std. error
Sørkedalite	4	7.7-20.2	11.8	12.8	6.0	46.7	3.0	2.2- 5.5	3.3	3.5	1.5	43.8	0.8
Monzodiorite[1]	7	5.6- 9.9	7.8	7.9	1.3	17.1	0.5	1.7- 2.5	2.2	2.2	0.3	12.3	0.1
Akerite - Kjelsåsite[2]	7	8.9-16.8	13.9	14.2	2.8	20.0	1.1	2.7- 4.9	3.9	4.0	0.8	20.6	0.3
Kjelsåsite - Larvikite	266	2.5-75.0	13.5	15.1	7.5	49.6	0.5	0.7-16.4	3.7	4.2	2.0	48.5	0.1
Tønsbergite	7	15.7-30.8	20.4	20.8	4.7	22.8	1.8	3.8- 7.1	5.0	5.1	1.0	20.4	0.4
Ditroite	8	14.4-75.0	30.7	34.8	19.5	56.1	6.9	4.8-13.4	8.7	9.1	2.9	31.6	1.0
Lardalite	12	13.0-43.8	24.2	25.5	8.5	33.4	2.5	3.5-11.9	6.5	6.9	2.4	34.4	0.7
Foyaite - Hedrumite	10	2.7-26.5	9.5	13.1	9.1	69.2	2.9	0.5- 7.6	2.3	3.4	2.6	76.7	0.8
Hedrumite ring-dike[3]	7	11.7-23.5	14.3	14.6	4.0	27.3	1.5	3.7- 6.0	4.2	4.2	0.8	19.2	0.3
Pulaskite	15	7.8-15.9	11.0	11.2	2.5	22.2	0.6	1.9- 7.2	2.9	3.0	1.3	44.0	0.3
Grefsen syenite	36	3.2-22.9	11.0	11.8	4.3	36.4	0.7	0.6- 5.6	2.5	2.7	1.0	36.6	0.2
Nordmarkite	196	0.7-78.6	13.5	16.5	11.7	70.7	0.8	0.2-14.5	2.4	2.9	2.2	74.1	0.2
Ekerite	108	7.4-87.1	25.5	29.0	15.5	53.3	1.5	1.1-27.6	5.2	6.3	4.3	67.9	0.4
Drammen granite	109	13.4-63.9	25.9	27.1	8.8	32.5	0.8	2.0-11.6	4.4	4.7	1.7	35.6	0.2
Finnemarka granite	22	11.2-33.9	20.5	21.1	5.3	25.2	1.1	1.2- 8.7	4.3	4.6	1.7	36.7	0.4
Biotite granites	38	13.2-67.3	30.7	32.4	10.9	33.6	1.8	1.7-19.2	5.3	6.3	4.0	62.5	0.6

[1] Associated with the Finnemarka granite.
[2] Transitional rocks from Nordmarka, northern Oslo Region.
[3] Øyangen cauldron, northern Oslo Region.

TABLE 3. K and Th/U data for plutonic rocks of the Oslo Region.

		K (%)						Th/U					
Rock type	No. of samples	Range	Geom. mean	Arithm. mean	Std. dev.	Coeff. of var.	Std. error	Range	Geom. mean	Arithm. mean	Std. dev.	Coeff. of var.	Std. error
Sørkedalite	4	1.83-3.11	2.25	2.31	0.61	26.37	0.30	3.42-3.78	3.59	3.59	0.16	4.52	0.08
Monzodiorite[1]	7	1.76-4.40	2.56	2.66	0.84	31.66	0.32	2.92-4.50	3.53	3.57	0.54	15.24	0.21
Akerite - Kjelsåsite[2]	7	2.76-4.01	3.49	3.51	0.39	11.14	0.15	3.30-3.97	3.56	3.57	0.22	6.08	0.08
Kjelsåsite - Larvikite	266	1.73-5.56	3.45	3.50	0.57	16.28	0.03	2.62-5.57	3.60	3.63	0.42	11.63	0.03
Tønsbergite	7	3.20-4.99	3.81	3.84	0.55	14.39	0.21	3.65-4.87	4.10	4.12	0.42	10.12	0.16
Ditroite	8	3.06-5.32	4.55	4.62	0.79	17.17	0.28	1.78-8.62	3.53	3.93	2.09	53.33	0.74
Lardalite	12	3.38-4.23	3.87	3.88	0.25	6.40	0.07	3.53-3.95	3.72	3.72	0.10	2.75	0.03
Foyaite - Hedrumite	10	3.86-5.14	4.76	4.77	0.34	7.22	0.11	2.03-6.00	4.10	4.23	1.05	24.78	0.33
Hedrumite ring-dike[3]	7	3.97-5.14	4.68	4.69	0.36	7.63	0.14	3.10-3.92	3.43	3.44	0.32	9.27	0.12
Pulaskite	15	3.04-5.11	4.42	4.46	0.61	13.66	0.16	1.75-5.13	3.84	3.94	0.81	20.59	0.21
Grefsen syenite	36	3.56-5.59	4.56	4.58	0.46	10.11	0.08	3.23-5.50	4.34	4.39	0.62	14.25	0.10
Nordmarkite	196	2.88-5.49	4.74	4.76	0.38	8.09	0.03	1.98-27.33	5.56	5.92	2.52	42.67	0.18
Ekerite	108	3.47-4.74	4.10	4.11	0.28	6.83	0.03	1.27-18.35	4.94	5.47	2.65	48.48	0.26
Drammen granite	109	3.54-4.85	4.25	4.26	0.20	4.81	0.02	2.36-11.71	5.86	6.19	2.04	32.97	0.20
Finnemarka granite	22	3.60-4.06	3.85	3.86	0.12	3.07	0.03	3.03-12.29	4.83	5.13	2.12	41.40	0.45
Biotite granites	38	3.61-4.75	4.09	4.10	0.26	6.46	0.04	1.69-14.02	5.75	6.42	3.01	46.85	0.49

[1,2,3] footnotes, see Table 2.

the range 4.5-6.5, includes nordmarkite, ekerite and biotite granites. The constant Th/U ratios for the first group of rocks are reflected in very high positive correlation coefficients between Th and U, mostly >0.9 (Table 4).

6. DISCUSSION

Taking into consideration additional evidence from Rb-Sr isotopic work [6,7] and gravimetric studies [13], it is reasonable to assume that the rock types with uniform Th/U ratios originated

TABLE 4. Correlation coefficients between Th, U and K for plutonic rocks of the Oslo Region.

Rock type	No. of samples	Th *vs.* U	U *vs.* K	Th *vs.* K	Th/U *vs.* K
Sørkedalite	4	0.998	0.660	0.699	0.965
Monzodiorite[1]	7	0.474	0.196	0.895	0.846
Akerite - Kjelsåsite[2]	7	0.968	-0.031	-0.124	-0.408
Kjelsåsite - Larvikite	266	0.964	0.387	0.386	-0.020
Tønsbergite	7	0.922	0.919	0.985	0.291
Ditroite	8	0.337	-0.352	-0.340	-0.234
Lardalite	12	0.998	-0.463	-0.498	-0.169
Foyaite - Hedrumite	10	0.856	0.401	0.292	-0.156
Hedrumite ring-dike[3]	7	0.961	0.230	0.128	-0.174
Pulaskite	15	0.564	-0.087	-0.241	-0.047
Grefsen syenite	36	0.924	-0.162	-0.000	0.527
Nordmarkite	196	0.796	-0.207	-0.179	0.048
Ekerite	108	0.733	-0.188	0.090	0.405
Drammen granite	109	0.447	-0.145	-0.506	-0.298
Finnemarka granite	22	0.726	0.184	0.262	0.066
Biotite granites	38	0.424	-0.169	-0.493	0.006

[1,2,3] footnotes ,see Table 2.

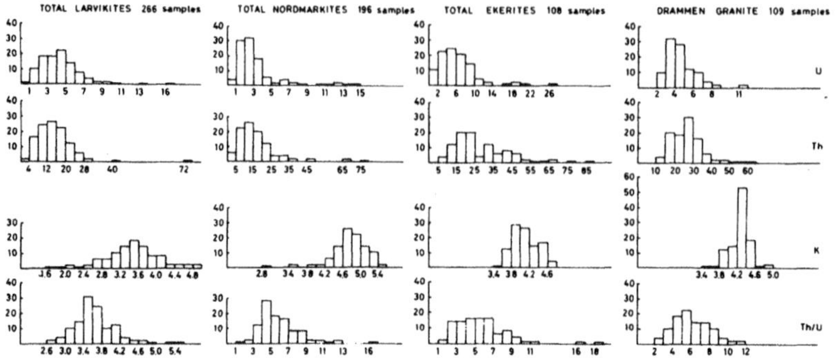

Fig. 2. Histograms showing U, Th, K and Th/U distributions (per cent frequency) for rock groups with more than 100 samples. Th and U values are in ppm, K in per cent. Note that the class intervals for one variable may not be uniform. There are two additional K values for larvikite at 5.03 and 5.56%, and one additional Th/U value at 27.33 for nordmarkite.

directly from a mantle-derived source-rock mainly by differentiation, retaining a primeval homogeneity in Th/U ratios.

For rocks with variable Th/U ratios a different origin and/or differentiation history must be assumed. Contamination from the overlying Precambrian and Cambro-Silurian rocks may explain the variability in Th/U ratios at least for some rock types. Differentiation at high levels in the crust with a concomitant selective loss of U by oxidation, due to a higher oxygen fugacity in the upper crust as compared to the lower crust, is perhaps a more likely explanation. This is consistent with the observation that when passing from older rocks with uniform Th/U ratios to younger rocks with variable ratios, the average U abundances are clearly going down, while the abundance of Th remains fairly constant or is only slightly less (Fig. 3). The magmas forming nordmarkite, ekerite and biotite granites were rich in volatiles, as attested by the presence of miarolitic cavities [11] and contact metasomatic deposits [5]. The fluid phases associated with these rocks may have played an important role in the development of variable and high Th/U ratios by acting as a transport medium for oxidized uranium.

Possible mutual relationships between the different plutonic rocks of the Oslo Region are presented diagramatically in Fig. 3, as deduced from a knowledge of Th/U ratios, petrographic similarities and field relations. However, a common magmatic evolution sequence for the whole Oslo Region is very unlikely, and several local trends of differentiation have certainly been of significance. It must therefore be strongly emphasized that Fig. 3 represents a generalization and oversimplification. Especially the rocks grouped as 'nordmarkites' are very inhomogeneous and may belong to different magmatic sequences.

Compositional modification of magmas by assimilation of bedrock seems to be far less common in the Oslo Region than previously assumed. For instance, the uniform and low Th/U ratios of sørkedalite would favour the interpretation that it represents an early cumulate of heavy minerals from a kjelsåsitic magma [14,3], rather than being of hybrid origin [2]. The same argument can be applied to the akeritic transitional rocks from Nordmarka, which were considered to be a contaminated border facies of kjelsåsite by Sæther [14], but which rather represent a quartz-rich member of a differentiation series akerite-kjelsåsite. A hybrid origin for monzodiorite which is associated with the Finnemarka granite is, for the same reason, highly improbable, and the complex was probably formed by multiple intrusion [13], although not necessarily of comagmatic origin as proposed by Czamanske [4].

The Grefsen syenite and various biotite granites (Drammen granite, Finnemarka granite, etc.) are not easily related to the other rock types. The biotite granites might be accounted for by

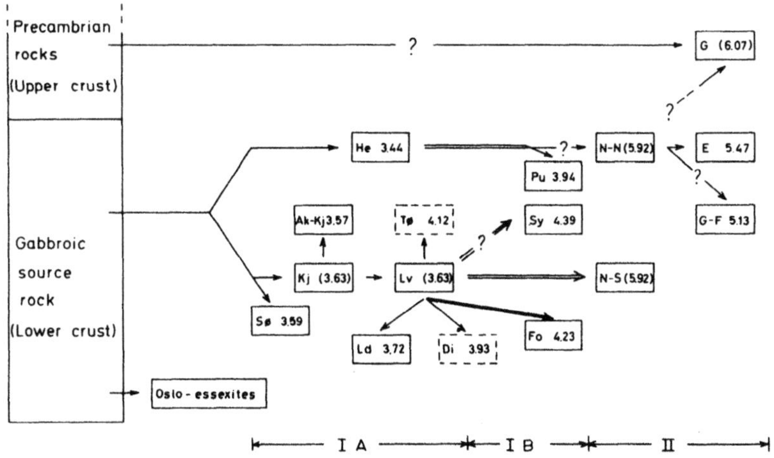

Fig. 3. Simplified evolution scheme for the principal plutonic rock types of the Oslo Region (see text). Double arrows indicate decrease in average U abundance. Stippled boxes are used for rocks of metasomatic origin. IA/IB and II refer to rocks with uniform and variable Th/U ratios, respectively. Average Th/U ratios are quoted besides the rock type abbreviations (SØ: sørkedalite, Kj: kjelsåsite, Lv: larvikite, Ak-Kj: akerite transitional rocks, TØ: tønsbergite, Ld: lardalite, Di: ditroite, Fo: foyaite, Sy: Grefsen syenite, N-S: 'nordmarkite' with augite, N-N: nordmarkite sensu stricto, He: hedrumite ring-dike, Pu: pulaskite, E: ekerite, G-F: Finnemarka alkali granite, G: biotite granites).

postulating a process of contamination from overlying Precambrian rocks, a theory also favoured by Killeen and Heier [8] who studied the radioelement distribution in adjacent Precambrian granites.

Two varieties of larvikite, the red-coloured tønsbergite and the gneissic nepheline-bearing ditroite, are clearly of metasomatic origin and have higher average Th, U and K contents than larvikite. The strong positive correlation between U-K and Th-K for tønsbergite should be noted (Table 4). This rock also contains secondary quartz [12]. The metasomatic addition of SiO_2, K_2O, Th, U, etc. to form the variety tønsbergite most likely represents mobilization of a late silica-rich phase of the larvikite magma, although it could also be associated with the emplacement of a deep-seated, granitic magma of younger age, or preferably a rock type with higher density than granite [13].

ACKNOWLEDGEMENT

Financial support for the investigation was provided by Norges Almenvitenskapelige Forskningsråd.

REFERENCES

1. J.S. Adams and P. Gasparini, Gamma-ray spectrometry of rocks, Elsevier Publishing Company, Amsterdam, 1970.
2. T.F.W. Barth, Skr. Norske Vidensk.-Akad., I. Mat.-Naturv. Klasse, No. 9, 1944.
3. M.K. Bose, Skr. Norske Vidensk.-Akad., I. Mat.-Naturv. Klasse, Ny Serie No. 27, 1969.
4. G.K. Czamanske, J. Geol., 73, 293, 1965.
5. V.M. Goldschmidt, Skr. Videnskapsselsk. Kristiania [Oslo], I. Mat.-Naturv. Klasse, No. 1, 1911.
6. K.S. Heier and W. Compston, Lithos, 2, 133, 1969.
7. S.B. Jacobsen and G. Raade, Norsk Geol. Tidsskr., 55, 171, 1975.
8. P.G. Killeen and K.S. Heier, Geochim. Cosmochim. Acta, 39, 1515, 1975.
9. P.G. Killeen and K.S. Heier, Skr. Norske Vidensk.-Akad., I. Mat.-Naturv. Klasse, Ny Serie No. 35, 1975.
10. R.F. Link and G.S. Koch, Jr., Mathem. Geol., 7, 117, 1975.
11. G. Raade, Mineral. Record, 3, 7, 1972.
12. G. Raade, Distribution of radioactive elements in the plutonic rocks of the Oslo Region, unpublished cand.real. thesis, University of Oslo, 1973.
13. I.B. Ramberg, Norges Geol. Unders., 325, 1976.
14. E. Sæther, Skr. Norske Vidensk.-Akad., I. Mat.-Naturv. Klasse, Ny Serie No. 1, 1962.

ORIGIN OF THE MAGMAS OF THE VESTFOLD LAVA PLATEAU

C.Oftedahl

Department of Geology, Norwegian Institute of Technology, Trondheim, Norway

ABSTRACT. The Vestfold lava plateau is the largest (about 600 km^2) area of volcanic rocks preserved in the Oslo Paleorift. It also contains the longest, stratigraphically-continuous record of eruptions. The upper half is unique because of its diverse lithologies and petrochemistries: Between rhomb porphyry flows RP11 to RP26 occur four basaltic flows (B3-B6) and four trachytic flows, ranging from latitic to rhyolitic in composition. A genetic model to explain this compositional diversity must be constrained by the Ramberg [2] gravity anomaly analysis, which pictures a lens-shaped gabbro body, ten km thick, under the major part of the region and some 22 km below the surface. A corresponding basaltic magma chamber must have had a number of differentiation trends, to produce the complex Vestfold lava plateau and later plutonics. More than 100 complete wet silicate analyses and XRF trace element analyses (Rb, Sr, Y, Zr, and Nb) illuminate the complicated magmatic relations. The excluded elements Zr, Nb, and Y indicate:

1) Rhomb porphyries make up a rather concentrated cluster and are clearly different from the basalt group lineage, so that the surface basalts do not seem to represent the mother magmas for the rhomb porphyries. 2) Trachytes make up four groups. T3 and T4, rich in Zr and Nb, may be derived from the RP magmas by fractionation. A second group with very low contents, can not have developed by fractionation from basaltic magmas of compositions similar to the surface basalts, but need a more primitive mother magma, perhaps like the Kolsaas B1 tholeiite of Weigand [1]. A third group with aberrant values may represent anatectic magmas, and a fourth group with values comparable to those of rhomb porphyry flows is of uncertain parentage.

These conclusions are supported by computer mixing models of possible crystal fractionation paths.

E.-R. Neumann and I.B. Ramberg (eds.), Petrology and Geochemistry of Continental Rifts, 193–208.

Table 1. Stratigraphy of the Vestfold lava plateau.

Upper part

Symbol	Thickness (m) Aver.	Max.	Plagioclase phenocrysts (xx)	No. of analyses
Ig 1		250	Ignimbrite (outflow sheets), Ram.cauld.	1
B6	10	10	Aphyric basalt	2
RP26	200	300	Many RP flows, more or less like RP1	3
T4	100	200	Trachyte porph. flows, rhyol. ignim.	8
RP25	45	45	RP2 type, scattered 1 cm and smaller xx	1
RP24	50	70	Up to 8 flows. Rhomb xx, but scattered	4
T3	50	155	Trachyte porphyry lava flows (ignimbr.)	11
RP23	50	70	Typical RP1	5
B5	10	30	Aphyric andesitic basalt	5
RP22	10	10	Small xx (5 mm), very scattered	3
RP21	15	15	5-10 mm xx, scattered	4
RP20	15	15	Thin rhomb xx (1-2 cm), scattered	1
RP19	35	100	RP1 type, but rectangular xx and rhombs	7
RP18b	15	25	Small and scattered xx	2
B4/T2	15	25	Aphyric basalt in S, trach.porph. in N	4/6
RP18a	5	15	Small and scattered xx.	2
RP17	35	45	Typical RP, packed with rhomb xx, 2-3 cm	3
RP16	25	40	RP2-type, 1 cm irregular xx	4
RP15	35	35	Similar to RP1, but thick, boat-shaped xx	2
RP14	20	35	Ideal RP1	1
RP13b	3	3	Scattered, large xx (2-4 cm)	1
RP13	35	150	Rect.porph. 1-1,5 cm xx, rect. or quadr.	1
T1c	20	35	Latite porph., 0,5 - 1,0 cm rectangular xx	5
T1b	40	150	Trach. porph., many flows, up to 1 cm xx	8
T1a	10	20	Trachyte porphyry, 2-5 mm xx	-
B3	20	30	Complex basaltic unit	8
	868	1878		102

Lower part, type names and no. of analyses.

23. Rønneberg		15. Sukke		7. Kiste A, B	
22. Ende		14. RP6		6. Allum	
21. Rød		13. Stuaas		5. RP4	
20. Hegg	2	12. Rykaas		4. RP2	
19. Lakjell		11. Uleaas	1	3. RP1	1
18. Greaker A-E	2	10. Korsgaard A,B	1	2. Latite (To)	1
17. Løvald		9. Adal	1	1. B1	
16. Bjørnaas		8. RP5		Perm.sed.	
				Sil.s.st.	

1. INTRODUCTION

The main portion of the Oslo Paleorift consists of a downfaulted block which contains small areas of Precambrian gneisses, Cambro-Silurian sedimentary rocks, and mostly Permian igneous rocks. Recent summaries of the Permian geology with survey maps have been presented by Oftedahl (Vol. II) and Ramberg [2] and Vol. II.

The volcanic rocks which originally made up a lava plateau extending over the whole of the Oslo graben and most likely outside it, occur now in two larger areas and several cauldrons or cauldron remnants, due to the stoping action of later plutonic bodies and subsequent erosion. The two larger areas are the Krokskogen plateau west of Oslo and the Vestfold plateau to be treated below. The Krokskogen area, located in the middle of the 200 km long rift, contains a complete stratigraphy for the lower part of the lava plateau, comprising basalts B1, B2, and B3 and rhomb porphyries RP1-RP12. In neighbouring cauldrons direct continuity of the units B3 and the rectangle porphyry RP13 is obvious. The lower flows RP1, RP2, RP4, RP5, and RP6 may have been of regional scale and covered most of the graben area. The flows up to RP11 may have covered most of the central area, the 3000 km^2 around Oslo. A survey of the complete volcanic stratigraphy is presented elsewhere by Oftedahl (Table II, Vol. II).

In the Vestfold area the development was different (Table 1). The regional scale flows from RP4 on started to become mixed up with a great number of small and local flows, most likely deposited in small downfaulted areas in narrow, local grabens [3]. In all 40 different flows have been recognized among the flows corresponding to the interval RP4-RP12. The author has mapped the higher flows, mostly occurring between the Hillestad cauldron in the north and the Ramnes cauldron in the south. This area was especially chosen because of the wish to revise and improve the published stratigraphy [4] and because the rhomb porphyries here contain at least four basalt units and four trachyte units. Thus, this volcanic pile represents an extremely interesting case of interlayering of rocks of diverse compositions from basalts through latite and trachyte to rhyolite, when counting the last ignimbrite Ig 1. Since the eruptions must have covered a very short time span, geologically speaking, the different magma types must have existed at essentially the same time at various levels in the crust and below. Therefore investigations of this lava pile are exceptionally well suited to attack problems of magma formation in this area.

With the establishment of a very specific three dimensional model by Ramberg [2] on the basis of an extensive and detailed gravity anomaly analysis, it seems very likely that the monzonitic magmas developed on top of a basaltic magma pillow at the base of the crust at about 20 km depth.

With this model as a background, the major problems are then:

1. Did the intermediate and acidic magmas develop from one or more basic magmas essentially by fractional crystallization?

2. If their development was not wholely by fractional crystallization, which magmas are fractionated magmas but contaminated by crustal material and which magmas may represent partial or complete anatexis?

2. THE VESTFOLD LAVA PLATEAU

The central area of the Vestfold lava plateau from the Hillestad cauldron southwards to the Ramnes cauldron was mapped in the years 1961 to 1975 by the author, with Mr. Roar Anton Bruun as field assistant in 1958. The lower stratigraphy east of the central area and extending from Holmestrand southwards past Horten towards Tønsberg was mapped by Mr. Henrik Heyer in the years 1967-69, 1971, and 1974. Part of the volcanic rocks within the Ramnes cauldron was mapped by Mr. Erik Schou Jensen in 1963 and 1964 and by the author, but most of the cauldron as well as much of the area to the south and west was mapped by Mr. Rolf Sørensen during 1966-69. The ignimbrite sequence in the key area within the Ramnes cauldron was described by Schou Jensen [5]. A revised stratigraphy of the lava plateau as well as preliminary information on the Hillestad and the Ramnes cauldrons was presented by Oftedahl [6], who also discussed possible magma formation processes, based on some 60 major element analyses, but without coming to any definite conclusion.

3. STRATIGRAPHY

The Vestfold lava plateau can be divided into three major groups with different lava types or lava association:

3. The Vivestad group. Basalt B3, trachyte T1, and rhomb porphyries RP13-RP26, with intercalated flows of basalts and trachytes, and starting with B3.
2. The Horten group, the lower rhomb porphyries starting with RP1 and comprising all local flows up to the base of B3 or trachyte T1.
1. The Holmestrand group. B1 basalts, at least 20 single basalt flows of many petrographic types.

Figure 1 shows the central part of the Vestfold lava plateau, cut into a number of fault blocks by antithetic step faults. The detailed stratigraphy appears in Table 1, which also gives thicknesses, a short characterization of each flow, and number of analyzed samples. The most important facts about the Vestfold lava plateau as compared to the Krokskogen lava plateau at Oslo and

Fig.1 Central part of the Vestfold lava plateau. For location, see Oftedahl, this volume, Fig.1. Symbols: 1, 2, 25, 26 - RP flows; B1, etc. - basalts; T1, etc. - trachytes; A1, Ki, etc. - lower, local RP flows, only sketchily shown.

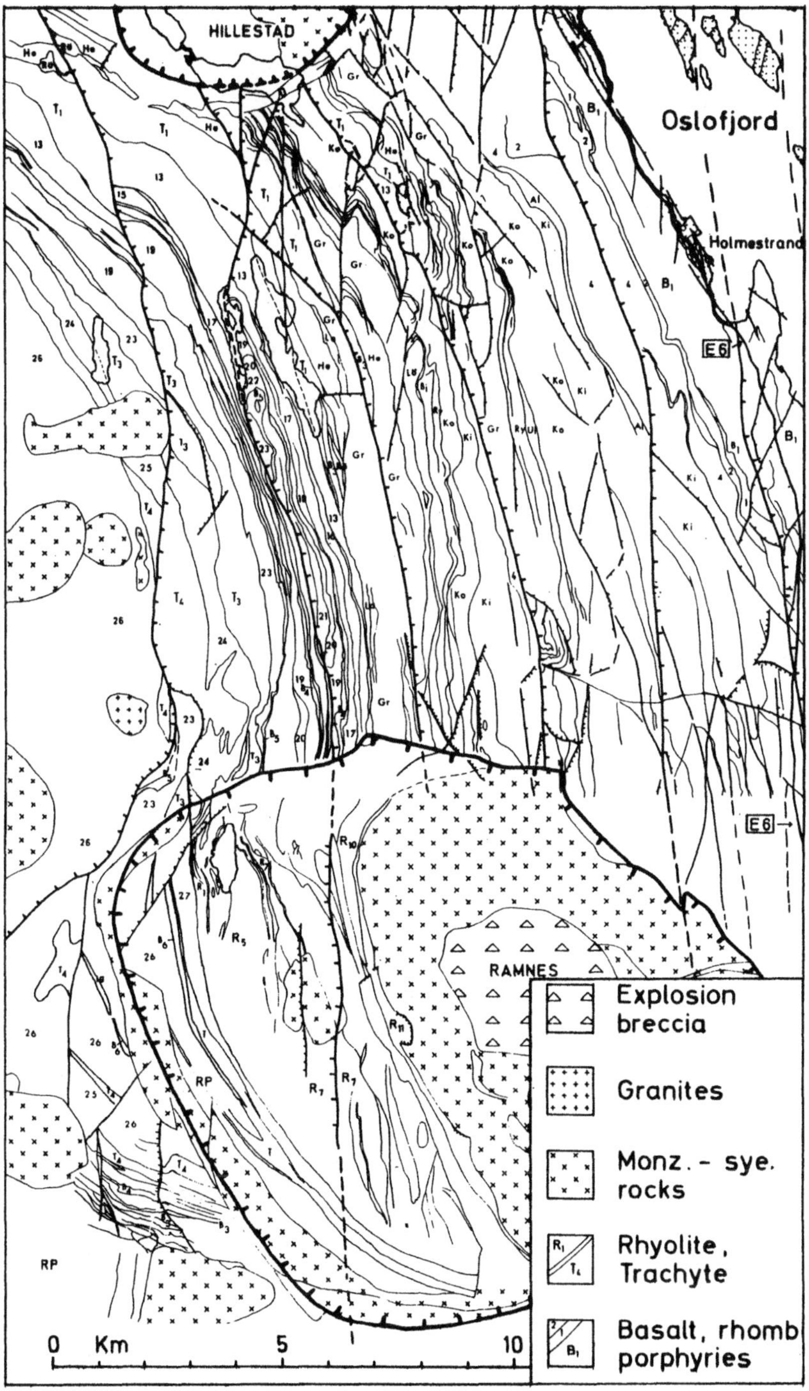
HILLESTAD
Oslofjord
Holmestrand
E6
RAMNES
0 Km 5 10
Explosion breccia
Granites
Monz. - sye. rocks
Rhyolite, Trachyte
Basalt, rhomb porphyries

surrounding cauldrons are as follows.

The plateau starts with a basalt formation B1 resembling those in Krokskogen and in the cauldrons, followed by the well-known rhomb porphyry types RP1, RP2, and RP4. RP3 appears to be a local flow for the Krokskogen area. The Krokskogen types RP5 and RP6 can also be recognized in some small areas in Vestfold, otherwise the rhomb porphyry types from RP4 and through the Horten group appear to be local for the area, whereas the Krokskogen stratigraphy has a much more regional flavour, in that the Krokskogen types up to RP12 and B3 also appear in a number of the cauldrons [4, 7]. The most interesting part of the Vestfold plateau starts with the uppermost flows of the Horten group, among others the Hegg type, which petrographically is the least altered in the Oslo region and contains pyroxene grains in the well crystallized groundmass.

In the Vivestad group some flows like RP19 and RP20 contain pyroxene microphenocrysts. Otherwise the rhomb porphyries are more or less altered. The more-or-less rhomb-shaped phenocrysts consist of a faintly to strongly altered plagioclase of a very specific composition [8]. The rhomb porphyry lava flows obviously contained the rhomb crystals as intratelluric phenocrysts, and some had in addition microphenocrysts of iron ore, augite, and possibly olivine. The lava flows crystallized by formation of more or less well developed tablets of anorthoclase in a glass. This groundmass glass later devitrified into what now appears as a light greenish chlorite (smectite).

Among the trachytes T1 consists of as many as 10 or more single trachyte lava flows, distinguished from the rhomb porphyries by their small and rectangular phenocrysts of somewhat varying size and frequency. A marked flow banding is developed in some of these trachytes. Chemical analyses showed, however, that the youngest and most widespread flows looking like the other trachyte porphyries really are latites. These are included in the T1 unit in the description below. Most likely trachytes T2 and T3 also represent lava flows, but the formation T4 seems to be composite, consisting of some trachytic to rhyolitic lava flows and some ignimbrites. Only in the south-western part of the plateau are found ignimbrites of classical appearance (Ig 1).

All the basalts are thin units and may represent single flows. At Rød near Revovann the B3 flow is unique in containing small areas of another porphyritic lava type with 1 cm long rectangular and heavily altered feldspar phenocrysts and 0,5 cm quartz phenocrysts. This so-called quartz porphyry has been analyzed in detail and is described below.

4. MAJOR ELEMENTS

More than 100 major element analyses have been performed on rocks from the lava plateau, essentially from the Vivestad group with some analyses from the Horten group. Harker plots of the major oxides against SiO_2 seem to show that the basalts make up one major band (in some cases several), whereas the rhomb porphyries and the trachytes each make up very broad and fairly well separated groups. The trend through the major band of basalts can be seen as continuing to the rhomb porphyries and then again to the field of trachytes. This observation suggests that the rocks are all genetically related by one or more magma generating processes which eventually overlap in a smooth way. The diagram also points to several problems. The most obvious one is the fact that both rhomb porphyries and trachytes make up quite wide fields in the Harker plots. Points from several analyses of one lava flow may also show a big spread. Thus the seven analyses from RP19 nearly cover the whole RP area for many oxides. This fact accentuated the problem: How much of change in major elements resulted from the alteration which is evident in all thin sections? This alteration clearly occurred within a relatively high temperature hydrothermal regime, because the vesicle fillings in all lavas consist of quartz, calsite, epidote, and chlorite, never zeolites. Obviously the lava plateau was totally cooked through after its formation, although the alteration has produced surprisingly different results in the different lava flows. Therefore the peak of alteration is assumed to be connected with the congealing of each separate flow. The trace elements give more information about this process.

As has already been noted [6], there is a clear tendency toward increase of SiO_2 in rhomb porphyries near the top of the plateau. Thus one RP24 and one RP26 analysis lie in the trachyte field. Nearly all of the lower rhomb porphyries fall on the left side of the RP field. Surprisingly the most well crystallized of all, the Hegg RP type located around RP10-RP11, is the most basic of them all. When examined in detail, the points for successive RP flows jump back and forth within the RP field in a very irregular manner, suggesting that in detail the different RP types all have their special story, at least during the last phase before eruption.

Norm calculations of 110 major element analyses have been run on the Center for Volcanology norm computer program. The relations saturation/undersaturation and a few other points are commented on below.

Undersaturation is the main rule for the basalts. They usually contain a little normative nepheline and 1-6% olivine, with 13,5% as a maximum. Of 19 analyzed basalts only five contain normative quartz.

Rhomb porphyries fall fairly well close to the neutral line, with a little quartz being somewhat more common than a little nepheline and olivine in the norm. Of 53 analyzed rhomb porphyries, 19 are undersaturated and 34 oversaturated. Of the latter 16 analyses contain 5-10% quartz and one 14,8%, a strongly altered RP13, rich in secondary calcite and quartz. Actually most of the 14,8% may reflect vesicular fillings. Of the undersaturated, four analyses contain between 5 and 10% nepheline and only one more than 5% olivine (RP13b with 6,5% ol). This general picture is in accordance with what was found by Barth [9] for larvikite specimens, which in part are faintly saturated, in part faintly undersaturated.

Within the four trachyte formations the quartz content varies considerably. Thus the latitic T1 runs from 0,3-9% quartz. The trachytic T1b falls in the interval 9-15% quartz. Five trachytes within T2 fall in the same range (10,6-14,7%) with one T2 rhyolite at 21,1%. Within the 11 analyses of T3, four are low (4-5%), five are medium (9-11%), and two prove to be rhyolites (27 and 31% quartz). Also T4 analyses fall in a low group of 6-8%, a medium group of 13-17%, and show one rhyolite with 21,6% quartz.

Normative minerals like rutile and perovskite appear often, owing to the highly oxidized state of the lavas which mostly contain little ferrous iron. The catanorms contain only one trachyte with acmite. Corundum is present in most of the norms in quantities up to a few per cent, being clearly more abundant in rocks with normative quartz. This trend is interpreted as being due to leaching of alkalies (mostly soda?) and deposition of quartz during deuteric alteration.

5. TRACE ELEMENTS

All the Vestfold volcanic rocks which have been analyzed for major elements have also been analyzed for the trace elements Zr, Nb, Sr, Rb, and Y, in the X-ray lab of the Dept. of Geology, University of Oregon. The Nb-Zr plot proved to be the best for indicating magmatic relationships. These two elements are excluded during fractional crystallization, and they show the widest variations. In the Zr-Nb plot (Fig.2), the analytical points roughly make up a broad band which is fairly straight and trends through the origin. Ideally, this would happen if a basaltic magma differentiated through the intermediate to the acid range by fractional crystallization in a very regular way. In detail, however, the present diagram shows no simple relationship, with the basalts in one group, grading into rhomb porphyries, then trachytes. The rather confused picture has the following major features:

1. The points for the trachytes show a surprising degree of dispersion throughout the diagram. Thus the points for T1 lie between the points for the basalts and the rhomb porphyries. The points for T2 occur concentrated within the RP field, whereas T3 and T4 follow clearly after the rhomb porphyries. But all of the four units have anomalous points all over the diagram. Some T3 and T4 samples have surprisingly high Nb values.

2. The basalts make up one group with Zr- and Nb-values, well separated from the main cluster of the rhomb porphyries which have much higher values.

The basalts concentrate around a line which does not point into the RP field but has a steeper slope. One may therefore conclude that the rhomb porphyry magmas did not develop from this path of evolution, which ends so abruptly that it can hardly have occurred in the big basaltic pillow. A solution to this problem may be found in one of the results of the detailed gravity study of Ramberg [2]. He found that some local highs could be explained only by shallow and heavy bodies. This could be either heavy Precambrian rocks just below the sub-Cambrian peneplane or a Permian sill of basaltic composition. Since the last possibility was preferred, I conclude that the basalts B1-B3 and the Vestfold B4-B6 may have come from such local magma chambers, with some exceptions. The names and volumes of these chambers are as follows [2, p.116 and 128]:

Horten-Tønsberg high:	2415 km^3,	ca.(50x10x4) km^3.	3,9-5,1 km depth	
Asker-Lier " :	250 "	" (20x5x2,5) "	1,4-1,9 "	"
Narrefjell(?) " :	457 "	" (35x15x1) "	3,5-4,7 "	"

At present it seems an attractive hypothesis to postulate that these gravity highs represent crystallized magma chambers from which came the basalts, - in the Vestfold area from the Horten-Tønsberg chamber, in the Krokskogen area from the Asker-Lier chamber, with the doubtful Narrefjell high as a possible magma chamber for the western lava flows which have completely disappeared owing to later emplacement of plutonic bodies. The emplacement of these basalt lenses would be by saucer-like subsidence along small faults, really a sort of place-trading with the corresponding volume of older rocks that were initially on top of the basalt pillow. Within these sills a certain differentiation within upper local domes of the chamber must be postulated. With fissures penetrating various parts of the magma chamber, locally different flows may have formed, such as B4-B6 in the Vestfold area. The anomalous Skien basalts [10], being nephelinitic melilitites, presumably came directly from the upper mantle below the big basaltic pillow.

Such a tectonically-induced tapping also explains why there is no progression in Zr and Nb according to stratigraphy. Fig.2 shows a low value cluster which consists of B1, B3, some B4, B5, and B6 points. High values are shown by most B4 and some special B3 rocks. But most of B3 in the Vestfold area seems to be within the normal range of 200-400 ppm Zr with some exceptions. In the first place the basalts southwest of the Ramnes cauldron must be something special, with their high Zr and Nb values. Since samples from the supposed B3 unit range from an andesitic basalt into trachyte composition, the B3 flows here obviously were melts strongly differentiated or otherwise modified in composition. A similar magma pocket may be indicated by a thin basalt just south of Hillestadvann with 1054 ppm Zr. The very special basalt B3 with included quartz porphyry is commented on below.

The data of Baker et al. [11] indicate a rather constant Zr-Nb ratio around 3,0 for their southern Gregory Rift lava series, or within the range 3,25-2,75. The bulk crystal/liquid distribution coefficient of Zr is shown to increase from nearly zero for basalts to 0,4 and 0,6 for the benmoreite and the trachyte range, respectively. Accordingly the bulk distribution coefficients increase proportionally for Nb. The same may be the case for the general

Oslo trend, as shown by Fig.2, with a general Zr/Nb ratio around 5,75. But special trends run steeper, such as the basalts with a Zr/Nb ratio of about 3,75-4,00, which is not too far from the results of Baker et al.

Explanations for this double trend may be sought in a factor such as depth of fractionation. Flat lines (or high ratios) may indicate fractionation at greater depth and higher temperature, steep lines or low ratios may indicate shallow level fractionation as postulated for the B1-B6 basalts. Also, southern Gregory Rift lavas [11] are assumed to have undergone relatively shallow fractionation. Contamination by bedrock is preliminarily assumed to be unimportant.

The rhomb porphyries make up a cluster in the middle of the diagram, but there is a clear tendency to a concentration along a broad band through the origin. A number of analyses fall on or close to a line through the origin with Zr/Nb = 5,6. The central part of this cluster consists of points representing melts which can hardly have formed from any of the surface basalts, because the points of the central basalt cluster comprising B1-B3 represent magmas with Zr and Nb contents that are much too low. Straightforward olivine-augite-plagioclase fractionation of melts corresponding to this cluster would produce latitic melts with Zr and Nb contents much lower than those for the central part of the RP cluster. Therefore the conclusion seems inescapable: If the latitic rhomb porphyries are melts formed by fractional crystallization of basalts, these basalts are not seen at the present surface. Thus under the assumption of fractionation, a mother basalt liquid for these latites must be postulated to have differentiated out of the original basalt pillow magma by fractionation, producing a basalt which was considerably higher in excluded trace elements. Exactly this trend was found in the East-African basalts studied by Baker et al. [11].

Within the whole group of rhomb porphyries a few rocks show up as special cases. In the first place the Hegg sample stands out because it has much lower Zr and Nb contents than does any other RP sample. This RP flow is stratigraphically located a little below B3 and was unique in some special properties, resulting in this porphyry being the most well crystallized of all RP flows in the Oslo region. The feldspar phases both in rhombs and groundmass are fairly clear and the rest of the groundmass is well crystallized, with minerals like augite, brown hornblende, biotite, and finally some chlorite. Thus this rhomb porphyry which is the most primitive one also possibly had the highest content of volatile components to permit the most complete crystallization. Next again in Zr-Nb contents come RP19 samples and RP6 of Krokskogen. The RP19 flow

Fig.2 Plots of Nb, Y,and Rb against Zr (all in ppm) in basalts, → rhomb porphyries (dots), and trachytes in 121 samples from the Vestfold lava plateau and 8 rhomb porphyries from the Krokskogen plateau (see Fig.3). The association quartz porphyry and surrounding B3 is specially indicated.

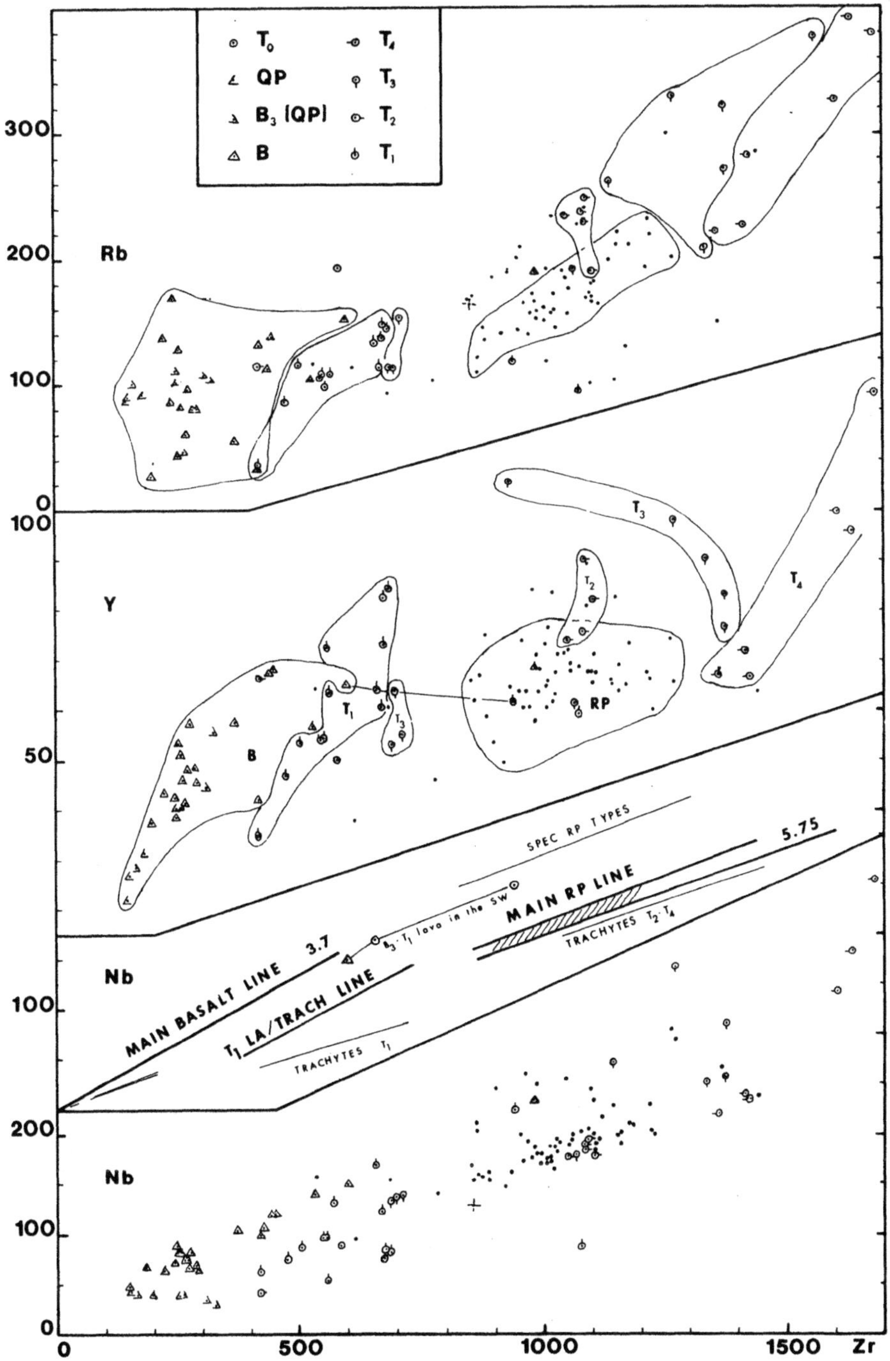

T0
QP
B3 (QP)
B
T4
T3
T2
T1
Rb
Y
Nb
RP
SPEC RP TYPES
MAIN RP LINE
5.75
TRACHYTES T2-T4
B3-T1 lava in the SW
3.7
MAIN BASALT LINE
T1 LA/TRACH LINE
TRACHYTES T1
Zr

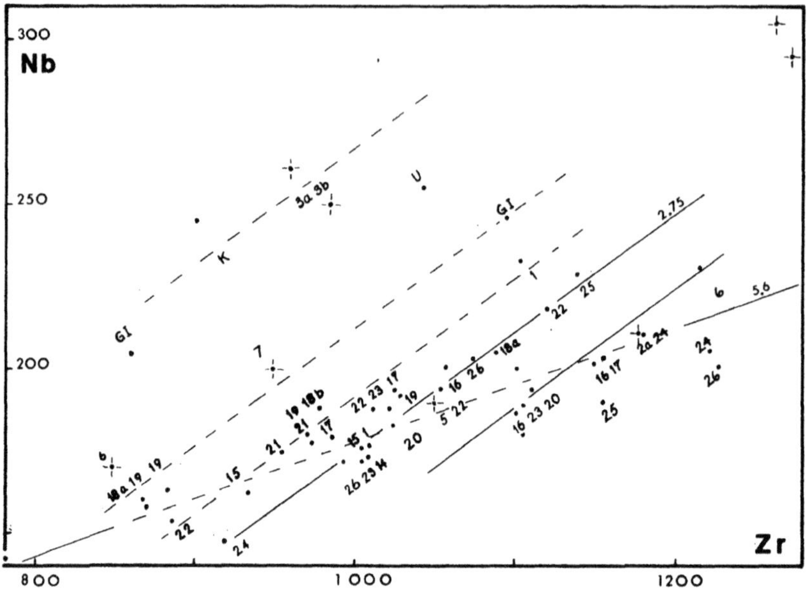

Fig.3 Zr-Nb plot of rhomb porphyries, with stratigraphic position indicated (16 = RP16, etc.). Letters: rock from lower half, see Table 1. Stars: 8 Krokskogen samples. Possible genetic lineages are mentioned in text.

is distinguished as one of the few which carry abundant augite microphenocrysts. A few rhomb porphyries deviate from the main trend in being higher in Nb, the Uleaas type and the Krokskogen types RP2b, 3a, and 3b. Among these the Uleaas rock is special in containing 5,96% K_2O against 3,85 Na_2O. However, high contents of Nb and K_2O together is not a rule, since a Vestfold RP6 contains 7,14% K_2O and still lies on the main trend for Fig.3.

The Zr-Nb relations appear in detail in a blow-up of Fig.2 (Fig.3). Firstly, this diagram shows that there is no stratigraphic progression in RP evolution, since the RP numbers jump back and forth, the only certain rule being that there is a clear tendency for the older RP flows to be more primitive, i.e., lower both in Zr and Nb as well as in SiO_2. Secondly the points show groupings that could define steeper evolution lines (Zr/Nb around 2,75).

Thus the RP flows may represent magmas which have undergone a two-stage fractionation, a deep one and a surface-near one. Obviously a number of different, near-surface chambers may have contributed to the flows, each being cut by fissures at different places to tap magmas of different stages of fractionation.

Among the T3 samples two are so rich in Nb that they fall above the top line in Fig.2. These two rocks were sampled within 2 km of the Ramnes cauldron ring dike. At least for one sample (from Kjønnerød quarry), a marked fumarolic alteration is obvious. Thus this alteration may well involve more than iron oxidation and might include enrichment in Nb from below, most likely by means of a ring fault intrusion not seen in the present surface. The reason for the existence of the three-sample group poor in Zr and Nb is very uncertain. A separate lava flow from a separate source looks most probable but the field geology does not support this hypothesis.

The Zr-Y diagram shows the same general trend as with the Zr-Nb diagram, but is not so telling as the latter (see Fig.2). The basalts and T1 samples again fall along trends deviating from the trends of the other rocks. After the basalts follow the trachytes T1, then the rhomb porphyries with a sprinkle of different trachytes. Quite surprisingly the T2 group is rich in Y. Most of the T3-T4 samples fall beyond the RP field. Most of the trachytes anomalously rich in Nb are also anomalously rich in Y. Among the rhomb porphyries some cases of the inverse are found. Some RP-types rich in Nb are correspondingly poor in Y. For trachytes T1b the opposite is the case. Finally a few rocks which have normal Nb contents are anomalously rich in Y.

The Zr-Rb diagram mirrors the Zr-Nb diagram in a general way, with only a few small differences. Thus the B3-quartzporphyry group is a little richer in Rb relative to other basalts, just as the T2 group is relatively richer than the main RP cluster. It is to be noted that Rb contents continue to rise with Zr through T4, without levelling out or falling, as is the case with K against Zr or total alkalies against Zr (not shown). The Zr/Rb ratio varies from basalts (ca. 3,9), T1 (ca. 5,0), to the RP cluster around 6,3 (5,0-7,4).

The K-Rb diagram (not shown) has a trend very similar to that of the Zr-Nb diagram, with a broad-band progression from low K and Rb values (basalts, then T1), then medium values (the RP group), to high values (T3 and T4). In all there is a surprising analogy with the Zr-Nb diagram, indicating that K and Rb behave more or less as excluded elements, in analogy with the two first-mentioned.

6. THE QUARTZ PORPHYRY - B3 OCCURRENCE

In the southern part of the long B3 exposure (at Rød), a quartz porphyry occurs as part of basalt lava (areas 50-100 m long), and as smaller fragments (5-10 cm long). One well-rounded ellipsoid (5,0 x 2,5 x 1,75) cm^3 of basalt was found in the bigger area of quartz porphyry. The latter rock as well as the adjacent basalt are seen to be more heavily altered in thin sections. Analyses of pairs from 1) adjacent rocks, 2) samples each 2 m from the contact, 3) quartz porphyry fragment and surrounding basalt, and 4) half the ellipsoid and surrounding quartz porphyry have been performed. Preliminary results are as follows.

The quartz porphyry has roughly a rhyolitic chemistry but, as seen from Fig.2, contains very little Zr, Nb, etc. Therefore it is certainly not any differentiate formed during evolution of the Permian magmas. The most likely explanation seems to be that this

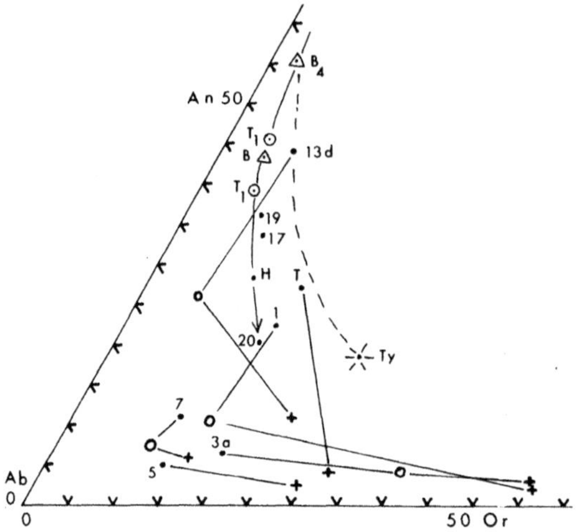

Fig.4 Analyses of feldspar compositions by microprobe: 1, 3a, 5, 7, 13d, and T - phenocryst cores (dots) and rims (rings), and groundmass feldspar (crosses) of RP1, RP3a, RP5, RP7, RP13d, and Tyveholmen dike in Oslo from Harnik [8, Fig.14]; B4, T1, B, T1, 19, 17, 20 - new analyses of plagioclase phenocryst cores in Vestfold B4, etc., to RP20. Ty - wet chemical analysis of Tyveholmen phenocryst core by Oftedahl [12, p.50]

quartz porphyry was formed from fragments of the roof of a basaltic magma chamber. Such fragments included in a basalt magma could undergo a partial melting but not mix with the basalt owing to their much higher viscosity when melted. The low trace element contents can be due to the fragments coming from paragneisses.

7. FRACTIONAL CRYSTALLIZATION MODELS

The computer program at the Center for Volcanology, University of Oregon, for imitating fractional crystallization has been used for testing if the analyzed rocks from the Vestfold area could have been produced by crystal fractionation. It proved impossible to produce a basic RP (Hegg type) from a normal B1 pyroxene basalt (Weigand No.49). On the other side Hegg RP could fractionate to give a latitic T1, which again is an impossible process according to the Zr contents. An attempt to produce the first RP19 that could have originated from Hegg-like magma according to the Zr-Nb plot, gave a complicated fractionation model as a possible path: 101,5 Hegg - 5,4 Ol + 6,8 rhomb feldspar + accessory minerals = 100,6 RP19. Such precipitation of heavier minerals and addition of rhomb-feldspar crystals seems possible, even if not very probable. Still further along, an RP23 sample may have been derived from an RP19 melt by simple fractionation: 100 RP19 - 43,1 rhomb feldspar ± (Ol, Py, accessory minerals) = 61,6 RP23.

One fractionation course which seems really probable from the trace element plots is the fractionation of rhomb porphyries to produce T3 and T4 magmas. Such attempts gave fairly good computer results, as for instance: 100 RP19 - 34,62 RP19 rhomb feldspar = 62,92 T3.

The bases for these fractionation computations are microprobe analyses of B and RP augites and feldspar phenocrysts (Fig.4). Compositions of rhomb feldspar crystals are shown to vary considerably, making it very difficult to get fractionation results with acceptably small errors (sum of squares of residuals below 1,0). In the cited examples the sums of squares of residuals are 1,9 - 5. The rhomb crystal analyses of Harnik confirm this picture; in his investigated samples every crystal, rim, and groundmass feldspar is of markedly different composition (Fig.4).

8. CONCLUSIONS

I here base my conclusions on the Ramberg [2] hypothesis of a 10 km thick basalt pillow under the Oslo region, located at depth of about 20-30 km, and on differentiation of this basalt as the major source for all non-basaltic rocks. The Vestfold basalts originated from shallow magma chambers undergoing local, high-level differentiation. Locally, differentiation produced trachytic magmas (T1, part of T3). Most of the rhomb porphyry lavas did not originate from any magma like the B1 to B6 basalts but from a top layer on the big basalt pillow, where differentiation already had produced a melt rich in Zr and other excluded elements. The high contents of Zr, Nb, Y, and REE in the RP flows make it unlikely that assimilation of crustal roof rock has played any important role during generation of the RP (and larvikite) magmas. Early rise in these magmas to high levels (2-4 km depth) along a number of fissures and channels furnished the sources for numerous RP flow types, each with their different local story of rhomb crystal and trace element development. On top of a number of such chambers, more or less flat-topped, cauldron subsidences developed through tectonically induced loss of magma support for the thin, circular roof lids. Continued differentiation within such chambers produced trachytic to rhyolitic top layers, yielding flows T3 and T4. Finally the caldera collapse ignimbrites of the Ramnes cauldron, as well as its ring dike and central intrusion, represent magmas that appear to be much more primitive in their excluded trace element contents. Although roof rock contamination is assumed for the B3-quartz porphyry association and is thought possible for the B3-T1 transition in the SW, no well-founded support for contamination or anatexis as important, general factors has been found.

ACKNOWLEDGEMENTS

The work reported here was done during a one year stay at the Center for Volcanology, University of Oregon. I want to thank Dr. G.G.Goles for friendly help during the stay, for many and stimulating discussions, and for suggesting improvements in the present paper. I am also indebted to Dr. B.H. Baker for suggestive discussions, to Dr. D.F. Weill for the opportunity to carry out microprobe analyses and to Dr. G.A. McKay for performing them, to Mr. G.Cunningham for help with XRF analyses, and to Mr. T.L.Robyn and Mr. L.N.Peterson for help with mixing models. The work received financial aid from the Norwegian Research Council for Science and the Humanities and from the Fulbright-Hays travel fund.

REFERENCES

1. P.W.Weigand, Geochemistry of the Oslo basaltic rocks. Skr. Norske Vid.-Akad. i Oslo, I. Mat.-Naturv. Kl. Ny serie 34, 1975.
2. I.B.Ramberg, Gravity interpretation of the Oslo Graben and associated igneous rocks, Norges geol.Unders. 325, 1976.
3. H.Heyer, Rombeporfyr-stratigrafi vest for Holmestrand. Norges geol. Unders. 213, 86-96, 1968.
4. C.Oftedahl, The lavas, Skr. Norske Vid.-Akad. i Oslo, I.Mat.-Naturv. Kl.3, 1952.
5. E.Schou Jensen, Geologisk undersøgelse af Ramnes vulkanfelt, nordvestlig del. Unpubl. thesis, København Univ. 1964.
6. C.Oftedahl, Magmen-Entstehung nach Lava-Stratigraphie im südlichen Oslo-Gebiete, Geol. Rundschau 57, 203-218, 1967.
7. C.Oftedahl, The cauldrons, Skr. Norske Vid.-Akad. i Oslo, I. Mat.-Naturv. Kl.3, 1953.
8. A.B. Harnik, Strukturelle Zustände in den Anorthoklasen der Rhombenporphyre des Oslogebietes, Schweiz.Min.Petr.Mitt. 49, 509-567, 1969.
9. T.F.W.Barth, Systematic petrography of the plutonic rocks. Skr. Norske Vid.-Akad. i Oslo, I. Mat.-Naturv. Kl.9, 1945.
10. T.V.Segalstad, Skien og Nevlunghavn basalters geologi og geokjemi, Unpubl. thesis, Univ. of Oslo, 1976.
11. B.H.Baker, G.G.Goles, W.P.Leeman, and M.M.Lindstrom, Geochemistry and petrogenesis of a basalt-benmoreite-trachyte suite from the southern part of the Gregory Rift, Kenya. Submitted to Contributions to Min. & Petr., 1977.
12. C.Oftedahl, The feldspars, Skr. Norske Vid.-Akad. i Oslo, I. Mat.-Naturv. Kl.3, 1948.

PETROLOGY OF THE SKIEN BASALTIC ROCKS AND THE EARLY BASALTIC (B_1) VOLCANISM OF THE PERMIAN OSLO RIFT

T.V. Segalstad

Institute of Geology, University of Oslo
P.O. Box 1047 Blindern, Oslo 3, Norway.

ABSTRACT. A pile of Permian continental basaltic rocks at least 1500 m thick occurs near Skien, some 100 km SW of Oslo. The pile consists of numerous thin (1-5 m) flows dipping towards the NE. Most flows are porphyric, and amygdules are abundant. Flows of melanite-ankaramite form the lower two third of the basalt pile, and flows of basanite form the upper one third. The basaltic rocks evolved from an assumed olivine-nephelinitic magma at great depths, principally by clinopyroxene crystal fractionation accompanied by later olivine and perhaps melilite reactions at moderate pressure in minor, shallow magma chambers. Chemical analyses of the Oslo B_1 basalts show increasing differentiation/evolution with time, from SW towards NE. Assuming one primary magma source, this can be explained by the rise and movement of magma along the rift during its development, or by mantle plume activity associated with movement of the lithosphere.

1. OCCURRENCE

Overlying the Precambrian gneisses we find a sequence of marine Cambro-Silurian shale and limestone ending with the Ringerike continental sandstone. The sedimentary rocks are dipping towards NE from 18° in W to 36° in E. Concordant red coloured Permian sandstone, between 110 and 125 m thick, overlies the Ringerike sandstone. The B_1 basaltic rocks [1], here called basalts for convenience, lie above a quartz conglomerate. The basalts are overlain by latite (RP) but the border is always tectonic. Geological data [2] and geophysical data (magnetometry [2] and gravimetry [3]) support the theory that the basalts are cut by ring faults with ring dikes belonging to two cauldrons, now parti-

E.-R. Neumann and I.B. Ramberg (eds.), Petrology and Geochemistry of Continental Rifts, 209–216.

ally obliterated by later intrusions of larvikite and nordmarkite plutons.

A detailed study of the basalts along the road between Skien and Siljan showed the minimum aggregate thickness of the basalts to be 1500 m, with a constant 36° dip towards the NE. Numerous thin flows, usually 1-5 m thick, indicate an origin close to central volcanoes. Tuffs occur sparsely, and sediment horizons are even less common. Most flows are rich in amygdules. Ultramafic nodules have not been found, despite intensive searches.

2. PETROGRAPHY AND MINERALOGY

The lower 1000 m of basalts consist of picritic lavas with clinopyroxene + olivine phenocrysts, and $\pm$ melanite (6 wt.% TiO_2)$\pm$ nepheline $\pm$ apatite $\pm$ magnetite in the groundmass. The rocks are classified as ankaramites or *melanite-ankaramites*. *Melanite-nephelinite* occurs in the middle part of the basalt pile. This rock contains the same minerals, without olivine, but with chlorite + possibly melilite + increased amounts of nepheline in the groundmass. The upper 500 m consists of *basanites* with plagioclase (An 20-30) + clinopyroxene $\pm$ kaersutite (7 wt.% TiO_2) phenocrysts, and plagioclase + chlorite in the groundmass.

The clinopyroxene is generally strongly zoned, showing hourglass zoning and resorbed cores. Microprobe analyses show the pyroxenes to have diopside cores, titaniferous salite intermediate zones, and titansalite rims. The olivine (Fa 12) may display a corona reaction rim.

3. PETROLOGY

The major element geochemistry is characterized by low SiO_2 (37-46 wt.%), high TiO_2 (2.5-4 wt.%), high CaO (8-19 wt.%), high P_2O_5 (0.5-1.8 wt.%), and relatively high alkali contents (Na_2O 1.5-4.5 wt.%; K_2O 1-3.5 wt.%). Both the geochemistry and the petrography indicate that these basaltic rocks must have originated from a silica-undersaturated magma.

Relevant experimental work [4,5] and plots in the diopside - nepheline - albite [6], the silica - alkali, and the AFM [1] diagrams indicate that the rocks generally fall on a differentiation trend which may be explained as follows. Starting with an initial magma with a composition represented by olivine + nepheline ("olivine nephelinite"), clinopyroxene + olivine fractionation will lead to melts of ankaramitic compositions. Olivine reaction with the melt will then occur, and at lower temperatures melilite can separate. When all the olivine is consumed, nepheline is the next mineral to form. To finally form plagioclase, melilite must react with the melt.

The fractionation theory was tested with the help of a computer mixing program, which showed that ankaramitic melts fractionating 30-40% clinopyroxene would yield the basanitic melts. Magma differentiation between ankaramites would involve 15-20% clinopyroxene fractionation. Other mineral phases would participate only to a lesser extent in the fractionation processes.

Experimental work suggests that the strongly undersaturated basaltic rocks like the Skien basalts have originated at very great depths [7-9]. Olivine nephelinites are believed to originate after very small degrees of partial melting at some 100 km depth, in the mantle's low velocity zone [10,11]. Ascent and extrusion of these basaltic magmas may occur when this zone is tapped rather directly by major fault systems, like the ones which may have initiated the formation of the Oslo rift.

Silica undersaturated lavas are most often accompanied by explosive extrusions of pyroclastic material [7] and contain ultramafic nodules of deep origin. No ultramafic nodules have been found in the Skien basalts. Ultramafic nodules even seem to be lacking in the whole East African province [12], the most extensive known region of nephelinite volcanism. Nephelinitic magmas are expected to come rather directly from the upper mantle to the surface [10]. It seems possible that temporary residence and differentiation in shallow magma chambers at higher pressure (deduced from stratigraphical geochemical variations [1]) have caused the peculiar petrography of the Skien basalts. Clinopyroxene fractionation was thus the dominant process of evolution, followed by an olivine reaction, in contrast to the lower pressure olivine fractionation more common in basalts of such composition. Any ultramafic nodules would sink during the shallow-level differentiation; if they remained, the olivine-rich nodules would react with the melt. This may explain the absence of ultramafic nodules in these rocks.

4. EVOLUTION OF THE EARLY OSLO BASALT VOLCANISM

Available chemical analyses of the Oslo basalts [1,13,14] were plotted in different variation diagrams in order to see potential *trends* in the evolution of the basaltic magmas in the Oslo rift. The most striking features of the plots in the silica - alkali diagram is the scatter of points across the whole basalt area of the diagram. A grouping of samples within geographical districts seems to demonstrate the following paths within the B_1 basalts. The Skien basalts in the SW are the most undersaturated ones, with ankaramite trending towards basanite. Towards N the basalts around Vestfold and Drammen show differentiation from alkali-basalt towards basa-

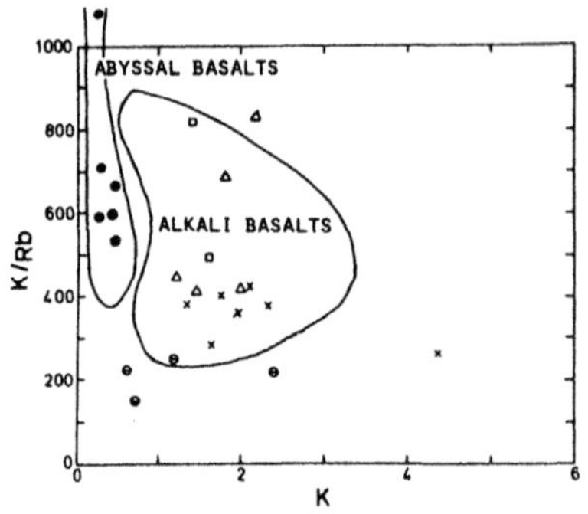

Fig. 1. K/Rb versus K (wt.% oxide) diagram slightly modified after Weigand [14] showing plots of the Oslo basalts. All samples from Bærum (filled circles) plot within the abyssal basalt field while the other Oslo basalts plot within or near the alkali basalts field [15].

nite. Going eastwards to Jeløya the basalts are even more differentiated basanites, enriched in alkalis and silica. On the other hand, the Bærum B_1 basalts in the north are distinctly different from other basalts in the Oslo rift [14]. Their chemistry shows characteristics tending towards the oceanic tholeiites (Fig. 1) while all other samples from the Oslo rift S of Oslo show alkali basalt affiliations [14].

From the geochemical observations reported above, a simplified map of the Oslo rift (southern part) was constructed (Fig. 2). If one assumes that the basalts have formed from one primary magma, the conspicuous trend of silica enrichment (from W towards E) and of alkali-enrichment (from NW towards SE), are shown by thinner arrows, while the differentiation trends are shown with thicker arrows. As the increasing differentiation also may be taken as an expression of time, the following is to be taken as a proposal for the evolution of the early basalt volcanism in the Oslo rift.

The volcanism started with a B_1 alkaline trend in the SW, with chemically very primitive magmas evolving at high pressure and in the presence of CO_2-rich volatiles [16-18]. With increasing differentiation the magmatism evolved towards the north and then towards the east. As a result of crustal thinning [3], shallow near-surface differentiation at low pressure and in the presence of H_2O-rich volatiles [16-18] and possibly contamination of magma from partial melting of the crust, tholeiites were erupted in Bærum. Further differentiation of the same major magma mass may also have produced the later rhomb porphyries. (The later B_2 stage seems to have originated from a new mass of magma, but more differentiated than the first one which produced the B_1 basalts. The B_3 stage seems also to have been produced by a new

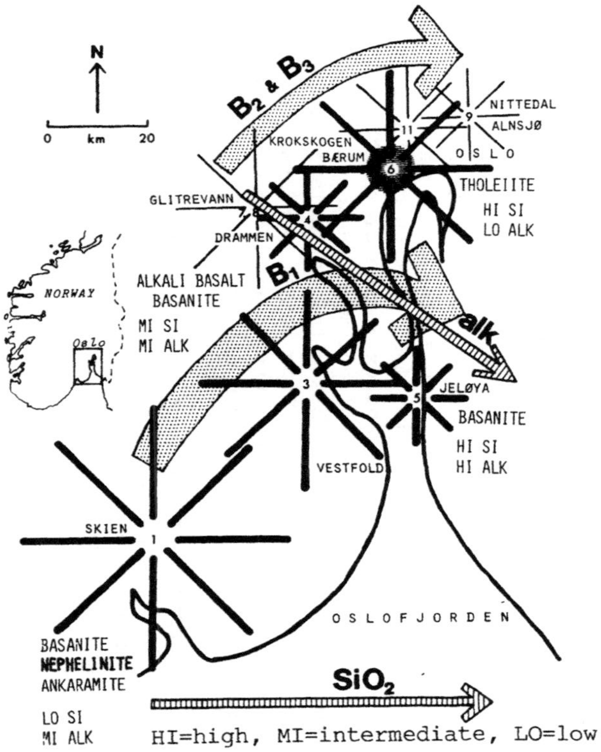

Fig. 2. Sketch map showing major centres of early basalt volcanism in the southern part of the Oslo rift. * thick lines: B_1. * thin lines: B_2 & B_3. Shaded: central tholeiite area. See text.

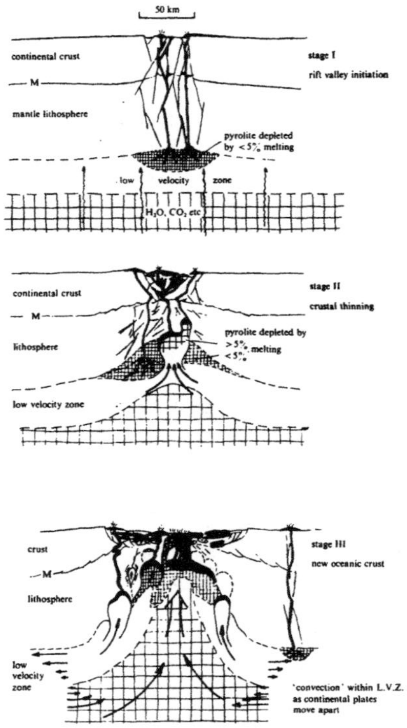

Fig. 3. A three-stage model after Green [11] illustrating the possible processes relating rift valley volcanism to continental rifting and the formation of a mid-oceanic rift system.

mass of magma, further differentiated than the B_2 magma. To produce these trends a model with magma chambers at different depths, all differentiating at different rates seems appropriate. Both the B_2 and B_3 volcanism seem to have started in the NW and moved towards NE with time and differentiation, probably differentiating into the later rhomb porphyries.

5. COMPARISON BETWEEN THE EARLY VOLCANISM AND TECTONISM OF THE OSLO RIFT AND THE KENYA RIFT VALLEY

The suggested evolution of the early stages of rifting in the Oslo rift is quite similar to that found in the Kenya rift valley [19]. Table 1 gives a comparison between the volcanic and tectonic events for these two continental rifts. Both rifts were preceded by downwarping and deposition of sediments, followed by eruption of undersaturated lavas. After a possible up-doming there were eruptions of alkali basalts, followed by step faulting. After this period of basaltic volcanism, extensive fissure eruptions of latite (Oslo) and phonolite (Kenya) took place, covering a wide area. The character of the volcanism is very similar, though the African volcanism seems to be even richer in alkalis, and thereby richer in nepheline.

6. TECTONIC MODELS FOR THE EARLY FORMATION OF THE SOUTHERN OSLO RIFT

Green's [11] theory for the relation between rift valley volcanism and continental rifting (Fig. 3) seems to correspond with the observed volcanism in the Oslo rift. Stage I involved extrusion from a very deep level (low volocity zone) of primitive derivatives of olivine nephelinite magma now found at Skien. Stage II involved crustal thinning under tension [3], and extrusion of the basanites now found in Vestfold. Stage III involved rifting of the continental crust and mantle lithosphere, and extrusion of the tholeiites now found in Bærum. The basaltic volcanism is supposed to have proceeded from ankaramites through basanites to tholeiites, from SSW to NNE, relative to the crust. This can be performed by the rise and movement of magma along the rift during its development in the NNE direction, if the crust was not moving.

However, the opposite might be a possibility; i.e. the volcanism could have been produced by mantle plume activity at one spot, while the lithosphere moved in a SSW direction over this hot spot [20]. A constant upwelling of magma from this centre, accompanied by increasing differentiation, could explain the existence of progressively older volcanic centres towards the SSW.

A third possibility is a relative movement of the western crustal block towards the SSW. The NNW-SSW direction of movement

OSLO RIFT		KENYA RIFT VALLEY	
VOLCANIC EVENTS	TECTONIC EVENTS	VOLCANIC EVENTS	TECTONIC EVENTS
EXTENSIVE FISSURE ERUPTIONS OF LATITE (RP) OVER A WIDE AREA	Major and local faulting	EXTENSIVE FISSURE ERUPTIONS OF PHONOLITE OVER A WIDE AREA	Major and local faulting
B_1 ERUPTION OF THOLEIITE (Bærum)			
	Step faulting		Gentle down-warping of rift zone
B_1 ERUPTION OF ALKALI BASALTS AND BASANITES FROM VOLCANIC CENTRES		ERUPTION OF ALKALI BASALTS AND TUFFS FROM MANY SMALL CENTRES	
	Up-doming?		Up-doming
B_1 ERUPTION OF ANKARAMITES, NEPHELINITES, BASANITES (Skien)		ERUPTION OF NEPHELINITES, EMPLACEMENT OF ALKALINE IGNEOUS ROCKS AND CARBONATITES (Kavirondo Gulf area)	
	Deposition of sediments (Permian)		Deposition of sediments (Miocene)
	Down-warping?		Down-warping?

Table 1. Comparison between the initial phases of the formation of the Oslo rift and the Kenya rift valley. The right half of the table is modified after Bailey [19]. Age decreases upwards.

would then imply a boundary transform fault [21], not consistent with the normal kind of spreading perpendicular to the NNE-SSW rift axis. (This theory was already suggested by Cloos [22] in 1928, who emphasized the NNW-SSE direction of so many dikes and faults, and concluded that they comprised a shear pattern (*fiederspalten*) caused by the southward movement of the precambrian block west of the rift, relative to the eastern block. This theory was quoted and tentatively supported by Oftedahl [23] in his remarks on faulting in the Oslo Graben).

Further work, including field work, radiometric and paleomagnetic age determinations, and paleomagnetic pole analyses, may enlarge our knowledge of the tectonism of the Oslo rift.

REFERENCES

1. T.V. Segalstad, Petrology of the Skien basaltic rocks, southwestern Oslo Region, Norway, Lithos (in prep.), 1977.
2. T.V. Segalstad, Nor. Geol. Tidsskr., 55, 321, 1975.
3. I.B. Ramberg, Nor. Geol. Unders.,325, 1, 1976.
4. J.F. Schairer, and H.S. Yoder, Jr., Carnegie Inst. Wash. Yr. Bk., 63, 65, 1964.
5. J.F. Schairer, C.E. Tilley, and M.A. Brown, Carnegie Inst. Wash. Yr. Bk., 66, 467, 1968.
6. J.F. Schairer, and H.S. Yoder, Jr., Am. J. Sci., 258-A, 273, 1960.
7. M.J. O'Hara, Scot. J. Geol., 1, 19, 1965.
8. D.H. Green, Tectonophysics, 7, 409, 1969.
9. D.H. Green, Phys. Earth Planet. Interiors, 3, 221, 1970.
10. D.H. Green, Trans. Leichester Lit. Phil. Soc., 64, 28, 1970.
11. D.H. Green, Phil. Trans. Roy. Soc. London, A 268, 707, 1971.
12. I.S.E. Carmichael, F.J. Turner, and J. Verhoogen, Igneous petrology, McGraw-Hill, 500, 1974.
13. W.C. Brøgger, Skr. Nor. Vidensk. Akad. i Oslo, Mat.-Naturv. Kl., 1933.
14. P.W. Weigand, Skr. Nor. Vidensk. Akad. i Oslo, Mat.-Naturv. Kl., 1975.
15. P.W. Gast, Geochim. Cosmochim. Acta, 32, 1057, 1968.
16. B.O. Mysen, Carnegie Inst. Wash. Yr. Bk., 74, 454, 1975.
17. B.O. Mysen, and A.L. Boettcher, J. Petr., 16, 549, 1975.
18. H.S. Yoder, Jr., Generation of basaltic magma, Nat. Acad. Sci., Wash., 1976.
19. D.K. Bailey, Continental rifting and alkaline magmatism, in: The alkaline rocks, ed. by H. Sørensen, J. Wiley & Sons, Inc., 148, 1974.
20. J.T. Wilson, Sci. Am., 208, 86, 1963.
21. W.N. Gilliland, and G.P. Meyer, Geol. Soc. Am. Bull., 87, 1127, 1976.
22. H. Cloos, Fortschr. Geol. u. Pal., 21, 233, 1928.
23. C. Oftedahl, Skr. Nor. Vidensk. Akad. i Oslo, Mat.-Naturv. Kl., 1952.

COMPOSITE PLUTONIC RING-COMPLEXES: A STRUCTURAL CHARACTERISTIC OF RIFT-ZONE PLUTONISM

Jon Steen Petersen

Geologisk Institut, Aarhus Universitet, Denmark

ABSTRACT. Composite plutonic ring-complexes are found in a number of places over the world and characteristically form by non-orogenic igneous processes. They occur in distinct structural belts which, according to global tectonic models, define major continental rupture lineaments associated with former triple junctions. Their origin therefore seems to be related to processes of crustal rifting.

Many rifts develop into active spreading zones and ultimately form new continental margins. Accordingly, plutonic rock suites or remnants of these suites that occur along continental margins may reflect early stages of a continental break up. "Failed rifts" on the other hand, preserve the relations between initial rifting and related igneous processes and consequently display more complete information about the beginning stages of a continental rupture. The analysis of rift-related plutonism may therefore provide criteria for the recognition of older and more deeply eroded rift systems along present, as well as former, continental margins.

The common structural and compositional characteristics of rift-related plutonic ring-complexes, including newly recognized composite ring-complexes in the southern Oslo province are summarized. These features are compared with marginal igneous complexes of uncertain origin, such as the massif-type anorthosite and rapakivi suites in Scandinavia, which show remarkably similar compositional and structural characters.

These similarities suggest a petrogenetic analogy between the formation of anorthosite rock kindres and rift-related magma series.

E.-R. Neumann and I.B. Ramberg (eds.), Petrology and Geochemistry of Continental Rifts, 217–229.

1. RIFT ZONE PLUTONISM

Plutonic ring complexes form composite, anorogenic occurences in many well preserved paleorift systems such as the Oslo province [1], the Niger-Nigeria province [2], the New Hampshire igneous series [3], the Monteregian plutons [4], the Gardar alkaline province [5], and the Marysville igneous complex [6]. Their structural characters include cauldron formations, associated volcanic series, dike swarm injection and fault-bordered sediment troughs together with the plutonic complexes. They occur in distinct structural belts which are often transverse to the regional structures, supporting their cratonic origin. The many similarities with recent rift systems; exposed at more elevated levels, suggest an origin as former rift structures. The term "failed rift" has been designated many of these provinces since rifting did not lead to the formation of a continental margin. A "failed rift" may therefore preserve evidence of the preceeding rift episodes and the analysis of plutonism in such regions may accordingly reveal evidence of rifting which becomes obliterated by later episodes in more highly evolved rift zones.

The igneous activity in these areas commonly displays at least two contrasting lines of magmatic evolution, one acid and another basic to intermediate. They commonly show evidence of multiple intrusion that involve both series in single massifs, suggesting the contemporaneous existence of two magma series. Dike swarm injections, often lamphrophyric, typically form the final stages of magmatic activity in such provinces.

Since plutonic central complexes apparently represent characteristic structural features of the deeper levels in many rift systems, they form very important indicators of igneous processes related to rift formation and therefore provide criteria for the recognition of older, more deply eroded continental rupture linements.

1.1 Summary of events in rift formation

The development of crustal ruptures into spreading zones and subsequently into Atlantic-type, rifted continental margins, is characterized by a sequence of tectonic and magmatic episodes which describe progressive stages of evolution. Based on the interpretation of the tectonic development in embryonic rift-systems and young continental margins, schematic evolutionary models e.g. [7] can be outlined as follows:

A) The beginning stage of a continental break up is initial uplift and thinning of the continental crust associated with a marked increase in heat flow. The cause of uplift may possibly be related to mantle diapirism and thereby affect the position and characters of the rifting geometry [8] by formation of triple-rift junctions.

B) The subsequent initial stage is followed by linear volcanic activity, often of central extrusion type, and with graben formation. A rift valley of about 50 km width is formed [9,10], and the subsidence is associated with mainly basic, alkaline igneous activity and contemporaneous sedimentation in a normal faulted valley system.

C) Later, the rift system widens as the result of block subsidence between single rifts, and commonly a structural depression develops in the scale of 300-400 km across, bounded by complex patterns of rifts and horsts. This stage is associated with the most extensive graben collapse and basic alkaline volcanic activity as well as granitic (rhyolitic) intrusions. The igneous activity occurs mostly about the rift axis and incipient transform offsets in the rift, and rarely extends beyond the subsided regions.

D) The rift process may finally develop into a continental break up and the formation of new oceanic crust. It has been argued that the rifting and collapse stages in fact do not indicate true plate boundaries, as essentially no lithospheric displacement has occurred during the preceeding episodes [11] but are merely block subsidence associated with the igneous activity. It seems that the break up stage is connected with the extensive injection of igneous material along the marginal crust in the form of dike swarms [12].

A purpose of the present contribution is to point to some structural and compositional similarities between anorthosite - rapakivi suites in Scandinavia and rift related plutonism of the southern Oslo province, and to discuss evolutionary models which conform with the above patterns of rift formation.

2. THE OSLO PROVINCE

Recent geological mapping in the larvikite district south of Oslo has revealed characters of a large, composite ring complex that occupies a major part of the southern Oslo province [13]. The complex is bordered to the west by Permian sediments and volcanics resting unconformably on Cambro-Silurian sediments found in a series of progressively down-dipping faults between the graben structure and the Precambrian basement. Strong shearing along the larvikite borders is apparent from the gneissic and schistose appearance of the marginal zones, which are locally considerably more mafic than the central parts. Major pegmatites, famous for their content of rare minerals are found along these boundaries, and are often injected into the country rocks of volcanic and sedimentary origin.

The northern part of the larvikite complex is cut by another ring-complex, consisting of nordmarkite, larvikite and pulaskite among others, and towards the east the complex is partly overlain by, and partly cross-cut lava series of rhomb-porphyry (latite) [14].

The larvikite massif is composed of a sequence of cusp-shaped segments which are repeatedly cut by others in a manner suggesting structural younging towards west [13](Fig. 1).

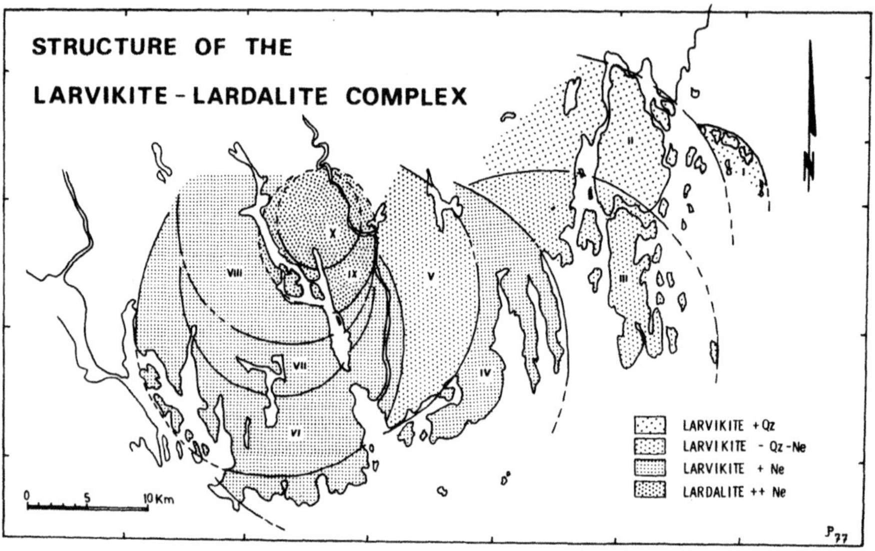

Fig. 1. Sketch map of the structural geometry in the larvikite (monzonite) massif in the southern Oslo province.

The complex covers almost the total width of the Oslo rift and the continuity of this structure reveals a progressive shift of centres of activity which can be viewed either as the result of sequential cauldron subsidence or multiple intrusion of circular plutons. The individual segments apparently show concentric structures defined by feldspar lamination dipping steeply inwards towards the center of each section. A tendency of slightly less inclined dip (40-50°) inwards has been observed. The feldspar lamination is occasionally associated with rhytmic layering, formed by the successive concentration of felsic and mafic phases, often in a graded manner. A large ilmenite-magnetite-apatite ore body in the Kodal area [15] apparently follows the direction of igneous lamination and shows a strongly asymmetrical graded border, suggesting structural relationship to the layering in the surrounding larvikite. Field relations are similar for other minor oxide ore-bodies in the disbrict and suggest a cumulative, segregational origin for such apatite-rich ores in the region.

The contact between individual segments is occasionally

characterized by fine grained and often porphyritic varieties which may represent chilled border facies grading into a normal, coarse grained larvikite type. These features suggest thermal contrast between some of the segments and favour an origin by multiple injection of magma, possibly derived from a common magma reservoir, triggered by successive rift subsidences towards the west.

This view is supported by the fact that the composition of larvikite changes systematically from Qz-bearing varieties through intermediate members towards increasingly undersaturated larvikite types towards the west [13,16]. The Bolærne and Nøtterøy segments which form the structurally most primitive variants are characterized by the presence of quartz as clearly interstitial grains. The following sections of Tjøme as well as the well known dark-type larvikite of the Klåstad segment consist of intermediate members which have neither quartz nor nepheline. The youngest segments in the western region contain interstitial and exsolved nepheline to varying degrees. The final intrusion in the complex is the lardalite massif which contain between 10% and 40% Ne. The compositional variation therefore suggests derivation from increasingly undersaturated and accordingly more evolved source magma in the course of time.

The larvikite area is dissected by numerous NNW-SSE trending camptonite dikes as well as a few rhomb-porphyry dikes that form the youngest events in the igneous activity of the region. This direction follows a major trend of subsidence in the graben structure. Other plutonic ring complexes in the Oslo province are found in the Skien district [19] where ring complexes, composed of larvikite and nordmarkite, are associated with several large cauldron structures [20]. In the Nordmarka district north of Oslo the occurrences of kjelsåsite-larvikite form composite ring-complexes that apparently suggest structural younging towards north [21]. The structural development in this area, involving complex successive cauldron subsidences, makes interpretation of the structural characters difficult. The sørkedalite [22] which is a cumulate of ol-gabbroic composition forms part of a marginal facies that dips steeply towards the interior of the complex.

The larvikites (=monzonite) form part of a basic suite that includes ol-diorite, kjelsåsite and larvikite, all characterized by low initial Sr-ratios and Th/U ratios of about 4 [16]. Another major igneous suite in the Oslo province is formed by acid rocks which include the nordmarkite-ekerite suite and the Drammen granite suite. The latter shows low Sr-ratios similar to the larvikite series and displays Th/U ratios of 5-6 [16]. This grouping of Th/U values despite similar Sr-initial ratios is identical to that found in the Rogaland mangerite-charnockite and related hbl-bi granite suites [17,18] in spite of considerably different absolute abundances, which, however, can be ascribed to different depth of crystallization [18]. The distribution of rocks belonging to the so-called basic and acid suite in the Oslo region suggests the presence of both magmas throughout the period of evolution and supports the existence of contemporaneous, contrasting magmas in the source area. This is similar to rift-related igneous provinces elsewhere, e.g. the New Hampshire province and the Niger-Nigeria province.

The oldest igneous events in the Oslo region were apparently the emplacement of a number of essexite plugs which occur exclusively outside the igneous bodies and are situated along a N-S trending fault across the present rift structure. This line also defines the position of the Drammen biotite-granite which apparently intruded immediately before the main igneous episodes and was subjected to cauldron subsidence [1]. The major plutonic activity includes intrusion of the larvikite series rocks as well as the nordmarkite-ekerite series. Slightly younger bi-granites and syenites of variable compositions transgress the early ring-complexes and often show highly irregular boundaries against these. Igneous activity in the province was terminated by the intrusion of lamprophyric dikes, essentially along NNW-SSE trending lines.

3. THE ROGALAND PROVINCE, SW NORWAY

The Rogaland igneous province forms a composite plutonic complex which almost totally covers the coast region between Egersund and Mandal in SW Norway. The trend is NW-SE and is clearly transverse to the regional fold pattern which trends roughly N-S [23,24]. Country rocks are high-grade metamorphic gneisses, where isograds of increasing metamorphic state (low-pressure pyroxene-granulite facies) apparently follow the borders of some of the plutons [25]. Similar features have been reported from anorthosite-rapakivi kindreds in SW Finland [26] and S-Greenland [27] where they are interpreted as contact aureoles.

The complex is cross cut by a swarm of WNW-ESE trending dolerite dikes. They occur essentially within the igneous rock members and point towards a relationship with magmatic activity in the region. The occurrences are similar to ol-dolerites of the Lofoten anorthosite-mangerite province, where isotope dating suggests injection closely following the youngest members of the mangerite suite. Chemically, these dolerites are comparable with alkali basalts typical of continental margins [28] and are suggested to have been emplaced in a similar structural environement. The Rogaland dolerites are grouped into two types, one of which is similar to the Lofoten dikes [23].

The plutonic complex is made up of several oval-shaped plutons of the anorthosite kindred, arranged in a manner suggesting sequential structural younging towards ESE (Fig. 2). The most pristine

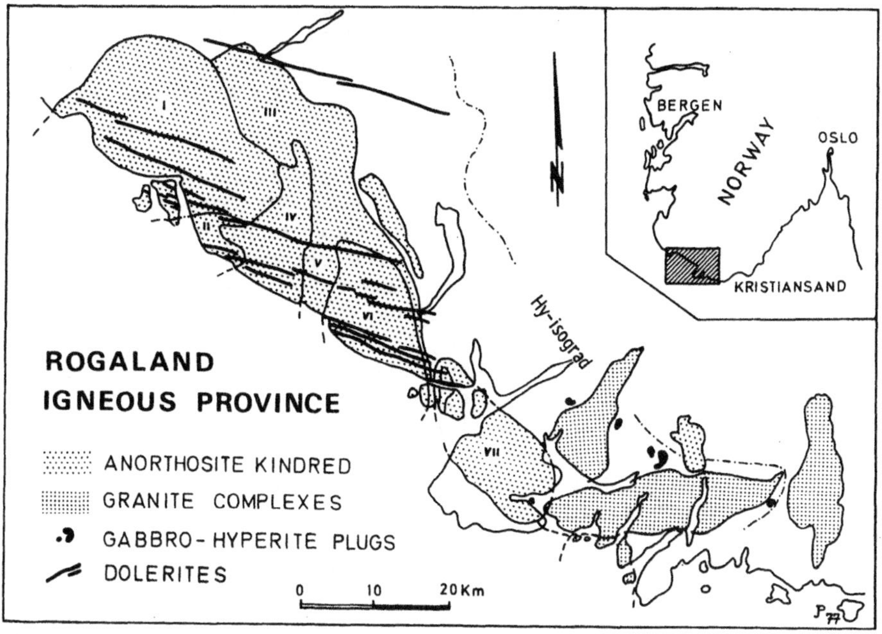

Fig. 2. Simplified geological map of the Rogaland igneous province, southwest Norway.

body, the Egersund-Ogna pluton (I) is composed of massif anorthosite with a strongly sheared marginal facies of leuconorite and norite. The pluton cross cuts regional structures, including a large recumbent fold [29]. This anorthosite dome is invaded to the south by the Haaland massif (II), which shows a partly brecciated border zone to the Egersund body [30]. To the east the Egersund pluton is partly invaded by the Bjerkheim-Sogndal layered lopolith (III). The province plutons are then cut by the structurally complex Helleren massif (IV). The Eia-Rekefjord mangero-norite (V) that transgresses the Helleren massif is a thin, sheet-like intrusion. Then the Åna-Sira anorthosite massif forms a roughly oval body, where the distribution of schillerized feldspars and the orientation of conformable ilmenite-magnetite deposits suggest a concentric structure (VI), elongated in NW-SE direction [31]. In the SE part the similarly oval shaped Farsund charnockite occurs (VII). The NW corner of this is occupied by the Hidra anortho-norite which grades from pure anorthosite in the core through norite to mangerite and a clear relation with the surrounding charnockite is suggested.

The Farsund charnockite cross-cut members of the Farsund plutonic complex which consists of a complex of several biotite and hornblende granites occurring to the SE of the anorthosite series and which shows similar absolute ages about 900 m.y. [32, 33].

The granitic series includes the Holum granite to the N of Mandal [34]. Chemically, strong relations between the charnockites and anorthosite appear evident whereas a co-magmatic origin for the granites seems less simple [35,33] despite similarly low Sr-initial ratios.

On a WNW-ESE trending line in the southern part of the region several small gabbroic plugs are situated. The compositions vary between bahiaite [23] through hyperite and Qz-norite to Qz-diorite and they are clearly postkinematic. The presence of primary as well as inverted pigeonite suggests moderate to high Fe/Mg ratios and at least marginally quenched magma, that may have derived from iron-enriched, anorthosite-forming liquids. Their absolute age is not known, but the fact that they occur exclusively outside the plutonic members of the province, similarly to the Oslo essexites, may possibly be taken as an indication of their older age.

The strucutral characters of the Rogaland province show many features in common with plutonic rocks in the Oslo-region and may be interpreted as follows:

A) Mantle diapirism and increased heat-flow lead to crustal doming and large-scale magma generation in the upper mantle. The beginning of separation of feldspars in an upper mantle reservoir and injection of more primitive, though clearly evolved basic plutons (gabbro plugs) along lines of crustal weakness (WNW-ESE) may have been the earliest events in this period.

B) The major plutonic episode involves roughly contemporaneous intrusions of granitic magma and more basic plutons of the anorthosite kindred. The latter derived as crystal mush from a deeper magma reservoir. The intrusion of anorthosite suggests a sequential shift of centres of activity, possibly reflecting migration of zones of subsidence, roughly analogous to the Larvikite-model [13] described previously. No systematic change in composition is apparent although the youngest rocks of the anorthosite suite seem to be the more differentiated, acid members, like the Qz-mangerite and charnockite. This period would correspond to the major collapse stage of an evolving rift system.

C) Finally the complex is transgressed by numerous alkali dolerites along the axis of magmatic and tectonic activity and this may correspond to the terminal stage of rift evolution where the true plate boundary is formed followed by injection of fissure basalts of continent marginal type. Further igneous activity then takes place away from the continental margin in an active ocean spreading zone.

4. THE RAPAKIVI PROVINCE, SW FINLAND

The anorogenic plutonic rock province in SW Finland and Sweden consists essentially of rapakivi granites and associated rocks of the gabbro-anorthosite kindred. Besides these there are numerous small granite-monzonite bodies situated along the coastal sections of central Sweden and south Finland [36,37]. Diabases and other minor basic plutons are associated with almost all rapakivi intrusions, as well as transitions to porphyric, sub-volcanic varieties.

The beginning of cratonization in the rapakivi district is marked by dolerite swarm intrusion along NW-SE trending zones [38] in the Häne district (S. central Finland) as well as the Satakunta area north of Åbo and in the Åland islands district. These early dolerites contain abundant inclusions of large plagioclase megacrysts of intermediate composition (An50-60). Euhedral crystals up to 30 cm across as well as broken fragments are described [38]. These findings are similar to dolerites of the alkalic Gardar province of SW Greenland, where the anorthosites are interpreted as top accumulates in an early stage of developing Gardar magmas [39].

The intrusion of the major rapakivi plutons succeeds the period of dolerite dike intrusion, and form the bodies of Wiborg, Ahvenisto and Salmi in eastern Finland and Russia; the Åland islands, Vehmaa and Laitali massifs in SW Finland and the Nordingrå, Rødo and Ragunda areas as well as other anorogenic granite varieties occurring along the coasts of Sweden and south Finland [36,37]. The spatial relations between individual bodies, however, is not so close as that in the Rogaland or the Oslo provinces and the pattern resembles that found in e.g. the New Hampshire and Nigerian provinces, where numerous separate and composite bodies are confined to a more or less well defined structural belt.

Single massifs, however, show evidence of multiple injection; the Ahvenisto complex e.g. forms a slightly off-centred ring-complex in which early anorthosite-gabbro associations form a horseshoe-shaped pluton, due to the transgression of the central rapakivi body which shows brecciated borders against the anorthosite-gabbro segment [40]. The latter invariably shows chilled borders and furthermore evidence of high temperature contact metamorphic recrystallization of country rock gneisses [26]. - The Nordingrå massif likewise shows associations between gabbro-anorthosite and rapakivi granite originated by multiple injection including ol-diabase. The Ragunda complex forms a sheetlike, shallow intrusion of roughly contemporaneous Qz-gabbro and rapakivi granite.

The rapakivi suite is cut by younger lamprophyres and dolerite dikes particularly in the SW region of the Åland islands and the Satakunata district, which follows another principal direction of fracture in the region trending NE-SW and apparently marks a final stage of igneous activity in the region.

The sequence of episodes in the Finnish rapakivi district closely follows that found for the Rogaland and Oslo provinces, and at least three distinct periods can be recognized:

A) Early cratonic activity succeeding the Svecofennian orogenic episode is associated with emplacement of olivine-dolerites along NW-SE trending zones of rupture. Relations between these dolerites and the later rapakivi intrusions have often been stressed and the abundant feldspar inclusions may have been derived from a deeper magma reservoir that may possibly relate to the rapakivi-forming series.

B) Intrusion of anorthosite-rapakivi kindred rocks as well as other related granite series form the major period of igneous activity in the region. This period is associated with graben subsidence and contemporaneous sedimentation in graben troughs often found associated with the rapakivi suite.

C) Finally, faulting along another set of fractures, which may form part of a conjungate fracture system associated with the rift evolution, occurs. The intrusion of lamprophyric dikes particularly in the Åland island district and diabases of the Satakunata area are evidences of this activity.

5. DISCUSSION

The association of anorthosite-rapakivi kindred and rift related tectonism has been pointed out in a number of recent discussions including the north American anorthosites [39,41], the south Greenland area [42,43] as well as in the Lofoten area [44]. Relation between plutonic ring complexes in paleorift systems and rocks of the anorthosite kindred are reported from the Niger province [45] where gabbro-anorthosite forms one line of a bimodal magmatic evolution. In the Gardar alkalic province numerous anorthosite fragments are found in associated dolerite dikes and are suggested to form early members of developing Gardar magmas crystallized at depth.

In a discussion by Wynne-Edwards [41] involving formation of anorthosite suites in the Greenville province as the result of ductile plate tectonics, a similar sequence of events to that proposed here is suggested for the beginning stages of evolution. These include crustal thinning and mantle diapirism followed by graben formation and basic igneous activity. The subsequent spreading period of Wynne-Edwards is comparable with the major rift-collapse stage suggested here and includes the main igneous activity that comprises intrusion of the anorthosites and related monzonite - charnockites as well as of granitic plutons.

The problem of contemporaneous basic and acid magma series in these provinces may, following the discussion by Yoder [46], be considered as evidence of progressive partial melting in the upper mantle, derived at separate invariant points, thus resulting in formation of contrasting magma compositions. The process of melting may result from adiabatic decompression associated with mantle diapirism and/or crustal rupture, where hydrous mineral phases in the primary mantle assemblages promote the formation of acid melts derived directly by fusion of ultramafic mantle source as suggested by Kushiro [47].

The initial acid melts would dehydrate the source rock and presumably possess high concentration of "volatile", incompatible elements such as K, Rb, U, Th, F, Li, Bi, Sn, Mo, Cu, Pb among others, and the resulting plutons and their environments therefore constitute considerable economic potential. The melt itself may be subsaturated with water and thus be capable of dehydrating country rocks (pyroxene-granulite aureoles) despite a hydrous mineralogy of the resulting pluton.

The lower density contrast with the surrounding material of the basic fraction may, following Ramberg [48], cause it to rest in a reservoir of a lower crustal level than the acid melts, and eventually intrude successively along zones of weakness, associated with crustal subsidence, and thereby form series of related, slightly evolving plutons. The consistency of Sr-isotope data for many rocks of both suites supports an origin from a common mantle source material.

REFERENCES

1. C. Oftedahl, Permian rocks and structures of the Oslo region, Nor. Geol. Unders., 208, 298, 1960.
2. R.E. Jacobson, W.N. Macleod and R. Black, Geol. Soc. Lond. Memoir, 1, 1958.
3. R. Greenwood, Geol. Soc. Am. Bull., 62, 1151, 1951.
4. A.R. Philpotts, The Monteregian province, in: The Alkaline Rocks, ed. by H. Sørensen, J. Wiley and Sons. 1974.
5. C.H. Emeleus and B.G.J. Upton, The Gardar period in south Greenland, in: Geology of Greenland, ed. by Escher and Watt, Geol. Unders., 1976.
6. E.S. Hills, Geol. Rundsch., 47, 543, 1958.
7. J.F. Dewey and J.M. Bird, J. Geophys.Res., 75, 257, 1970.
8. K. Burke and J.F. Dewey, J. Geol., 81, 406, 1973.
9. B.H. Baker, P.A. Mohr and L.A.J. Williams, Geol. Soc. Am. Spec. Paper, 136, 1972.
10. J.H. Illies, Graben tectonics as related to crust-mantle interaction. in: Graben Problems, ed. by J.H. Illies and St. Mueller, Schweitzerbartsches, Stuttgart, 1970.
11. D.A. Falvey, APEA J., 1, 95, 1974.
12. K. Burke, Tectonophysics, 36, 93, 1976.
13. J.S. Petersen, Structure of the larvikite-lardalite complex, Oslo region, Norway and its evolution. Geol. Rundsch., 67, (in press) 1977.
14. C. Oftedahl, Geol. Rundsch., 57, 203, 1967.

15. S. Bergstøl, Min. Deposita, 7, 233, 1972.
16. G. Raade, Distribution of radioactive elements in the plutonic rocks of the Oslo region. Thesis, Univ. Oslo, 1973.
17. P.G. Kileen and K.S. Heier, Chem. Geol., 15, 163, 1975.
18. P.A. Ormåsen and G. Raade, Heat generation versus depth of crystallization for Norwegian monzonitic rocks (in prep.)
19. T.V. Segalstad, Nor. Geol. Unders., 55, 321, 1975.
20. C. Oftedahl, Studies on the igneous rock complexes of the Oslo region XIII, Skr. Nor. Vid. Akad. Oslo, Mat-Naturv. 3, 1953.
21. E. Sæther, Studies on the igneous rock complex of the Oslo region XVIII, Skr. Nor. Vid. Akad. Oslo, Mat-Naturv. 1, 1962.
22. M.K. Bose, Studies on the igenous rock complex of the Oslo region XXI, Nor. Vid. Akad. Oslo Mat-Naturv. 27, 1969.
23. T.F.W. Barth and J.A. Dons, Nor. Geol. Unders. 208, 6, 1960.
24. T. Falkum, Nor. Geol. Unders., 242, 19, 1965.
25. J.S. Petersen, Nor. Geol. Tidsskr., 57, 65, 1977.
26. A. Vorma, Geol. Surv. Finl. Bull., 255, 1972.
27. D.B. Bridgewater, J. Sutton and J.S. Watterson, Rapp. Grønl. Geol. Unders., 11, 52, 1966.
28. S.N. Misra and W.L. Griffin, Nor. Geol. Tidsskr., 52, 409, 1972.
29. J. Michot and P. Michot, The problem of anorthosite: The South-Rogaland igneous complex, SW Norway, in: Origin of anorthosite and related rocks, ed. by Isachsen, Univ. State New York Mem. 18, 1968.
30. J. Michot, Nor. Geol. Tidsskr., 41, 151, 1961.
31. R. Zeino-Mahmalat and H. Krause, Nor. Geol. Tidsskr., 56, 51, 1976.
32. S. Pedersen and T. Falkum, Chem. Geol., 15, 97, 1975.
33. D. Demaiffe, J. Michot and P. Pasteels, Time relationship and strontium isotopic evolution in the magma of the anorthosite--charnockites of south Norway. Int. Geochron. Meet. Paris 1974.
34. J.R. Wilson, S. Pedersen, C.R. Berthelsen and B.M. Jacobsen, New light at the Precambrian Holum granite, south Norway. Nor. Geol. Tidsskr. (in press) 1977.
35. T. Falkum, J.R. Wilson, J.S. Petersen and H.D. Zimmermann, The intrusive granites of the Farsund area, south Norway. Nor. Geol. Tidsskr. (in press) 1977.
36. N.H. Magnusson, G. Lundquist and G. Regnéll, Sveriges Geologi, Scandinavian University Books, 1963.
37. A. Simonen, Bull.Comm.Geol.Finl., 191, 1960.
38. J. Laitakari, Bull. Comm. Geol. Finl., 241, 1969.
39. D. Bridgewater, Can. J. Earth Sci., 4, 995, 1967.
40. A. Savolahti, Bull. Comm. Geol. Finl., 174, 1956.
41. H.R. Wynne-Edwards, Am. J. Sci., 276, 927, 1976.
42. D. Bridgewater and B.F. Windley, Spec. Publ. Geol. Soc. S.-Africa, 3, 307, 1973.
43. D. Bridgewater, J. Sutton and F. Watterson, Tectonophysics, 21, 57, 1974.

44. O.A. Malm and D.E. Ormåsen, Mangerite-charnockite intrusions in the Lofoten-Vesterålen area, north Norway. Nor. Geol. Unders. (in prep) 1977.
45. R. Black, C.R. Akad. Sci. Paris, 260, 5824, 1965.
46. H.S. Yoder, Am. Min., 58, 153, 1973.
47. J. Kushiro, J. Petrol, 13, 311, 1972.
48. H. Ramberg, Gravity deformation and the earths crust. Acad. Press. London, 1967.

PETROGENESIS OF THE LARVIK RING-COMPLEX IN THE PERMIAN OSLO RIFT, NORWAY*

Else-Ragnhild Neumann

Mineralogical-Geological Museum
Sarsgt. 1, Oslo 5, Norway

ABSTRACT. Geochemical data indicate that all the rocks in the Larvik ring-complex (southwestern part of the Oslo Region), have been derived from a magma chamber undergoing continuous fractional crystallization, but which was fed periodically with new batches of magma. The lardalites represent cumulates from this magma chamber, formed at a late stage when the melts were markedly silica undersaturated.

1. INTRODUCTION

The Oslo rift plutonic rocks range from strongly silica oversaturated to strongly undersaturated types. Fractionation trends leading from silica saturated alkaline to oversaturated peralkaline rocks are well established from both field evidence and chemical relations [1-7].

Strongly undersaturated plutonic rocks such as lardalites (plagifoyaites) and foyaite/hedrumites (nepheline syenites), are found only in the Larvik ring-complex in the southwestern part of the Oslo Region. This complex consists of a number of semicircular zones with chilled contacts [8] (Fig. 1). The easternmost zones, which are believed to be the oldest, contain quartz-bearing larvikite (monzonite). There is a progressive decrease in silica activity westwards through zones with both silica saturated and nepheline-bearing larvikites, to zones with almost exclusively nepheline-bearing larvikite. The innermost zones consist of nepheline-rich lardalite, transgressed in the center by foyaite/hedrumite with variable nepheline contents.

*This work has been financially supported by A/S Norsk Varekrigsforsikrings Fond.

E.-R. Neumann and I.B. Ramberg (eds.), Petrology and Geochemistry of Continental Rifts, 231–236.

The ring complex (primarily the larvikite) is cut by dikes of different compositions. Among these are numerous pegmatites, nepheline syenite pegmatites to the west [9], and quartz-bearing syenite pegmatites to the east (G. Raade, personal communication). Nepheline porphyry has also been found in the westernmost part of the area [9].

Radiometric dating has given significantly younger ages for the lardalite than for the larvikites, 269±5 Ma and 277±3 Ma respectively (B. Sundvoll, personal communication).

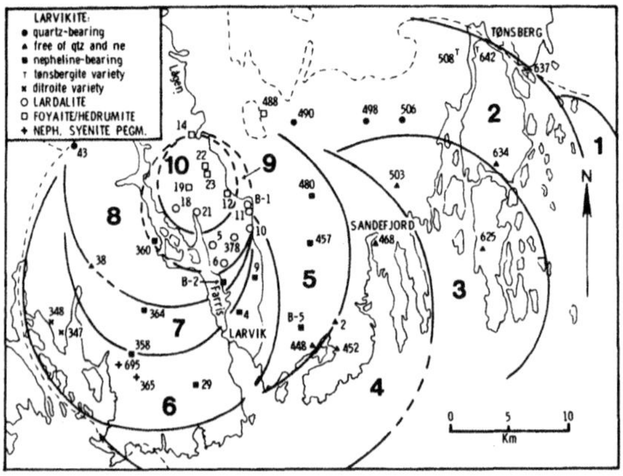

Fig. 1. Sketch map of the zones in the Larvik ring-complex with sample localitites. Zones 1-8 consist of larvikite, zone 9 of lardalite, and zone 10 of lardalite and foyaite/hedrumite.

The field relations of larvikite, lardalite and foyaite/hedrumite led Brøgger [10] to believe that the lardalite is a differentiation product of a larvikitic parent magma. This hypothesis has been accepted in later work [1,4,15,16].

The present study was undertaken in order to investigate whether geochemical data can throw any further light on the genetic relations a) between quartz- and nepheline-bearing larvikite, and b) between larvikite, lardalite and foyaite/hedrumite. This is a preliminary report; a full discussion of analytical methods and data will be presented later.

2. RESULTS

The analyzed larvikites have 0-6.4 percent normative quartz, or

0-7.3 percent nepheline, whereas the lardalites (but for 1 sample) have 19-25 percent and the foyaite/hedrumites 3-37.5 percent normative nepheline. Among the larvikites, silica activity cannot be correlated with other parameters such as differentiation index (D.I. = normative qtz + or + ab + ne) or $(Fe_2O_3+FeO)/(Fe_2O_3+FeO+MgO)$ ratios. The lardalites and foyaite/hedrumites have higher D.I.s, but lower $(Fe_2O_3+FeO)/(Fe_2O_3+FeO+MgO)$ ratios relative to D.I. than the larvikites (Fig. 2).

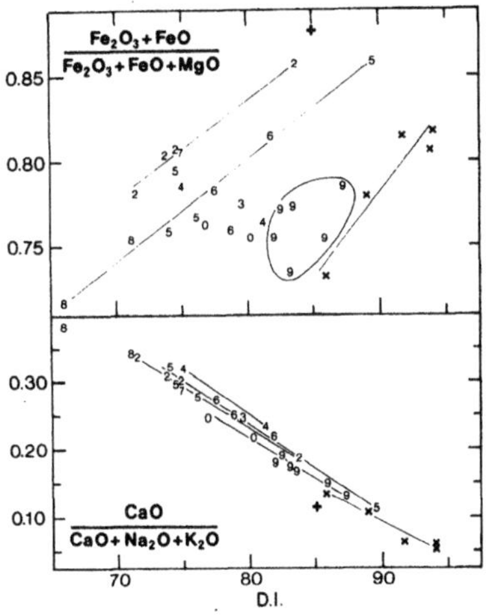

Fig. 2. Example of major element versus D.I. relations among larvikites, lardalites and foyaite/hedrumites. Numbers refer to zones in Fig. 1. 0 = lardalites in zone 10, x = foyaite/hedrumites in zone 10, + = foyaite/hedrumite outside zones 9 and 10.

The larvikites show weak correlations between major elements and D.I. (Fig. 2), and no general correlations between trace elements and D.I. The ratios between pairs of incompatible elements vary considerably, the Ta/Th ratios, for example, range from 0.4 to 1.1. REE (rare earth element) patterns are quite similar to those of other larvikites in the Oslo Region [7], with strong enrichment of REE relative to chondrites, and of light relative to heavy REE. The rocks are slightly enriched to slightly depleted in Eu relative to other REE. The registered Eu anomalies are believed to be related to settling of early-formed feldspar crystals [7]. There is no systematic change in the behavior of Eu in larvikites from east to west.

The lardalites, like the larvikites, show weak correlations

between major elements and D.I. (Fig. 2), and none between trace elements and D.I. They are strongly enriched in REE relative to chondrites and have La/Lu ratios between 140 and 230. Their REE patterns are characterized by major variations in the behavior of Eu. Some samples show no Eu-anomalies [7] while others show prominent positive Eu-anomalies, again suggesting accumulation of feldspar. There is no correlation between relative Eu-enrichment and D.I.

The foyaite/hedrumites in zone 10 show good correlations between both major and trace elements and D.I. (Fig. 2). Also these rocks are strongly enriched in REE relative to chondrites and have high (but variable) La/Lu ratios. They show increasing Rb/Sr ratios, decreasing Ba-content and increasing degree of Eu-depletion with increasing silica undersaturation, strongly suggesting that they are related through fractionation processes dominated by feldspar.

3. DISCUSSION

Three simplified models may be suggested to explain the genetic relations between the various larvikites:

1. The rocks are cogenetic and the zoned batholith is the result either of multiple intrusions from a single magma chamber at greater depth, or of a series of cauldron subsidence into one major magma chamber.
2. The zones represent intrusions of genetically unrelated magmas.
3. The zones represent multiple intrusions from, or multiple cauldron subsidences into a magma chamber which was undergoing continuous fractional crystallization, but which was periodically fed with a new batch of magma from depth, which then mixed with the magma in the chamber.

The compositional characteristics of the larvikites do not indicate a single line of fractionation from a common parent magma as in model 1, but are sufficiently similar to rule out model 2. The data agree reasonably well with compositional relations postulated by O'Hara [11] for model 3 (large variations in the concentrations of, and ratios between, incompatible elements with small variations in the concentrations of major elements).

With regard to the relationship between larvikite, lardalite, and foyaite/hedrumite, the major and trace element data at least make it abundantly clear that the lardalites cannot have been formed by fractionation from a larvikitic magma as has formerly been widely believed. On the basis of the chemistry three possible models of formation seem possible: a, partial melting; b, feldspar (+nepheline) accumulation; c, separation of an immiscible liquid.

a, The lardalites fall near the 1 bar temperature minimum in the quartz-nepheline-kalsilite diagram, suggesting that the rocks may have been formed by partial melting of silica under-saturated larvikites. Partial melting of a feldspar-rich source with fairly flat REE patterns will yield a melt with a positive Eu-anomaly. The lardalites, however, have similar or higher $MgO/(MgO+FeO+Fe_2O_3)$ ratios, similar Rb/Sr ratios and Ta, Th and U-contents to those of the larvikites, which is contrary to what may be expected from a partial melt relative to its source.

b, The lardalites have compositions similar to those of felsic ocelli as reported by Philpotts [12]. It is, however, difficult to test this hypothesis further as data are not yet available on the partitioning of trace elements between immiscible liquids.

c, The conclusions drawn in the preceeding discussion of the geochemistry was that the lardalites have geochemical characteristics of cumulate rocks, and the foyaite/hedrumites those of rocks related to each other through fractionation. With their field relations in mind it is therefore tempting to suggest that the foyaite/hedrumites and the lardalites represent residual liquids and cumulates (respectively) from a single fractionation process. In most diagrams, however, the lardalites plot well off the foyaite/hedrumite trend, indicating that the true story is more complicated. It is most likely that the lardalites represent a series of silica undersaturated melts. The residual liquids of these melts are represented by the group of highly evolved undersaturated rocks found in the area (nepheline syenite pegmatites, ditroite, and foyaite/hedrumite).

4. CONCLUSIONS

Geochemical data suggest that all the rocks in the ring-complex formed according to our model 3. The first melt(s) in the magma chamber was silica oversaturated, but injections of new undersaturated magmas gradually changed the overall composition towards silica undersaturation. This model requires the availability of several mafic magmas of differing compositions. The wide variety of basalts reported from the Oslo rift by Weigand [13] include both over- and undersaturated rocks, and tholeiitic and alkaline varieties. It thus seems likely that mafic magmas of suitable compositions were available as potential parent magmas.

The main magma chamber could have been situated at the top of the crust where the rocks are now found, implying that the zones formed by a series of cauldron collapses, or at a deeper level. The zones would then have formed as a series of intrusions. Gravity studies in the Oslo Region [16] has revealed no trace within the crust underneath the Larvik ring-complex of dense material left from the formation of the rocks in the ring-complex.

which, judging from their low initial $^{87}Sr/^{86}Sr$ ratio [17], probably originated from a mantle derived source. It is therefore most likely that the magmas that gave rise to the ring-complex formed in connection with pillow of dense material postulated by Ramberg and Smithson [14] to underlie the Oslo rift.

REFERENCES

1. T.F.W. Barth, Studies on the igneous rock complex of the Oslo Region. II, Skr. Norske Vidsk.-Akad. Oslo I. Mat.-Naturv. kl., 1945.
2. Chr. Oftedahl, Studies on the igneous rock complex of the Oslo Region. XIII, Skr. Norske Vidsk.-Akad. Oslo I. Mat.-Naturv. kl., 1953.
3. E. Sæther, Studies on the igneous rock complex of the Oslo Region. XVIII, Skr. Norske Vidsk.-Akad. Oslo I. Mat.-Naturv. kl., 1962.
4. G. Raade, Distribution of radioactive elements in the plutonic rocks of the Oslo Region, Unpubl. cand.real. thesis, University of Oslo, 1973.
5. J.P. Nystuen, Nor.Geol.Unders., 317, 1, 1975.
6. R. Sørensen, Nor.Geol.Unders., 321, 67, 1975.
7. E.-R. Neumann, A.O. Brunfelt and K.G. Finstad, Lithos, 10,311,1977.
8. J.S. Petersen, Structure of the larvikite-lardalite complex, Oslo Region, Norway and its evolution, Geol. Rundsch., 67, (in press), 1977.
9. W.C. Brøgger, Zeitschr. f. Kryst. und Min., 16, 1, 1890.
10. W.C. Brøgger, Die Eruptigesteine des Kristianiagebietes. III, Skr. Norske Vidsk.-Akad. Oslo I, Mat.-Naturv. kl., 1898.
11. M.J. O'Hara, Nature, 266, 503, 1977.
12. A.R. Philpotts, Am. J. Sci., 276, 1147, 1976.
13. P.W. Weigang, Skr. Norske Vidsk.-Akad. Oslo I. Mat.-Naturv. kl. Ny Serie, 34, 4, 1975.
14. I.B. Ramberg and S.B. Smithson, Tectonophysics, 11, 419, 1971.
15. Chr. Oftedahl, Studies on the igneous rock complex of the Oslo Region. IX, Skr. Norske Vidsk.-Akad. Oslo I. Mat.-Naturv. kl., 1948.
16. I.B. Ramberg, Norges Geol. Unders., 325, 1, 1976.
17. B. Sundvoll, Isotope- and trace-element chemistry, geochronology, in: The Oslo Paleorift Guidebook Part I, ed. by J.A. Dons and B.T. Larsen, Norges Geol. Unders. (in press), 1978.

PETROGENESIS OF THE HOLTERKOLLEN PLUTONIC COMPLEX, NORWAY.

Thomas R. Neff and S.O. Khalil

Department of Geography and Geology
Weber State College, Ogden, Utah 84403
Department of Geology, University of Alexandria, Egypt

ABSTRACT. The Holterkollen complex contains a biotite granite-quartz porphyry pluton, syenitic ring dikes and hornblende granite dikes. Chemical variations within the pluton show a series of "highs" and "lows" that have no relation to textural zones. The quartz porphyry apparently represents locations that were subject to sudden volatile loss whereas the chemical pattern probably is the result of multiple magma injection. The Holterkollen rock series probably was formed by fractionation of a mantle derived parent magma. However, the granite-quartz porphyry melt may have formed by anatexis.

1. FIELD RELATIONSHIPS

The Holterkollen pluton covers an area of approximately 12 km^2. It is composed of biotite granite and quartz porphyry. Internal contacts are gradational.

The Holterkollen pluton is cut by syenite and akerite* (biotite-syenite porphyry) ring dikes. Contacts of the ring dikes with plutonic and metamorphic wallrocks are very sharp with evident chilling of the akerite and syenite. Where the akerite and syenite ring dikes intersect (see fig. 1) the rocks appear to be intermediate between the two types and are therefore mapped as a hybrid variety.

Hornblende granite dikes not greater than 10 m in thickness

*The name "akerite" is retained in this paper to preserve continuity with earlier works.

E.-R. Neumann and I.B. Ramberg (eds.), Petrology and Geochemistry of Continental Rifts, 237–244.

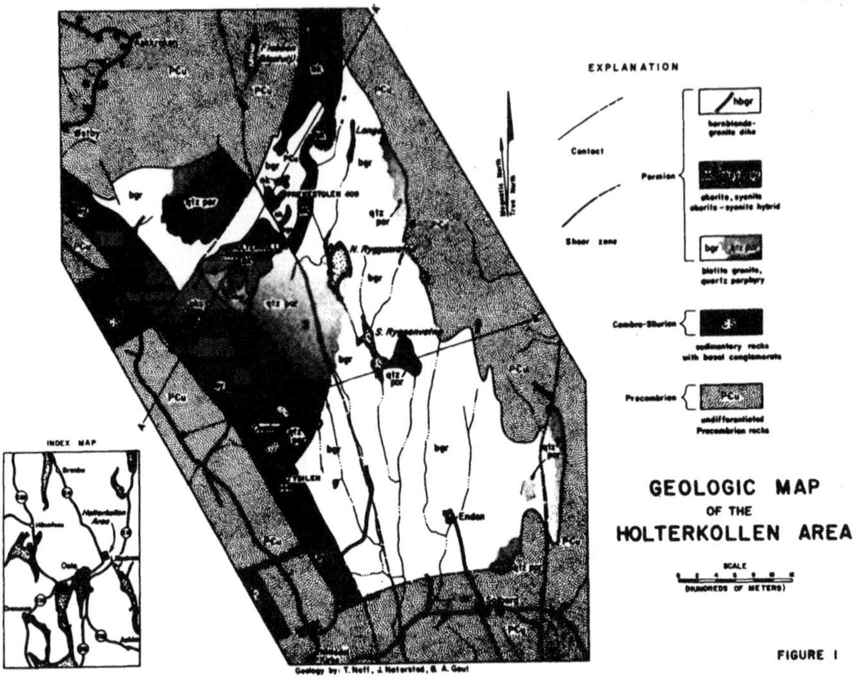

Fig. 1. Geologic map of the Holterkollen area.

cut the akerite, syenite and quartz porphyry. Contacts are sharp and borders in the hornblende granite show evidence of chilling.

2. PETROGRAPHY

2.1 Granite

The major rock type present in the Holterkollen pluton is biotite granite. Major mineral constituents of the granite are perthite (55-70%), plagioclase (An15-20) (10-15%) and quartz (15-30%). Accessory minerals are biotite (3-5%), magnetite (1-3%), and sphene (1%).

2.2 Quartz Porphyry

Quartz porphyry appears as a massive buff to pink aplite, studded with rounded, glassy, quartz phenocrysts ranging in size up to 1 cm in diameter. Sutured potassium feldspar and quartz make up 60-70% and 25-35% of the groundmass respectively with plagioclase (An10-15) constituting about 5%. Accessory minerals are magnetite, rutile and biotite.

Two generations of quartz commonly occur in quartz porphyry-1. the quartz phenocrysts which either show hexagonal crystal outlines or corrosion and 2. groundmass quartz existing as minute grains or in micrographic intergrowths. Phenocrysts of perthite, exhibiting multiple rimming, are also present in some cases.

2.3 Akerite

Akerite is a light to dark-gray porphyry whose groundmass is composed of orthoclase (60-75%), plagioclase (An_{5-20}) (10-20%), quartz (5-10%), and hornblende (5-10%). Accessory minerals are biotite (1-5%), magnetite (1-5%) and trace amounts of augite, zircon, rutile, sphene and apatite, (percentages given are for portions of the groundmass only). Phenocrysts make up from 10-60% of the rock and are composed of perthite (5-45%), plagioclase (An_{15-20}) (5-20%), biotite(1-5%), hornblende (1-5%), magnetite (.1-1%) and augite (1%).

Plagioclase and perthite both exhibit complex rimming. Two and sometimes three zones are recognizable in individual grains. Separate zones may be either plagioclase or perthite but identical compositions are never juxtaposed.

2.4 Syenite

Syenite is a dark-buff porphyry whose groundmass is composed of orthoclase (65-70%), plagioclase (untwinned) (10-15%), quartz (5-10%), hornblende (3-10%). Accessory minerals are magnetite (3-5%), biotite (.1-1%), augite (1%), sphene (<1%), and zircon (<1%). Phenocrysts which are mainly perthite and comprise 15-20% of the rock have single or multiple rims and corroded cores similar to those present in akerite.

2.5 Akerite - Syenite Hybrid

Samples from the akerite-syenite hybrid zone appear similar in hand specimen to akerite, but the number and size of biotite phenocrysts are reduced and at a few locations are entirely absent.

2.6 Hornblende Granite Dikes

Hornblende granite is fine grained, light buff in color, and even textured. Major constituents are perthite (65-70%), plagioclase (10-15%), and quartz (10-15). Accessory minerals are hornblende (3-5%), magnetite (1-2%), biotite (1%), zircon (1%), allanite (<1%) and pyrite (<1%).

Plagioclase crystals contain untwinned corroded cores surrounded by rims showing good polysynthetic twinning.

3. CHEMICAL RELATIONSHIPS

3.1 Chemical Trends

Figure 2 shows the separate major and minor elements plotted against (1/3 Si+K) - (Ca+Mg). Correlation between separate rock types shows that steadily decreasing trends exist for all major and minor elements except K. In addition there are no significant compositional differences between granite and quartz porphyry. Akerite and syenite are also chemically similar. The hornblende granite plots on the general trend for all major and minor elements except possibly for Fe++ and appears to be intermediate in composition between the ring dike rocks and those of the pluton.

All Holterkollen rocks have a differentiation index (normative Q+Ab+Or) greater than 80 and therefore can be treated as part of the "granite system" [1,5,6]. Granite system plots shown in figure 3 establish a clear path which includes hornblende granite. Moreover, the trend coincides approximately with the thermal trough at 0.5 Kb.

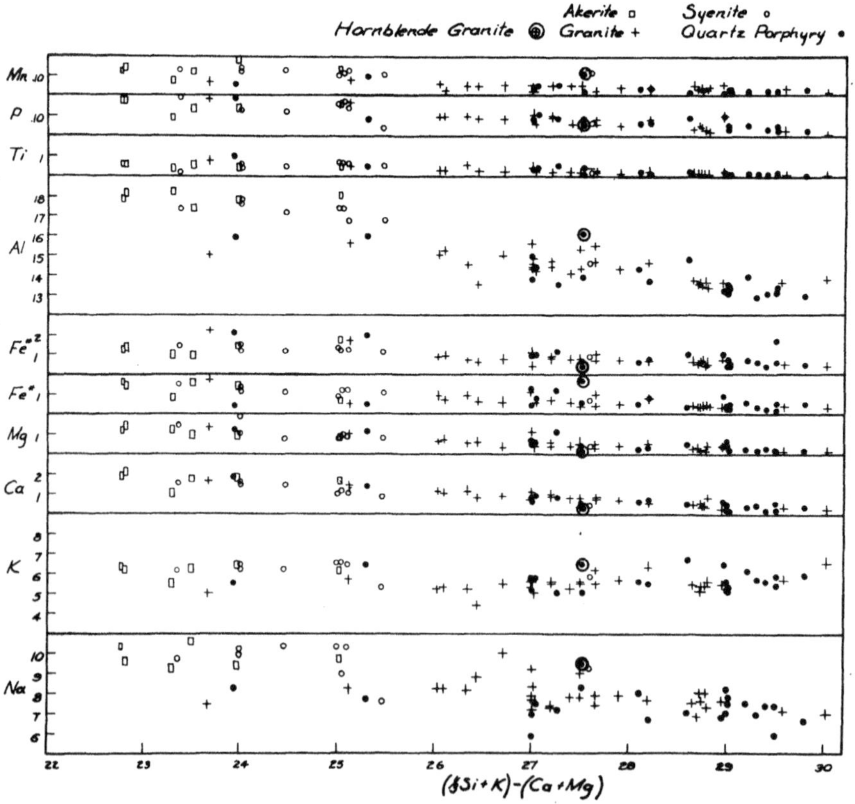

Fig. 2. Major and minor element variation diagram.

3.2 Spatial Variations

In an effort to determine whether the chemical variations noted in the granite and quartz porphyry analyses are systematic, a trend surface analysis of selected oxides, ratios and norms was carried out. Maps were drawn by computer using a program by O'Leary _et al_ [2].

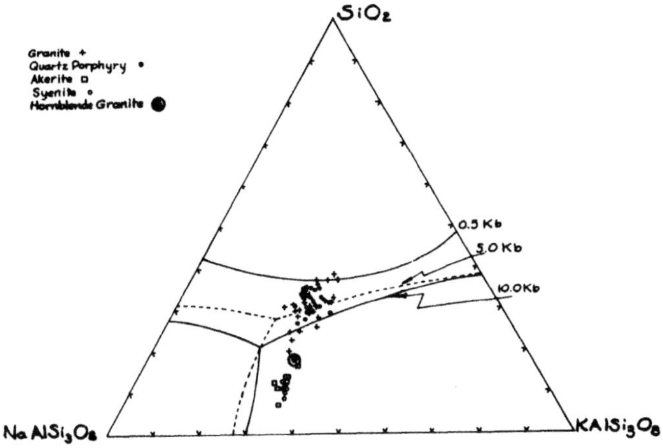

Fig. 3. Holterkollen rock compositions plotted in the "granite system".

The trend surface maps do not exhibit a simple quasi-concentric spacing of contours. Instead all maps have a series of "highs" and "lows" i.e., areas of closed contours.

Twelve maps were constructed using the following parameters: $Fe^{+++} + Fe^{++}/ Fe^{+++} + Fe^{++} + Mg^{++}$, Na_2O, normative An, normative quartz, MgO, CaO, Rb/Sr, K_2O, TiO_2, Al_2O_3 and P_2O_5. All variables are sensitive to chemical fractionation, especially Rb/Sr. One example is given here - $Fe^{+++} + Fe^{++}/ Fe^{+++} + Fe^{++} + Mg^{++}$ (fig. 4). All maps correlate well one with another.

4. DISCUSSION

4.1 Petrographic Phenomena

<u>Multiple Rimming of Feldspar-</u> All rock types contain feldspars exhibiting single or multiple rimming which indicates that they have had a complex petrogenetic history punctuated by periods of crystallization interrupted by stages of disequilibrium producing corrosion.

<u>Textural Variations Within the Pluton-</u> A comparison of the trend

surface maps (fig. 4) with the geologic map (fig. 1) shows that chemical zonation does not correlate with textural variations. The aplitic texture of the quartz porphyry is probably due to pressure release quenching due to sudden volatile loss. Vapor pressure when the granite and quartz porphyry commenced crystallization was approximately 0.5 Kb which corresponds roughly to the probable depth of emplacement - 2500 m. as calculated from the sedimentary and volcanic cover stacked above the pluton before erosion. When vapor rising above the pluton along fractures encountered the earth's surface, an explosion would occur suddenly, reducing the vapor pressure to 1 Atm in places directly under the fractures, thereby producing an aplitic texture in these zones.

Quartz Porphyry Phenocrysts- Apparently the phenocrysts were produced by primary crystallization from a melt. This supposition is supported by the existence of grains with hexagonal crystal outlines. It seems likely that the only way to produce such a shape is by unimpeded crystallization from a melt.

For quartz to be the first mineral to crystallize in quartz porphyry given their bulk compositions as plotted in the "granite

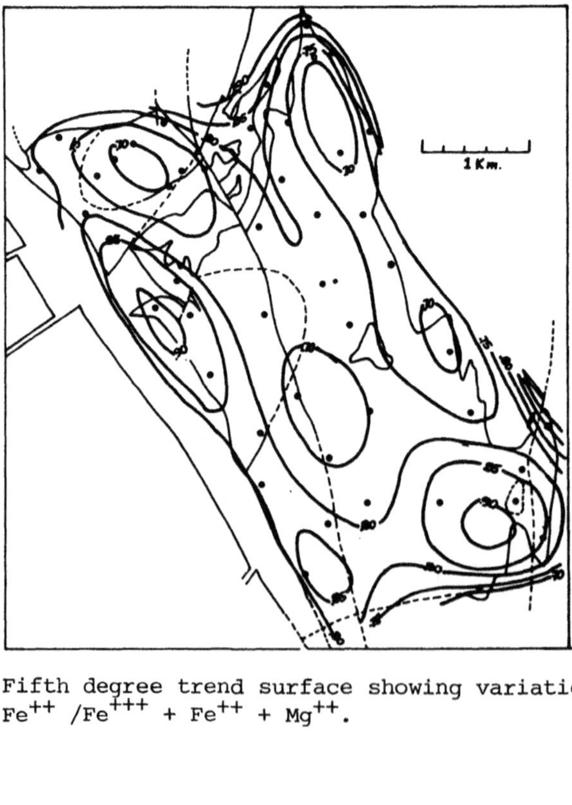

Fig. 4. Fifth degree trend surface showing variations in Fe^{+++}+ Fe^{++} /Fe^{+++} + Fe^{++} + Mg^{++}.

system" (fig. 3), the vapor pressure would have to be high enough to shift the phase boundary far enough toward the $NaAlSi_3O_8$ - $KAlSi_3O_8$ edge to allow the quartz porphyries to lie in the quartz phase field. Vapor pressures near 10 Kb are required for this to occur.

4.2 Spatial Variations in Chemistry

The pattern shown by the trend surface maps (fig. 4) appears on first glance to be compatible with an assimilation model. However, closer study shows that this does not appear to be a viable mechanism. If this process is assumed, then for example on the Fe^{++}+ Fe^{+++}/ Fe^{++} + Fe^{+++} + Mg^{++} Map (fig. 4), all centers (areas of closed contours) should be "lows" because the wallrocks are significantly higher in MgO than the plutonics. Instead, there are 3 "highs" and 4 "lows".

Centers on the trend surface maps (fig. 4) can be classed as either "salic" or "mafic". "Salic" centers are high in normative quartz, Fe^{++} + Fe^{+++}/ Fe^{++} + Fe^{+++} + Mg, Na_2O and Rb/Sr and low in normative An and TiO_2. The reverse is true for "mafic" centers. A comparison of the maps with one another shows that the "centers" correlate well in regard to location as well as type, which is best explained by multiple injection of various magma compositions fed upward from a fractionating chamber existing below the Holterkollen area.

4.3 Petrogenesis

A recent work by Ramberg [4] based mainly on geophysical evidence considers that the entire Oslo igneous rock series is most probably of mantle origin. The Oslo graben was produced by crustal extension and thinning over a spreading center. During this phase mafic, mantle derived magma invaded the lower crust. Differentiation of the mafic melt produced felsic magma which moved upward as diapiric masses to form the cauldron complexes and plutons. Ramberg, even though he favors a mantle origin for the biotite granite bodies, does accept the possibility that a minor amount of anatexis may have taken place in the lower crust and the biotite granite plutons, one of which is the Holterkollen, may have originated in this manner. Another recent work [3] based on U-Th distribution in Oslo area plutonic rocks supports the anatexis theory.

The major element data generated by this study seems to indicate with some reservations, that all of the Holterkollen rocks belong to the same fractionation series. This is shown for instance by the good negative correlations and overlaps on fig. 2, the major and minor element variation diagram. The granite system (fig. 3) shows a strong correlation of all rock types with the 0.5 Kb thermal trough - a relation suggestive of differentiation. The existence of feldspar rimming in quartz porphyry and a few granites also lends credence to the supposition that the dike and plutonic rocks are genetically related. Multiple rimmed K-feldspar and plagioclase which appear identical to those in akerite and syenite exist as minor phenocrysts in quartz porphyry.

The single magma hypothesis is clouded by other lines of evidence. The presence in the quartz porphyry of early formed quartz which probably crystallized at considerable depth supports a separate origin for the granite-quartz porphyry magma by anatexis. Production of a quartz rich melt at depths approaching the base of the crust by differentiation of a mantle derived mafic magma appears unlikely although possible.

Another support for the anatexis theory is the fact, as shown by chilled border relationships, that the granitic pluton was intruded before collapse of the Nittedal cauldron and injection of the ring dikes. This seems to indicate that the granitic melt was in existence before its supposed parent - the akerite-syenite-horneblende granite series. However, this supposition is only tenuous because the various Holterkollen magmas may have been tapped from differing levels within the same stratified chamber.

Due to the conflicting evidence given above a definitive conclusion cannot be drawn concerning whether or not all of the Holterkollen rocks are differentiates from the same parent magma. However, considering the information at hand it appears that a stronger case can be made at this time for the single parent differentiation model.

REFERENCES

1. W.C. Luth, R.H. Jahns, and O.F. Tuttle, Jour. Geophys. Res., 69, 759, 1964.
2. M. O'Leary, R.H. Lippert and O.T. Spitz, Computer Contrib. Kansas Geol. Surv., 3, 1, 1966.
3. G. Raade, Distribution of radioactive elements in the plutonic rocks of the Oslo Region, Unpub. Thesis, University of Oslo, 1973.
4. I.B. Ramberg, Norges Geol. Unders., 325, 1976.
5. J.C. Steiner, R.J. Jahns and W.C. Luth, Geol. Soc. America Bull, 86, 83, 1975.
6. O.F. Tuttle and N.L. Bowen, Geol. Soc. America Mem., 74, 1958.

ORE DEPOSITS RELATED TO THE KEWEENAWAN RIFT*

David I. Norman

Mineralogisk-Geologisk Museum, Oslo, Norway.

ABSTRACT. The Keweenawan-aged ore deposits in the Lake Superior region are described, and genesis briefly discussed, in terms of their relation with rift tectonics and associated rock types. Niobium-uranium-rare earth deposits, Cu-Mo porphyry-like deposits, and mineralized breccia pipes are associated with alkalic and felsic intrusives. Major alkalic and felsic intrusives appear to be located at the intersections of zones of rifting and transform faults. Ore deposits associated with mafic rocks are principally copper-bearing. These include disseminated Cu-Ni, native copper, chalcocite vein, and strataform deposits. Although basalt-gabbroic rock appears to be the source of copper for these deposits, the deposition of copper sulfides seems dependent on sources of sulfur other than the mafic rock.

1. INTRODUCTION

There has been considerable discussion of late concerning the relationship between ore deposits and plate tectonics. Most attention has been paid to convergent plate boundaries while only one paper [1] has dealt in detail with the occurrence of ore deposits in tensional environments. It is the purpose of this paper to summarize the deposits related to one intercontinental rift in order to illustrate the variety of ore deposits which can be associated with such tectonic environments, and to show the genetic relationship of the various deposits to rift tectonics. The

* This work has been supported in part by NSF grant GA 40544 and The Royal Norwegian Council for Scientific and Industrial Research. (NTNF).

E.-R. Neumann and I.B. Ramberg (eds.), Petrology and Geochemistry of Continental Rifts, 245–253.

Keweenawan event (1.2-1.0 AE) in central North America is well suited for this task because it is widely accepted as a paleorift [2,3] and has a sizeable number of well studies ore deposits.

2. RELATION OF ORE BEARING DEPOSITS TO RIFT TECTONICS

The major Keeweenawan-age, ore-bearing deposits are listed in Table 1. All are from the Lake Superior region, the area of principal Keweenawan exposure. The location of these deposits is illustrated in Fig. 1.

The various deposits can be divided according to their spatial relation with rift tectonics (Fig. 2) into: 1) deposits occurring within pre-Keweenawan age rocks; 2) those located within the pile of predominently basic volcanics, intrusives, and sediments; 3) those found in sediments overlying the volcanic pile.

2.1 Deposits in pre-Keweenawan age rock

Deposits in pre-Keweenawan age rock consist of niobium-uranium-rare earth deposits in alkalic complexes, Cu-Mo porphyries and breccia pipes associated with felsic intrusives, and native-silver, base-metal sulfide vein deposits.

The alkaline and felsic intrusives which are associated with ore deposits appear to represent an early phase in rift magmatism. Therefore their preservation, in part, depended on their being emplaced in areas adjacent to later extensive igneous acitivity or where rifting was minimal. The location of felsic and alkalic intrusives appears to have been influenced by the intersections of zones of rifting and offset or transform faults.

The plate tectonic model of the Keeweenawan rift proposed by Chase and Gilmer [3] may be extended to eastern Lake Superior (Fig. 3). There is geophysical and observational evidence of these proposed faults [8,15,16,17,18,19]. Each of the intersections of the proposed zones of rifting and offset faults is marked by felsic or alkalic intrusives. The same relation appears to hold for the model proposed by Chase and Gilmer. The largest accumulation of Keweenawan felsic intrusives and flows in western Lake Superior are found in the vicinity of Hoveland, Minn. and Mellon, Wisc.

The vein deposits occur, for the most part, in tensional faults which parallel the rift axis. Although not occurring within Keweenawan-age rock the veins are located in areas with numerous Keweenawanage dikes and sills and may have formed beneath an unknown thickness of volcanic cover.

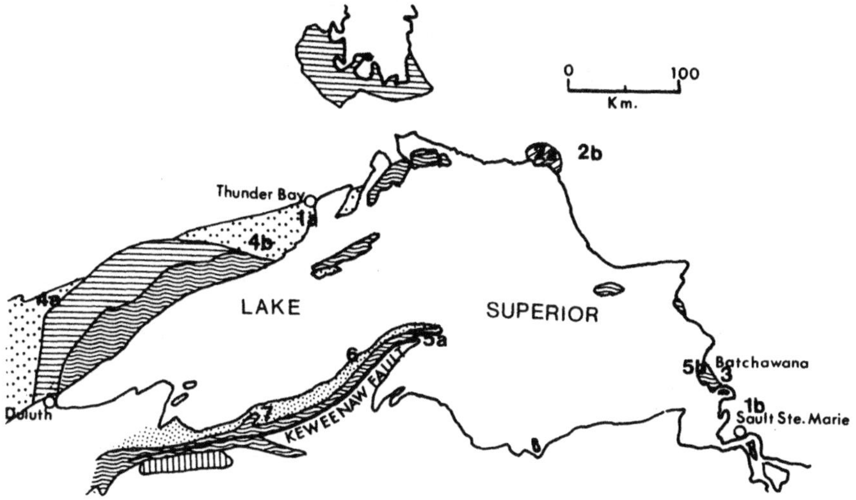

Figure 1. Generalized Keweenawan geology of the Lake Superior region. The numbers are keyed to the ore deposits in Table 1.

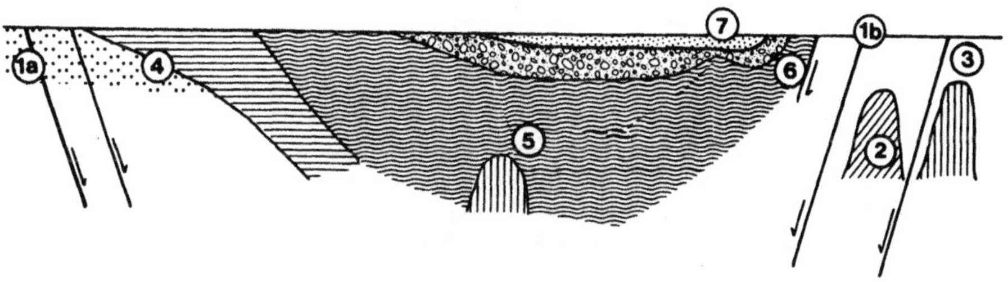

Figure 2. Schematic Keweenawan cross-section to illustrate the relationship of ore deposits listed in Table 1 to rift tectonics.

Table 1. Keweenawan-age Ore Deposits in the Lake Superior Region

	Type	Name or location	Ref
1.a	native-Ag, base metal sulfide veins	Thunder Bay	[4]
b	base metal sulfide veins	Sault Ste. Marie	[5]
2.a	niobium	Coldwell Complex	[6]
b	U-rare earths	Prairie Island Complex	[6]
3.	disseminated Cu-Mo	Batchawana	[7,8]
4.a	disseminated Cu-Ni	Buluth Complex	[9]
b	disseminated Cu-Ni	Crystal Lake Gabbro	[10]
5.a	chalcocite veins	Mt Bohemia	[11]
b	chalcocite veins	Coppercore	[12]
6.	native-copper	Northern Michigan	[13]
7.	stratiform copper	White Pine	[14]

2.2 Deposits within the volcanic pile

Three types of deposits occur within the volcanic pile; disseminated Cu-Ni sulfides, native-copper deposits and chalcocite veins.

The Cu-Ni sulfides occur as disseminations in the basal units of mafic intrusives. Economic concentrations of the sulfides occur principally where the intrusives are in contact with Middle Precambrian sulfide-bearing argillites.

Native-copper is found in small amounts, as vesical fillings, virtually throughout the volcanic pile. Economic concentrations occur in flow top breccias and interflow sediments near the southern basin margin where the pile rapidly thins, has been faulted and is steeply dipping [13]. Major chalcocite vein deposite occur within the volcanic pile in two localities. At both locations the deposits are associated with felsic intrusives into the pile.

2.3 Deposits in the overlying sediments

Along the Keeweenawan peninsula the volcanic pile is overlain by up to 1800 m of conglomerate which in turn is overlain by 200 m of sulfide bearing shales and siltstone, the Nonesuch Shale. Chalcocite with lesser amounts of native-copper are present throughout the basal units of the None such shale. Mineable concentrations occur over a 50 a.sq. km area near White Pine, Michigan. No controls of the ore zone have been determined other than the casual relationships between copper mineralization and favorable sulfiderich beds, and a thinning of the underlying conglomerate in the vicinity of the White Pine deposit.

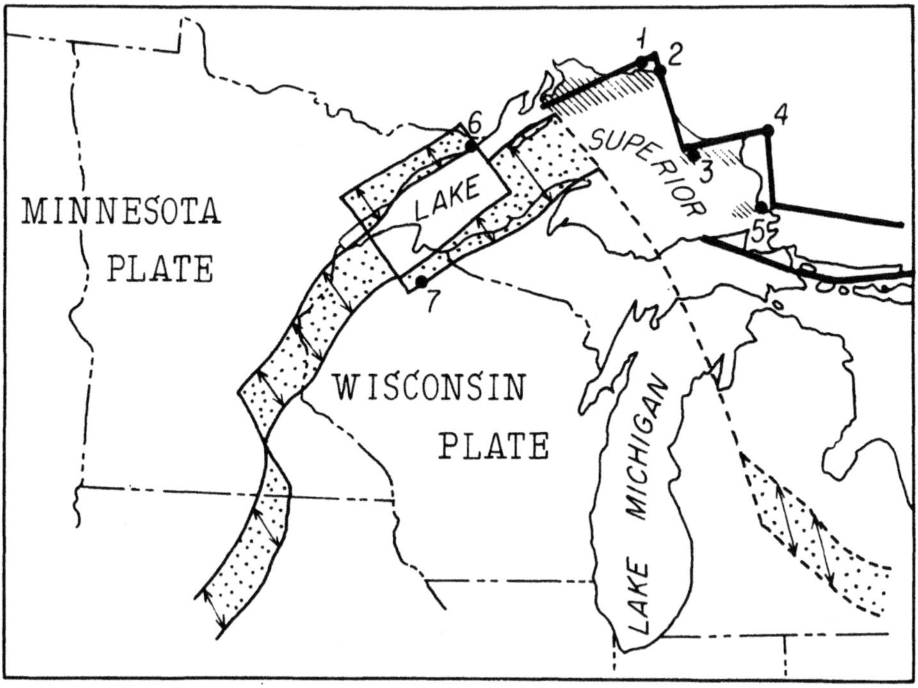

Figure 3. Proposed zones of Keweenawan rifting in eastern Lake Superior. Areas of gravity high [20], shown by hatchers, combined with the zone of predominently normal faulting (after Kumarapeli and Saull, [21]) shown by dark lines, suggest zones of rifting or protorifting separated by transform faults. The proposed rift zones and transform faults show the same general direction of relative motion as Chose and Gilmer's [3] model. Numbers refer to alkalic complexes and regions of extensive felsic magmatism which appear to be related to intersections of transform faulting and rifting. They are: 1. Prairie Lake complex, 2. Coldwell complex, 3. Michipicoten Island, 4. Fireside River complex, 5. Batchawana area, 6. Hoveland, Minnesota, and 7. Mellon, Wisconsin.

3. GENESIS OF THE DEPOSITS

The Keweenawan rift provided both areas of high heat flow for driving hydrothermal systems and various types of igneous rock as possible sources of metals. In this general aspect the Keweenawan rift appears different from convergent plate boundaries only in the prevalence of mafic magmatism. This supposition may explain why some of mafic magmatism. This supposition may explain why some of the Keweenawan ore deposits are similar to those commonly found in Cordilleran regions.

3.1 Deposits in pre-Keweenawan-age rock

Small, hydrothermal-vein deposits are present over large areas north of Sault Ste. Marie, Ontario and in the vicinity of Thunder Bay, Ontario. The broad distribution of these deposits perhaps can be most easily explained as being the result of local hydrothermal systems which acted during times of high heat flow, leaching metals from nearby rock and depositing them in veins. For example the native-silver-bearing veins near Thunder Bay occur in sulfide-bearing sediments. These deposits are similar to the native-silver veins at Cobalt, Ontario where evidence [22] strongly suggests the source of the silver to be the underlying sulfide-rich sediments.

There are four known disseminated Cu-Mo deposits in the Batchawan, Ontario area. The deposits are similar to Cu-Mo porphyry deposits in Cordilleran belts in their type of mineralization, high tonnage, lowgrade disseminations; and in their association with highly altered felsic porphyries and breccia pipes. Fluid inclusion studies of one of the deposits, Tribag [8], indicated similar fluid inclusion morphology and depth and temperature of mineralization as reported for Cordilleran porphyry deposits. Analyses of gases in fluid inclusions, temperature-salinity relationships of fluid inclusions, and stable isotope measurements from Tribag material indicated mineralization by fluids that had been in equilibrium with igneous rock at temperature greater than $600^{o}C$ [8]. The data strongly suggests the initial mineralization stage at Tribag was the result of magmatic-hydrothermal solutions, as has been proposed for other porphyry deposits [23].

The genesis of the niobium-uranium-rare earth deposits is straight forward in that they occur as primary constituents of alkaline intrusives.

3.2 Genesis of the deposits within the pile and overlying sediments

By far copper has been the single most important element mined from Keweenawan-aged ore deposits. To date, 6×10^6 m tons of copper have been produced with estimated reserves of 40×10^6 m tons. Almost all past production and reserves are from deposits either with in the pile of volcanics and mafic intrusives, or in the overlying sediments. The most logical source of this copper is the 127 ppm copper present, on the average, in basaltic rocks [24], of which the Keweenawan mafic rocks are no exception [1].

The disseminated Cu-Ni sulfide in the basal units of the Duluth complex only reach mineable grade where the complex intrudes the sulfide-bearing Rove argillites [25]. The Crystal Lake gabbro and many of the mafic dikes and sills which also intrude the Rove formation, have disseminated Cu-Ni sulfides along their contacts [10]. Mineralogic studies of the Duluth complex indicate an inward diffusion of H_2O, K, Si, and S into the magma after intrusion [26]. Diffusion or assimilation of sedimentary sulfur into the base of the Duluth complex was confirmed by stable isotope studies [27]. Furthermore, oxygen isotope studies of a smiliar Red Sea rift complex indicate the circulation of large volumes of meteoric water through the complex [28].

The indication that sulfur migrated into the Duluth complex, and the correlation of Cu-Ni ore deposits with intrusion into sulfur rich rocks, strongly suggests that the introduced sulfur was critical in the formation of the several disseminated Cu-Ni deposits. Cu-Ni deposits at the base of mafic massives have long been considered to be the result of the segregation and settling of liquid sulfides; however, this supposition does not negate the possibility of some of the sulfur being introduced by assimilation of country rock or by sulfide-bearing waters.

The native copper deposits have been explained by the leaching of Cu from the volcanics after deep burial, followed by upward migration of the fluids along porous channel ways (flow tops and interflow sediments) where the copper was precipitated [13].

The only notable occurrence of copper sulfides within the volcanics is near felsic intrusions [29]. This observation has been used as an argument against a magmatic source for the native copper deposits [13]. Analysis of fluid inclusion waters at Tribag [8] indicated the presence of reduced sulfur in hydrothermal fluids associated with a felsic intrusion. A possible origin for the copper sulfide deposits in the volcanic pile is the mixing of sulfide-bearing hydrothermal fluids derived from felsic intrusions with copper-bearing waters similar to those postulated to have deposited the native-copper deposits.

The White Pine deposit is considered to be syngenetic by some, epigenetic by others [14]. Ensign's et al [14] preferred model proposes mineralization by copper-bearing solutions which migrated through the underlying conglomerate, then moved upwards where the conglomerate thinned, whereupon the copper was precipitated by the sulfur-rich Nonesuch shale. Although the source of the copper is problematical in such deposits, a genetic link between basalt rock and stratiform copper deposits in indicated by trace element data [1]. The sulfur in the White Pine deposit on the other hand is indicated to be of biogenic origin by stable isotope studies [30].

In all three types of copper sulfide mineralization associated with basic rocks there is an indication that the origin of the sulfur was not the mafic rock itself. The reason for this is not clear.

4. CONCLUSION

Based on the example of the Keweenawan rift several observations concerning ore deposits in a rift environment may be made:

1) Cu-Mo porphyry like deposits may occur in rift environments.
2) Alkalic and felsic magmatism, which may have associated ore deposits, appear favored at the intersection of zones of rifting and transform faults.
3) Rift-related basic rocks are strongly associated with copper deposits, but the formation of copper sulfide deposits in particular appear to be related to an external source of sulfide.

REFERENCES

1. F. Sawkins, Jour. Geol., 84, 653, 1976.
2. R. King and I. Zietz, Geol. Soc. Am. Bull., 82, 2187, 1971.
3. C. Chase and T. Gilmer, Earth Planet. Sci. Lett., 21, 70, 1973.
4. M. Mudrey, Jr. and G. Morey, Geology of Minnesota: A Centennial Volume, Minn. Geol. Surv., Minneapolis, 1972.
5. Ontario Dept. of Mines. Sault Ste. Marie-Elliot Lake Sheet, Map 2108, 1966.
6. R. Douglas, Geology and Economic Minerals of Canada, Geol. Surv. Can., Ottawa, 1970.
7. M. Blecha, Can. Min. Jour., (Aug.), 71, 1974.
8. D. Norman, Geology and Geochemistry of the Tribag Mine, Batchawana Bay, Ontario (unpublished Ph.D Thesis), 1977.
9. B. Bonnichsen, Minn. Geol. Survey Inf. Circular, 10, 1974.
10. J. Geul, Ont. Dept. of Mines Geol. Rept., 87, 1970.
11. J. Pollock et al, 21st Int. Geol. Cong. (2), 20, 1960.
12. P. Giblin, Can. Min. Jour. (Ap.), 77, 1966.
13. W. White, Ore Deposits of the Western United States, 1933-1967, AIME, New York, 303, 1968.
14. C. Ensign et al, Ore Deposits of The Western United States, 1933-1967, AIME, New York, 460, 1968.
15. J. Hinz et al, Am. Geophy. Union Monogr., 10, 95, 1966.
16. R. Annells, Geol. Surv. Can. Bull., 218, 1974.
17. H. Halls and F. West, Proceedings 17th Ann. Inst. Lake Superior Geol., Duluth. Mn., 23, 1971.
18. E. Moore, 40th Ann. Rept. of Ont. Dept. of Mines, 40, 1, 1931.
19. L. Ayres, Ont. Dept. of Mines Geol. Rept., 69, 1969.
20. J. Weber and A. Goodacre, Am. Geophys. Union Mono., 10, 70, 1966.

21. P. Kumarapeli and V. Saull, Can. Jour. Earth Sci., 3, 639, 1966.
22. R. Boyle and A. Dass, Can. Mineral., 11, 414, 1971.
23. H. Taylor, Jr., Econ. Geol., 69, 843, 1974.
24. M. Prinz, Basalts, Interscience, New York, 1967.
25. P. Weiblen (personal communication)
26. P. Weiblen, EOS, 58, 516, 1977.
27. P. Mainwaring, The Petrology of a Sulfide-Bearing Layered Intrusion at the Base of the Duluth Complex, St. Louis County, Minnesota (unpublished Ph.D. Thesis), Univ. of Toronto, 1975.
28. H. Taylor, EOS, 58, 516, 1977.
29. B. Butler and W. Burbank, U.S.G.S. Prof. Paper, 154-A, 1929.
30. S. Burnie et al, Econ. Geol., 67, 895, 1972.

K/AR DATING OF CLAY-MINERAL ALTERATION ASSOCIATED WITH ORE DEPOSITION IN THE NORTHERN PART OF THE OSLO REGION

P.M. Ihlen, P.R. Ineson and J.G. Mitchell

Geol. Inst.
Univ. of Trondheim
Norway

Dept. of Geology
Univ. of Sheffield
England

School of Physics
The University
Newcastle upon Tyne
England

ABSTRACT. Twenty K/Ar isotopic ages are reported from 13 mines and prospects in the northern part of the Oslo region. Analyses of clay mineral concentrates from altered wallrocks give support for at least two distinct isotopic events. Upper Proterozoic ages are reported from Eidsvoll gold mines (663-788 Ma). Dates from the Midtskogen prospects located in the Precambrian basement and from altered igneous rocks at Skreia and Feiring reveal a distinct Permian event (237-277 Ma). Less well-marked events of Lower Ordovician and Triassic-Jurassic ages are, respectively, encountered in the Guldkis gold mine and in regional faults transecting Precambrian, Paleozoic, and Permian rocks.

1. INTRODUCTION

The applicability of K/Ar dating of clay minerals which appear to be the product of wall rock alteration associated with ore deposition has previously been discussed in papers by Dunham et al [2], Ineson and Mitchell [3,4,5] and Ineson et al [6]. Recently, Ineson et al [6] have applied this dating method to the problem of allocating emplacement ages for ore deposits genetically related to the Oslo Rift system. The present work can be looked upon as a continuation of the former, dealing with K/Ar dating of clay mineral assemblages connected with hydrothermal acitvity and ore mineralization in the Eidsvoll - Skreia area.

A complete description of the geology and metallogeny in this northern sector of the Oslo Region has already been given by one of the authors (P.M.I.) at this meeting.

E.-R. Neumann and I.B. Ramberg (eds.), Petrology and Geochemistry of Continental Rifts, 255-264.

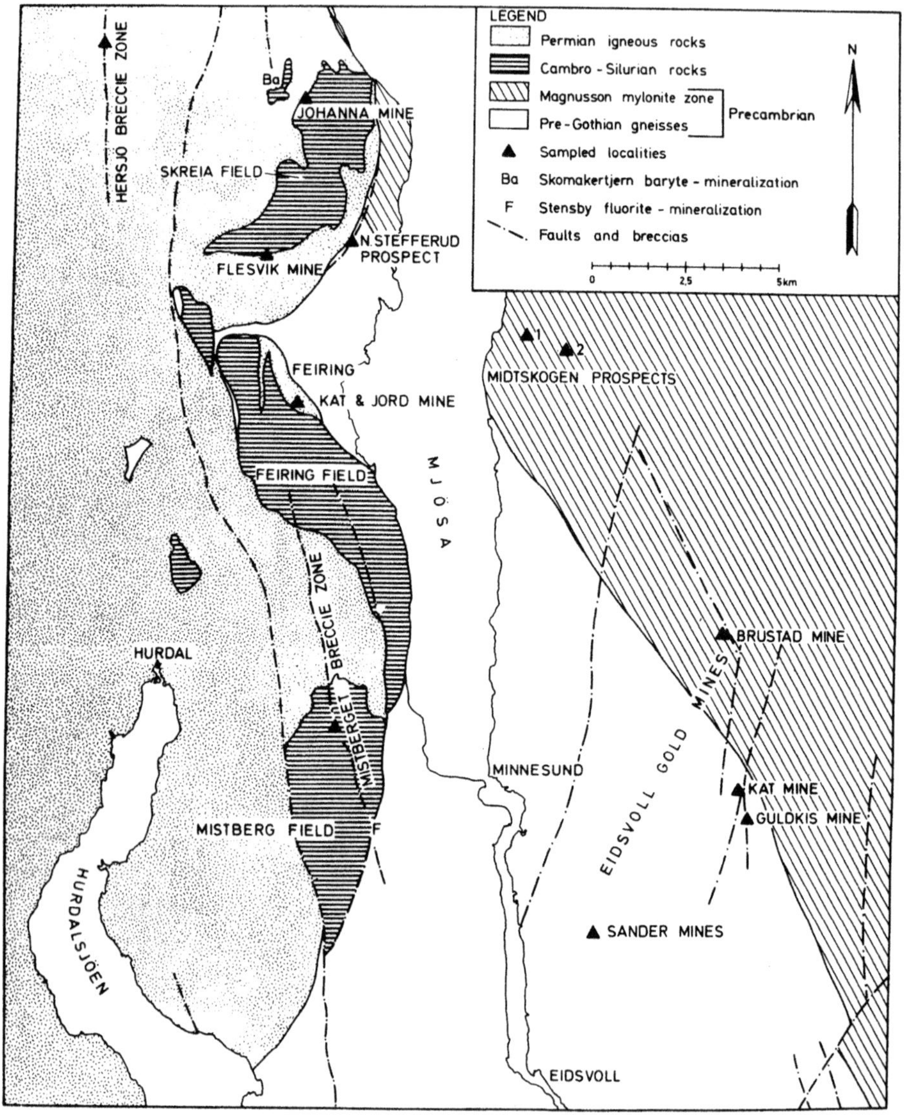

Figure 1. Generalized geological map of the Eidsvoll-Skreia area.

Table 1.

Reference	Sample locality	Field observation	Clay fraction^x	K_2O %	Argon content $v/m.mm^3 gm^{-1}.10^{-2}$	Atmospheric contamination %	Age (Ma)
	REGIONAL FAULT ZONES						
FZ1	Hersjø breccia	Alt.alkali-syenite	KAOLINITE-illite	0,415	2,750	55,7	191±2
FZ2	N.Stefferud prosp. A	Alt.plagioclase	ILLITE-kaolinite	5,45	4,120	9,6	216±3
	B	gneisses(Pre-Gothian)	KAOLINITE-illite	1,90	1,418	21,8	213±2
FZ3	Mistberget breccia	Alt.Silurian hornfelses	KAOLINITE	1,44	1,114	23,6	221±2
	FEIRING MINES						
F1	Jord mine	Alt.mænaite in breccia	ILLITE-kaolinite	6,85	5,710	7,1	237±3
F2	Kat mine A	Alt.mænaite with q-sulphide veins (spoil heap)	Illite-kaolinite-chlorite	2,37	2,100	50,5	250±3
F3	" " B	Alt.mænaite (spoil heap)	ILLITE-kaolinite	5,95	5,067	19,8	241±3
	SKREIA MINES						
S1	Flesvik mine	Alt.ekerite	ILLITE	6,97	5,980	11,7	243±3
S2	Johanna prosp.	Alt. q-syenite (spoil heap)	ILLITE	7,95	6,864	22,4	244±3
	MIDTSKOGEN PROSP.						
M1	Midtskogen prosp. no. 1A	Alt.Pre-cambrian mylonitic gneisses (spoil heap)	ILLITE	5,63	4,980	10,5	250±3
M2	Midtskogen prosp. no 2A	Alt.Pre-cambrian mylonitic gneisses (spoil heap)	Illite-kaolinite	4,89	4,630	8,4	267±3
M3	Midtskogen prosp. no 2B	" " " "	Illite-kaolinite	4,41	4,351	16,4	277±3
	EIDSVOLL MINES						
E1	Brustad mine A	Alt. augen gneisses	ILLITE-kaolinite	7,32	2,303	2,2	770±8
E2	" " B	" " "	ILLITE	4,02	1,277	2,8	776±8
E3	Kat mine A (Claim no.4)	Alt. granitic gneiss	ILLITE	8,93	2,347	2,6	663±7
E4	" " B	" " "	ILLITE	8,36	2,397	8,5	713±8
E5	New Sander mine	Alt. granitic gneiss (spoil heap)	ILLITE-kaolinite	6,95	2,003	3,1	770±8
E6	Old Sander mine	Alt. plag. gneiss (spoil heap)	ILLITE	7,72	2,498	2,6	776±8
E7	Guldkis mine A	Alt. augen gneisses	ILLITE-kaolinite	6,55	1,185	3,6	481±6
E8	" " B	" " "	ILLITE-kaolinite	7,19	1,337	14,5	492±6

Decay constants. $\lambda e = 0{,}584 \times 10^{-10}\ yr^{-1}$, $\lambda\beta = 4{,}92 \times 10^{-10} yr^{-1}$, $K/K = 1{,}19 \times 10^{-2}$ atom %

x) clay mineral constituents: Major component in capitals, minor component in lower case.

The aim of this discussion is to give further information regarding the minerallizing events related to the tecto-magmatic evolution of the Oslo paleorift and to demonstrate the applicability of this method to the problem of distinguishing Permian and Precambrian metallogenetic events.

2. K/AR ISOTOPIC AGE DETERMINATIONS

Twenty samples from highly altered wallrocks and spoil heaps were collected from 13 mines and prospects (Fig. 1). The potassium-argon ages, together with the mineralogical identification of separeted clay fractions, are reported in Table 1. The K/Ar ages were determined by standard isotope dilution techniques. An Omegatron mass spectrometer coupled to an on-line gas extraction system was employed. A flame photometer with a lithium internal standard determined the potassium contents of the clay concentrates. The analytical precision of all the reported results is better than ± 2 per cent. Argon analyses were performed once on each sample and potassium analyses in triplicate.

The clay fractions were run on a Phillips automatic XRD, model No. Pw 1130, using CuK_{α} and 35 kV/40 mA. The XRD runs revealed that the samples consists predominantly of illite with subordinate kaolinite and traces of chlorite. Quartz and carbonate minerals were usually detected as impurities.

The K_2O content of the analyzed samples reflects the ratios of illite to kaolinite. The presence of a small portion of unaltered or relatively unaltered mica or feldspar from the country rock would have appreciable effects on the K_2O content and corresponding K/Ar ages.

Microscope examination of argillitic and sericitic alteration in different deposits give evidence that most samples only consist of quartz and "white mica". Spectrograms of clay fractions from samples believed to carry partially altered feldspar show no or very weak lines for this mineral. The effect from these feldspar impurities on the K/Ar isotopic ages are, however, within the experimental error.

3. ORE DEPOSITS WITHIN THE PRECAMBRIAN BASEMENT

3.1 Eidsvoll gold mines

The Eidsvoll gold mines, including the Sander, Brustad, Kat and Guldkis mines are located 7-15 km east of the Permian igneous complex. In the past E-W and N-S trending quartz breccia veins carrying pyrite, chalcopyrite and subordinate amounts of gold, galena, sphalerite and galenobismuthinite have been worked. These lodes transect grey plagioclase gneisses and mylonitic gneisses of Precambrian age which are, respectively, part of the Pre-Gothian mega-unit and of the so-called Magnusson mylonite zone.

Samples were collected from altered wall rocks adjacent to the Brustad, Kat and Sander lodes, including one sample of altered gneisses constituting breccia fragments in the Old Sander vein. They yielded ages in the range 663 to 788 Ma with a mean age of 732 Ma. In view of the diverse location and petrology of the country rocks and the close grouping of isotopic ages, they may be interpreted as representing a single event of ore deposition in late Proterozoic.

Two samples of altered augen gneisses from the Guldkis mine gave the anomalous (Ordovician) ages of 481 ± 6 and 492 ± 6 Ma. Although old reports mention this lode as the site of the first gold discovery, it clearly deviates from the other gold bearing veins by being barren in sulphides.

3.2 Midtskogen prospects

Among the Midtskogen prospects occurring within mylonitic gneisses, 9 km east of the Oslo Region proper, two paragenetic types can be distinguished. The two southern vein mineralizations constitute mineral assemblages similar to the Eidsvoll gold mines. Prospects nos. 1 and 2, to the north, are situated on fracture zones infiltrated by quartz, ankerite and siderite veinlets carrying pyrite, sphalerite, galena and chalcopyrite with accessory hessite and Ag-Bi-Cu-sulphides. The clay alteration events in prospects nos. 1 and 2 have apparent ages of 250 ± 3 and 267 ± 3/277 ± 3 Ma respectively i.e. an undoubtedly Permian event.

3.3 The Northern Stefferud prospects

The sulphide mineralization at this locality occurs in a fault breccia located within Pre-Gothian gneisses along the contact of a Permian ekerite intrusion. The ore minerals which occur within a 2 m wide quartz veined zone, are pyrite and minor amount of sphalerite, galena and chalcopyrite. The three latter minerals are, however, strongly concentrated in small, transverse ankerite veins outside the main breccia. Two determinations on highly altered wallrocks in the breccia mineralization yielded ages of 216 ± 3 and 213 ± 2 Ma, apparently reflecting a Triassic hydrothermal event.

4. ORE DEPOSITS WITHIN THE OSLO REGION PROPER

4.1 Feiring mines

In the Feiring district sulphide bearing veins are encountered both within a Permian mænaite sill intruding Cambrian black shales a few meters above the sub-Cambrian peneplain, and in underlying Precambrian gneisses.

Microscope studies of veins transecting the mænaite sill have revealed two major phases of ore mineralization.

1) Pyrite and hematite were deposited along 1-20 cm wide quartz veins surrounded by a thin pyritic and sericitic alteration envelope.

2) Subsequent deposition mainly of sphalerite and galena occurred in calcite veins which are rimmed by aggregates of chlorite, sericite, illite and carbonates. The latter veins are usually found inside the early quartz veins. Accessory ore minerals usually found within these deposits are tellurides of Ag and Bi and Ni-Co-sulphides.

The age of argillitic alteration is represented by three samples of altered mænaite from the Kat and Jord mines which gave ages in the range 237 $\pm$ 3 to 250 $\pm$ 3 Ma with a mean age of 240 $\pm$ 3 Ma, i.e. an Upper Permian age.

4.2 Skreia Fe-mines

Within the Skreia-field, numerous contact-metasomatic iron deposits are encountered in altered Lower Paleozoic limestones. Two of these deposits; Johanna prospect and Flesvik mine, are situated at the immediate junction with Permian intrusives.

At the Johanna prospect magnetite bearing andradite skarn occurs in an originally Lower Silurian (Llandoverian) limestone which is transected by apophyses of a quartz-syenite intrusion. One sample of altered syenite from the spoil heap, showing quartz (partly amethyst) veins rimmed by strong argillitic alteration, gave an age of 244 + 3 Ma.

Similar post-magmatic or autometasomatic alteration has been observed in the Pauls mine, a few hundred meters to the west, where the same syenite intrude massive magnetite ore bodies and carry fragments of the latter. In addition, several mines along the eastern extension of this skarn zone contain larger areas where andradite and magnetite are decomposed to amphibole and chlorite, which also point to a late hydrothermal event.

One K/Ar determination on argillitic alteration inside the ekerite endocontact at the Flesvik mine resulted in an age of 243 $\pm$ 3 Ma. This mine is located in Upper Ordovician (Ashgillian) limestone which has undergone skarnification to an amphibole - calcite rock carrying irregular disseminations of hematite (specularite).

Combined with geological indications the nearly identical K/Ar ages may be taken to reflect post-magmatic hydrothermal activity related to the quartz-syenite intrusion which postdated the major phase of skarn alteration and ore deposition in this mining field.

4.3 Regional fault zones

During initial rifting and subsequent plutonic emplacement in the

Upper Paleozoic, a set of larger regional NNW-SSE striking faults and breccia zones were formed or reactivated. Within the outcrops of igneous rocks and Cambro-Silurian hornfelses these tectonic structures are traced along topographic depressions where the country rocks are fractured and brecciated and show argillitic alteration, features which indicate late stage movements and hydrothermal activity.

These reactivated fault lines are usually barren of mineral deposits. Exceptions are, however, local fluorite-calcite lenses in Cambrian shales along the Mistberget breccia and baryte veins following a quartz-breccia which transects alkali-syenites at Skomakertjern, east of the Skreia field.

Two samples of clay alteration in hornfelses and alkali-syenites were collected in the Mistberget and Hersjø breccia, respectively. K/Ar dating yielded: age of 221 $\pm$ 2 Ma for the former and 191 $\pm$ 2 Ma for the latter, indicating a Triaso-Jurrasic clay alteration event.

5. DISCUSSION

Summaries of the 20 analyses which are reported, are given in Fig. 2 and 3.

K/Ar dates from vein deposits are from field and microscope observations believed to represent true ages while the results from contact-metasomatic deposits must be interpreted as minimum ages.

The Upper Proterozoic (Riphean) event revealed by clay alteration in the Eidsvoll gold mines occurred more or less contemporaneously with the main chalcopyrite-gold mineralization which post-dates an early pyrite-(chalcopyrite) stage with subsequent fault movements. These gold lodes seem to be related to a Cu-vein belt following the main trend of the Magnusson mylonite zone down to L. Vänern in SW-Sweden.

Previous model lead isotopic dating on vein deposits transecting Dalslandian supracrustal and Pre-Gothian rocks in the latter area gave ages in the range 740 - 910 Ma, 850 - 890 Ma, and 530 - 700 Ma, (using the Houtermans, Russell - Farqhuar - Cummings 206/204 and Russel-Stanton - Farqhuar t_6 models respectively) [9]. These ages are consistent with recent Rb-Sr dating of Dalslandian slates yielding a seven point isochron of 1050 $\pm$ 40 Ma [10].

The ore deposition in the Eidsvoll gold mines is therefore probably associated with anorogenic block movements subsequent to

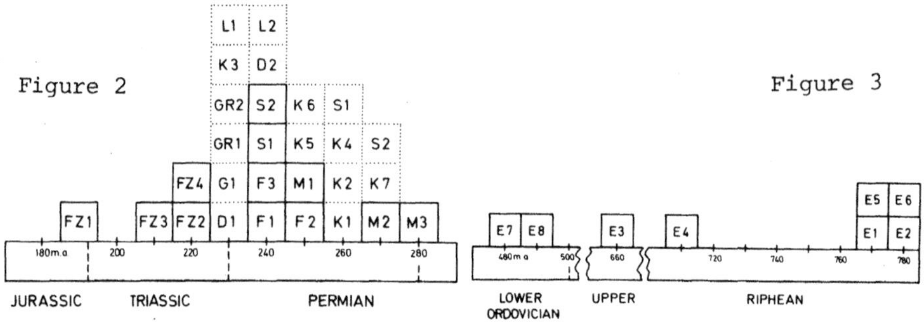

Figure 2. Histogram of Permian K/Ar isotopic ages. Dotted areas comprising previous results from the Oslo Region [6] (i.e. contact-metasomatic, (G, Gr, D) and Vein (K,L,S) deposits).

Figure 3. Histogram of Lower Paleozoic and Upper Riphean K/Ar isotopic ages.

the Sveco-Norwegian orogeny. The imprint of these movements are possibly represented by the dense pattern of N-S trending faults and quartz-breccias which dislocate blasto-mylonites along the southern border of the crush belt and on which the Kat and Guldkis mines are situated. The relation between the ore deposition and the block-faulting forming the sparagmite basins during the initial rifting and divergence of the proto-Atlantic [1,8] is less evident.

The anamalous ages of the Guldkis mine rocks can be explained in terms of reactivation of Precambrian structures in the Lower Ordovician (Arenigian). This is confirmed by previous work at the western margin of the Oslo Region in the Sandsvær district, where Rohr-Torp [7] has recognized block-faulting of Upper Cambrian to Lower Ordovician age. These movements possibly represent early subsidence of the Oslo Graben.

The Permian K/Ar ages encountered in the Midtskogen, Feiring and Skreia mines give no clear evidence for two separate mineralization events during the evolution of the Oslo igneous complex as recently proposed by Ineson et al [6] (Fig. 2). Instead the present work gives some indications that the ore deposition occurred continuously throughout the Permian.

In addition, the K/Ar dating on the Stefferud prospect, and

the Mistberget and Hersjø breccias points to a less marked event in the Triasso-Jurassic, representing reactivation of Permian tectonic structures and late hydrothermal activity with associated possible local mineralization of fluorite and baryte.

6. CONCLUSION

The major conclusions that can be drawn from the isotopic data obtained in the Eidsvoll-Skreia area are:

1) Vein mineralization commenced in the Upper Proterozoic (Riphean) with emplacement of sulphide and gold bearing quartz veins which are related to taphrogeny along the so-called Magnusson mylonite zone.

2) Reactivation of Upper Proterozoic tectonic structures occurred in the Lower Ordovician, possibly related to early subsidence of the Oslo Graben.

3) The present work supports previous results in confirming that the main period of ore deposition connected with the Oslo Rifting was of Permian age with a distinct clustering of ages in the range 237 - 250 Ma.
4) The hydrothermal activity continued until the Triassic and even Jurassic.

ACKNOWLEDGMENTS

The authors wish to express their appreciation of financial support and encouragement from several sources. The mineralogical and analytical work for the age determinations was carried out in the Geology and Geophysics Departments of the Universities of Sheffield and Newcastle-upon-Tyne and was financed by the research funds of these universities. The University of Trondheim, Norwegian Institute of Technology, A/S Norsk Hydro, A/S Sydvaranger and the Geological Survey of Norway provided funds for field-work and/or supplied information used in this report.

REFERENCES

1. K. Bjørlykke, A. Elvsborg and T. Høy, Nor.Geol. Tidsskr., 56, 233, 1976.
2. K.C. Dunham, F.H. Fitch, P.R. Ineson, J.A. Miller and J.G. Mitchell, Phil.Trans.R.Soc., 307 (A), 251, 1968.
3. P.R. Ineson and J.G. Mitchell, Geol.Mag., 109, 501, 1972.
4. P.R. Ineson and J.G. Mitchell, Trans.Instn.Min.Metall., 83, B13, 1974.

5. P.R. Ineson and J.G. Mitchell, Trans.Instn.Min.Metall., 84, B7, 1975.
6. P.R. Ineson, J.G. Mitchell and F.M. Vokes, Econ.Geol., 70, 1426, 1975.
7. E. Rohr-Torp, Norges Geol. Unders., 300, 53, 1973.
8. F.M. Vokes and G.H. Gale, Geol.Assoc.Canada, Spec. Paper, 14, 413, 1976.
9. F.E. Wickman, N.G. Blomquist, P. Geijer, A. Parwel, H.V. Ubisch and E. Welin, Ark. Mineral Geol., 11, 193, 1963.
10. T. Skiöld, Geol. Foren. Stockh. Förh., 98, 3, 1976.

FURTHER K/Ar DETERMINATIONS ON CLAY MINERAL ALTERATION ASSOCIATED WITH FLUORITE DEPOSITION IN SOUTHERN NORWAY

P.R. Ineson, J.G. Mitchell and F.M. Vokes

University of Sheffield, University of Newcastle-upon-Tyne, University of Trondheim

ABSTRACT. Further K/Ar ages on clay mineral (argillitic) alteration assemblages associated with fluorite deposition in southern Norway are reported and their geological consequences discussed.

Four determinations, including two already reported by Ineson et al [1] are available from the Lassedalen breccia deposit in the Precambrian rocks of the Kongsberg area, some distance west of the exposed Upper Palaeozoic rocks of the Oslo Region; the Gjerpen mining area north of Skien along the western border of the Oslo Region, is represented by nine determinations, while the Heskestad barite-fluorite deposit near Farsund in southwest Norway has yielded two clay mineral age determinations. In addition eight age determinations are reported from the Tveiten (Tveitstå) fluorite deposit in central Telemark. The dates from the Lassedalen and Gjerpen deposits, with three exceptions, are grouped in the range 270 to 230 Ma, with a distinct concentration (6 results) between 269 and 257 Ma (average 262$\pm$ Ma). The other four results lie in the range 231 to 242 Ma (average 237$\pm$5 Ma). These groupings agree well with those previously reported for other clay alteration ages in the same general area [1].

The results from the Gjerpen area show in addition, less well marked events both earlier (305, 297 Ma) and younger (195 Ma) than the above. Thus, the present work supports previous results in confirming that the main period of hydrothermal activity connected with the Oslo rifting was of Permian age. However, it also reinforces indications already obtained that hydrothermal events connected with the rifting took place over a relatively long age range, from the Late Carboniferous to the Triassic or even Jurassic.

E.-R. Neumann and I.B. Ramberg (eds.), Petrology and Geochemistry of Continental Rifts, 265–275.

The Heskestad deposit, situated some 200 km southwest of the exposed Oslo igneous rocks, has given clay alteration ages of 217 and 221 Ma (Triassic). This, and possibly similar deposits along the southern coast of Norway, seem to be related to faulting along the northwest boundary of the Skagerrak trough (rift).

Age determinations on argillic alteration associated with geologically similar fluorspar mineralization located in the southern Norwegian Sveconorwegian (Grenvillian) province at Tveiten in central Telemark show a spread of clay-alteration events ranging from about 700 to 300 Ma. This spread appears to be possibly explainable in terms of the repetition of hydrothermal events at periods later than that of the deposition of the fluorite in the area.

1. INTRODUCTION

Hydrothermal vein and/or breccia deposits of fluorite with or without quartz, calcite, barite, base metal sulphides and/or iron oxides are widespread and numerous, if economically not too important, in the southwestern Scandinavian region. The general setting of these deposits in respect to the rift structures and magmatic rocks of the Oslo Region has been presented in the Guide Book prepared for this Study Institute [2] while individual deposits have been dealt with in various published and unpublished reports, which will be acknowledged in the following text. Deposits of the type just outlined, of geologically similar character, are to be found at considerably varying distances from the exposed plutonic rocks of the Oslo Region. Their metallogenic affiliations are not always immediately obvious from an inspection of their geological features. Those in close proximity to the Oslo rocks and structures are plausibly referable to the events taking place during the Variscan taphrogeny which affected the region and which is the central theme of our present deliberations. Other deposits, often far removed from the present day Oslo Region, appear to be genetically related to large scale structures, which were active at least during the Variscan events. Yet others occur in parts of the southern Norwegian Precambrian province where any connection with the "Oslo events" is not immediately obvious (Fig. 1).

The geological grounds for assigning some of the southern Scandinavian fluorite-bearing deposits to the Variscan metallogeny have been previously presented, at various lengths, by Vokes [3], Ineson, Mitchell and Vokes [1], Vokes and Gale [4], and Ihlen and Vokes [2], while Ineson, Mitchell and Vokes [1] presented limited K/Ar isotopic evidence for a Late Permian event in connection with one of the deposits, situated a short distance from the western margin of the Oslo Igneous Province.

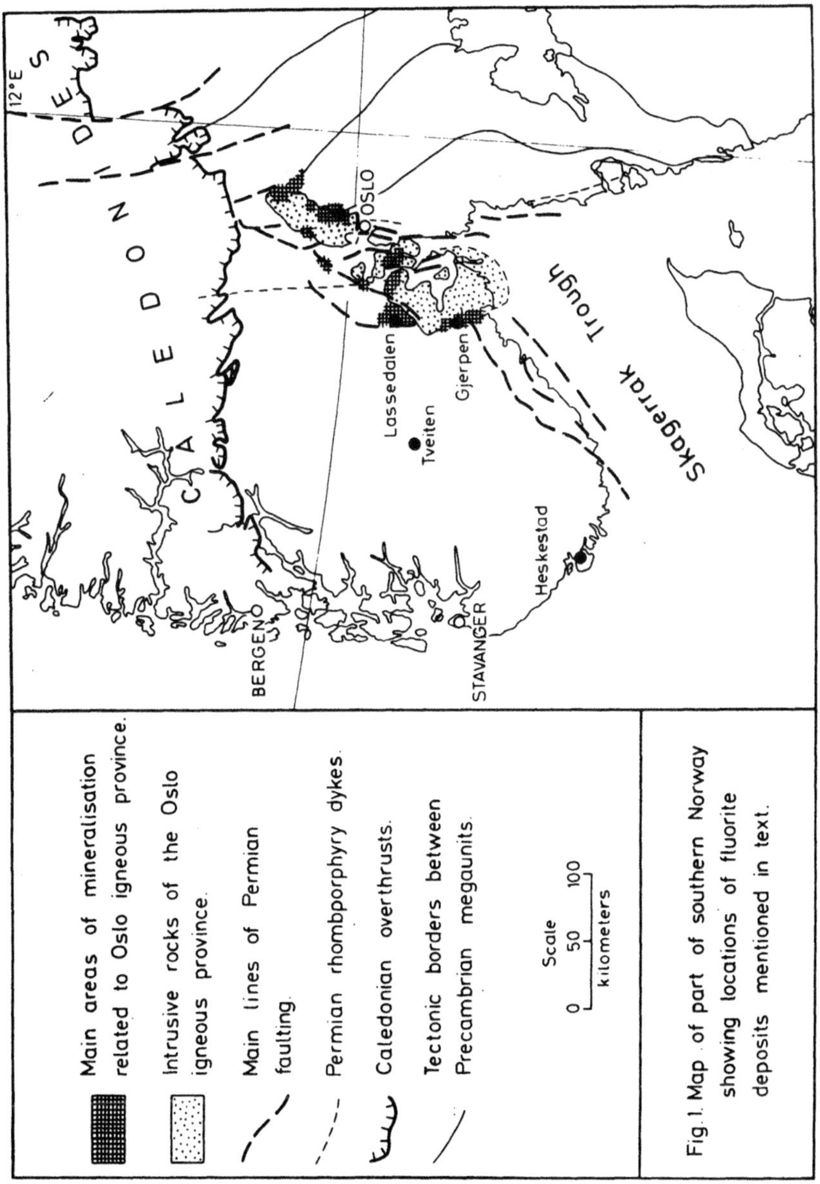

Fig. 1. Map of part of southern Norway showing locations of fluorite deposits mentioned in text.

This present contribution reports further isotopic evidence from a limited number of southern Norwegian fluorite-bearing deposits of the hydrothermal-epigenetic class outlined above, which confirms the wide regional spread of the Variscan events and serves to distinguish these from older events producing geologically similar deposits.

The dating method employed has been adequately described and discussed in a number of publications and interested readers are referred to Ineson, Mitchell and Vokes [1] for a review of these. Reference is also made to the contribution of Ihlen, Ineson and Mitchell [5] in the present volume for an account of the application of the method to the dating of clay minerals in the northern Oslo Region.

2. DEPOSITS INVESTIGATED

The present report covers isotopic geochronological results from four separate deposits or groups of deposits in the southwestern Sveconorwegian (Telemarkian) province of the Fennoscandian Shield. The generally Proterozoic rocks of this province became cratonized in an event of generally Grenvillian age, 1100-900 Ma ago, later tectonic events comprise almost exclusively taphrogenic movements culminating in the Oslo Region rifting in Permian times. Evidence is available that, following this culmination, faulting, mainly along old lines, continued in various parts of the region until at least Jurassic times, may be even later. The fluorite-bearing deposits examined in the present investigation occur in fault and breccia structures in this Precambrian Shield region, structures which appear to have resulted from the tectonic movements outlined above.

Two of the deposits, Lassedalen [6,7] in Buskerud and Gjerpen [8] in eastern Telemark, are located at distances of from, perhaps, 5 to 15 km from the nearest exposed Oslo plutonic rocks. A third deposit, Heskestad in Vest Agder is located some 200 km along the Skagerrak coast from the Oslo Region, but close to major fault structures associated with the Late Palaeozoic rifting. In contrast, perhaps, to the three above-mentioned deposits, those at Tveiten (Tveitstå) [6,9] in the Bandak area of central Telemark, have no obviously apparent relation to either magmatic rocks or major structures related to the Late Palaeozoic events. Geological consensus would place these central Telemark deposits in relation to the late orogenic granitic (magmatic or granitization) events of the area at around 800-1000 Ma ago [6,10,11].

3. RESULTS OF AGE DETERMINATIONS

The results of the K/Ar isotopic age determinations on clay minerals from the four deposits or districts are given in Table 1 and shown graphically in Fig. 2. There is a considerable spread in the apparent ages reported, from 712$\pm$3 to 195$\pm$3 Ma, though with one exception, all the ages older than 300 Ma refer to material collected from the Tveiten deposit, which geological reasoning points to as being probably of Precambrian age. The three other deposits

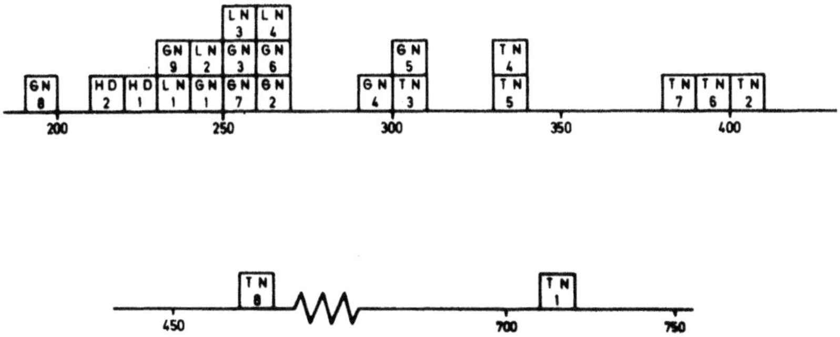

Figure 2. Distribution of K/Ar isotopic age determinations on clay mineral assemblages from southern Norwegian fluorite deposits.
(See Table I for locations of samples).

show, collectively, a range of ages from 305 to 195 Ma; 67 percent of the determinations falling in the time span 230 to 270 Ma and 40 percent in the range 250-270 Ma. Such results would seem unequivocally to confirm the Variscan (Pfalzian-Saalian) age of the events producing the clay alteration of the deposits in question. In the case of the Lassedalen and Gjerpen deposits, the hydrothermal activity producing the alteration was apparently active between 270 and 230 Ma ago and it is also reasonable to conclude that the deposition of fluorite and other veinfilling minerals occurred during this time span - possibly near its beginning. Ihlen et al. [5] have demonstrated that in the northern Oslo Region, clay alteration dates tend to give a minimum age for the ore deposition in the skarn type deposits and the same relation may possibly hold in the vein deposits under discussion here.

Occurrences of fluorite in the intrusive igneous rocks of the

TABLE 1.

Sample No.	Sample Locality	Separated Mineralogy		K_2O content (%)	Argon Content	Atmospheric contamination (%)	AGE (m.a.)
		Major	Minor				
LN1	Lassedalen, Buskerud dump material	Quartz, Illite	Chlorite, Kaolinite	1.70	$1.40mm^3gm^{-1}x10^{-2}$	32.4	233 ± 3
LN2	Lassedalen Buskerud dump material	Quartz, Chlorite Illite	Kaolinite	1.54	$1.32mm^3gm^{-1}x10^{-2}$	27.3	242 ± 3
LN3	Lassedalen, Buskerud Underground, main level	Quartz, Illite, Feldspar	Chlorite	7.29	$67.0mm^3gm^{-4}x10^{-3}$	10.4	259 ± 4
LN4	Lassedalen, Buskerud Lower Lassedalen Mine	Quartz, Illite Feldspar	Chlorite	7.91	$74.1mm^3gm^{-4}x10^{-3}$	8.5	264 ± 4
GNI	Gjerpen district, Telemark DDH153, 45-46m	Quartz, Chlorite	Epidote, Feldspars $\pm$ Illite	1.64	$14.01mm^3gm^{-4}x10^{-3}$	18.5	242 ± 3
GN2	Gjerpen district, Telemark DDH 157, 79.8-80.0m	Illite, Chlorite Quartz, var.mixed Layer Micas	Fluorite	8.23	$78.80mm^3gm^{-4}x10^{-3}$	9.4	269 ± 4
GN3	Gjerpen district, Telemark DDH 123/60, 59.5-63m	Quartz, Fluorite Illite	Calcite	2.04	$18.6mm^3gm^{-4}x10^{-3}$	24.6	257 ± 3
GN4	Gjerpen district, Telemark DDH 123/60, 63-67m	Illite, Quartz	Calcite, Chlorite $\pm$ Feldspars	5.26	$56.0mm^3gm^{-4}x10^{-3}$	8.0	297 ± 3
GN5	Gjerpen district, Telemark DDH 126/60, 82.9-83.0m	Quartz, Chlorite Fluorite	Illite, Calcite	1.24	$13.58mm^3gm^{-4}x10^{-3}$	18.9	305 ± 4
GN6	Gjerpen district, Telemark DDH 37, 100.0-101.4m	Quartz, Illite	? Chlorite Kaolinite	3.82	$36.3mm^3gm^{-4}x10^{-3}$	8.5	267 ± 4
GN7	Gjerpen district, Telemark DDH 37, 102.7-103.0m	Illite, Quartz Chlorite, Galena	Fluorite, Chalcopyrite ? Sphalerite, Feldspars	5.04	$49.8mm^3gm^{-4}x10^{-3}$	9.2	260 ± 3
GN8	Gjerpen district, Telemark Stulen prospect, outcrop	Chlorite, Quartz		0.93	$6.33mm^3gm^{-4}x10^{-3}$	51.0	195 ± 3
GN9	Gjerpen district, Telemark Stulen prospect, outcrop	Illite, Chlorite Kaolinite$\pm$ Feldspars	Fluorite	7.36	$59.8mm^3gm^{-4}x10^{-3}$	14.0	231 ± 3

HD1	Heskestad, Vest Agder DDH 5, 25-27m	Quartz, Feldspar (var) Illite	Chlorite, Kaolinite	6.46	$50.1 mm^3 gm^{-4} x10^{-3}$	13.4	221±3
HD2	Heskestad, Vest Agder DDH8, 30-40m	Quartz, Illite	Feldspar (var.), minor traces of various ore minerals	4.24	$32.3 mm^3 gm^{-4} x10^{-3}$	23.7	217±4
TN1	Tveiten, Telemark DDH1, 113.05-113.30m	Quartz, Illite Chlorite	Calcite	3.26	$93.4 mm^3 gm^{-4} x10^{-3}$	4.3	712±8
TN2	Tveiten, Telemark Upper Mine, Adit Portal	Quartz, Illite	Kaolinite, Feldspar	4.76	$71.7 mm^3 gm^{-4} x10^{-3}$	11.1	408±5
TN3	Tveiten, Telemark DDH 8, 53-55m	Quartz, Illite	Chlorite, Calcite	4.035	$4.403 mm^3 gm^{-1} x10^{-2}$	11.9	304±4
TN4	Tveiten, Telemark DDH8, 86-87m	Quartz, Illite Chlorite, Calcite	Kaolinite	5.564	$6.664 mm^3 gm^{-1} x10^{-2}$	8.7	331±5
TN5	Tveiten, Telemark DDH 9, 89m	Quartz, Illite	Kaolinite, Chlorite Calcite	4.821	$5.792 mm^3 gm^{-1} x10^{-2}$	8.7	332±5
TN6	Tveiten, Telemark DDH 9, 40-50m	Quartz, Chlorite Illite, Calcite	Kaolinite	2.680	$3.916 mm^3 gm^{-1} x10^{-2}$	11.3	397±5
TN7	Tveiten, Telemark DDH 10, 95-105m	Quartz, Illite Chlorite	Calcite, Mica Feldspar (?)	5.784	$8.076 mm^3 gm^{-1} x10^{-2}$	8.0	381±5
TN8	Tveiten, Telemark DDH 10, 70-90m	Quartz, Chlorite Illite	Kaolinite	3.233	$5.810 mm^3 gm^{-1} x10^{-2}$	7.7	477±6

Ramnes Cauldron in the southern part of the Oslo Region provide additional indications of possible ages of deposition. Sørensen [12] reports the presence of "veins and crystals of fluorite" in a felsitic variety of alkali granite to syenite in the central area of the cauldron. (He quotes Hysingjord [13] as estimating up to 20 per cent of fluorite in these rocks).

T. Vrålstad (pers. comm., 1977) who had inspected the locality, stressed the introduced nature of the fluorite in the Ramnes rocks. According to Sørensen [12] the alkali granite - alkali syenite series at Ramnes is of a later age than the kjelsåsite-larvikite series with which it is spatially associated. The larvikite has been dated by Heier and Compston [14] at about 276 Ma. The Ramnes fluorite deposition is thus demonstrably younger than the age of the larvikite by two geological events - the intrusion of the alkali granite body and the hydrothermal introduction of the fluorite. It is not possible on this evidence to arrive at a firm age for the fluorite deposition, but it is reasonable to suppose that it is of the same period as that indicated above for the Gjerpen and Lassedalen mineralizations, i.e. later than 270 Ma ago.

While the present results demonstrate the essential contemporaneity of the activity producing the Lassedalen and Gjerpen clay mineral alteration, there would seem to be some indication that it began rather earlier and finished somewhat later in the latter district.

The results seem also to indicate that the activity at Heskestad was rather later than at the first two localities. If these, admittedly sparse, data (only two determinations) can be relied upon, they seem to confirm evidence from other investigations that the rifting and accompanying magmatic and hydrothermal activity were of a generally later age in the western and southwestern coastal regions of Norway than in the vicinity of the Oslo Region per se. Faerseth, McIntyre and Naterstad [15], for example, have shown that fracturing and dyke intrusion, which they relate to rifting and differential movements along the margin of the North Sea, occurred in Sunnhordland between 275 and 169 Ma ago. These authors also report later calcite deposition in some of these structures, while Naterstad (pers. comm. 1977) mentions the presence of minor amounts of fluorite accompanying some of the dykes.

Unpublished results of the present authors on clay alteration associated with fluorite deposition in the same general region show an event at 185 ± 2 Ma.

If the above results can be taken at their face value, they would indicate progressively younger rifting and mineralizing events along the west of the Norwegian land mass as the opening of the present Atlantic Ocean continued in Mesozoic times. (cf. also

Oftedahl's demonstration of Mid-Jurassic tectonics in the coastal region of Trøndelag [16]).

When one turns to the clay mineral ages determined from the Tveiten deposit, the picture appears much less clear. As already stated, geological consensus and specific investigations would link the fluorite and related metallic mineral deposition in this area of central Telemark to granitic activity of the order of age of 900 Ma ago. Willms [6] reports the presence of fluorite replacing plagioclase feldspar in the Bessefjell granite, northwest of Tveiten. This granite has been dated at 933 Ma by the Rb/Sr method [10]. The oldest clay alteration age recorded from the Tveiten deposit in this study is 712±8 Ma, with other ages ranging up to 304±4 Ma.

If the general age of the mineralization is correct, the results of the clay-mineral dating could be taken to imply repeated hydrothermal (alteration) events at later periods along the same structures which controlled the original deposition.

It may possibly be only a question of coincidence, but some of the groupings of later clay-mineral ages at Tveiten appear to reflect the ages of orogenic events affecting the southwest Scandinavian Region as a whole; for example, 470-480 (Sardinian, Grampian) and 380-410 Ma (Ardennian or main Caledonian). The authors suggest, therefore, that we are seeing the results of re-equilibration of K/Ar isotopic composition in the clay minerals caused by hydrothermal surges produced in the Precambrian basement by the major tectonic phases mentioned above. The 304 Ma age at Tveiten may in this view represent an effect of the initial Variscan taphrogeny of the clay mineral assemblage at the deposit. No evidence has, however, come to hand indicating that the later Variscan movements affected the area; admittedly a weak point in the above argument.

4. RELATION TO GENERAL EPIGENETIC MINERALIZATION IN SOUTHERN NORWAY

The K/Ar clay-mineral age determinations presented here, together with those previously reported [1] and those being reported at this meeting [5], comprising almost 60 individual determinations from some 15 deposits or groups of deposits in Southern Norway, are beginning to give a clearer picture of the age spread of alteration connected with epigenetic mineralization in the southern Norwegian Precambrian areas on both sides of the Oslo Region per se.

The results show that clay mineral-forming hydrothermal events occurred over a time span of at least 600 Ma, from almost 800 until about 180 Ma ago. Of these ages, over 60 per cent fall in the time interval 280-210 Ma. Peaks are visible in the distribution of ages at roughly 265 Ma and 235 Ma, as previously pointed out [1], though the geological differentiation of the deposits showing these two age groupings is not by any means as clear as was previously thought to be the case. Further work is in progress and a review of the results achieved to date will be prepared in the near future.

ACKNOWLEDGMENTS

The authors wish to express their appreciation of financial support and encouragement from several sources. The mineralogical and analytical work for the age determinations was carried out in the Geology and Geophysics Departments of the Universities of Sheffield and Newcastle-upon-Tyne and was financed by the research funds of these universities. The University of Trondheim, Norwegian Institute of Technology, A/S Norsk Hydro, A/S Sydvaranger and the Geological Survey of Norway provided funds for field-work and/or supplied information used in this report.

REFERENCES

1. P.R. Ineson, J.G. Mitchell and F.M. Vokes, Econ. Geol., 70, 1426, 1975.
2. P.M. Ihlen and F.M. Vokes, Metallogeny associated with the Oslo rifting. NATO Advanced Study Institute, Sundvollen, 1977.
3. F.M. Vokes, Metallogeny possibly related to continental break-up in southwest Scandinavia, in: Implications of Continental drift to the Earth Sciences, ed. by D.H. Tarling and S.K. Runcorn, London, Academic Press, 1973.
4. F.M. Vokes and G.H. Gale, Geol. Assoc. Canada, Spec. Paper, 14, 413, 1976.
5. P.M. Ihlen, P.R. Ineson and J.G. Mitchell, K-Ar dating of clay-mineral alteration associated with ore deposition in the northern part of the Oslo Region. NATO Advanced Study Institute, Sundvollen. Contributions to discussions (metallogeny), 1977.
6. J. Willms, Fluorite-Vorkommen in Telemark, Süd-Norwegen, Diplomarbeit, Universität, Hamburg, (unpubl.), 1975.
7. T. Vrålstad, Lassedalen flusspat forekomst, Unpubl. report, A/S Norsk Hydro, 1976.
8. P.O. Kaspersen, En malmgeologisk undersøkelse av flusspatmineraliseringen i Gjerpenfeltet nord for Skien. Unpubl. Degree thesis, NTH, Trondheim, 1976.
9. T. Vrålstad, Flusspatprospektering i Tveiten CaF_2-felt, Tokke og Øvre Telemark forøvrig. Unpubl. report A/S Norsk Hydro, 1976.
10. P.G. Killeen and K.S. Heier, Norges geol. unders., 319, 59, 1975.
11. J.A. Dons, Norges geol. unders., 216, 1963.
12. R. Sørensen, Norges geol. unders., 321, 67, 1975.
13. J. Hysingjord, Geokjemisk prospektering i Oslo-feltet. Norges geol. unders., Rapport 1104 (Unpubl.), 1971.

14. K.S. Heier and W. Compston, Lithos, 2, 133, 1969.
15. R.B. Faerseth, R.C. MacIntyre and J. Naterstad, Lithos, 9, 331, 1976.
16. C. Oftedahl, Middle Jurassic graben tectonics in Mid-Norway. Procs. Jur. northern North Sea Symp., Norw. Petrol. Soc., Stavanger 1975, JNNSS/21, 1976.

ORE DEPOSITS IN THE NORTH-EASTERN PART OF THE OSLO REGION AND IN THE ADJACENT PRECAMBRIAN AREAS

Peter M. Ihlen

Geologisk Institutt, Universitetet i Trondheim
Norges Tekniske høgskole, Norway

ABSTRACT. The different ore deposits in the Eidsvoll-Skreia area are described and their relationship to the Permian magmatic activity is discussed. In all four main classes of ore mineralization can be distinguished: 1) Ortho-magmatic deposits of Fe-Ti-oxides and MoS_2, 2) vein deposits of Fe-Ti-oxides and barite, 3) contact-matasomatic deposits of Fe-W-oxides and minor base metal sulphides, and 4) perimagmatic vein deposits of mainly base metal sulphides and minor beryl and gold. Most of these are believed to be of Permian age and related to the tecto-magmatic evolution of the Oslo paleo-rift. Exceptions are, however, the vein deposits east of L. Mjøsa which have an uncertain position with reference to the Permian metallogenesis.

1. INTRODUCTION

The area under consideration is located at the north-eastern margin of the Oslo Region, comprising the districts around the southern end of L. Mjøsa (Fig. 1). Vogt [9-11], Goldschmidt [2] and Foslie [1] have earlier described various aspects of the ore deposition in this northern sector of the Oslo Region, and the subject has also recently been treated in a number of reviews of the Oslo Region metallogeny [6,7,12,13].

The ore deposits include oxide deposits of Fe, Ti, and W, sulphide deposits of Fe, Zn, Pb, Cu and Mo, deposits of gold and local mineralizations of fluorite, barite and beryl. The metallogeny is, however, dominated by contact-metasomatic Fe- and W-deposits and quartz-vein deposits carrying base-metal sulphides and minor gold. Liquid-magmatic Fe-Ti-deposits, intra-magmatic

E.-R. Neumann and I.B. Ramberg (eds.), Petrology and Geochemistry of Continental Rifts, 277-286.

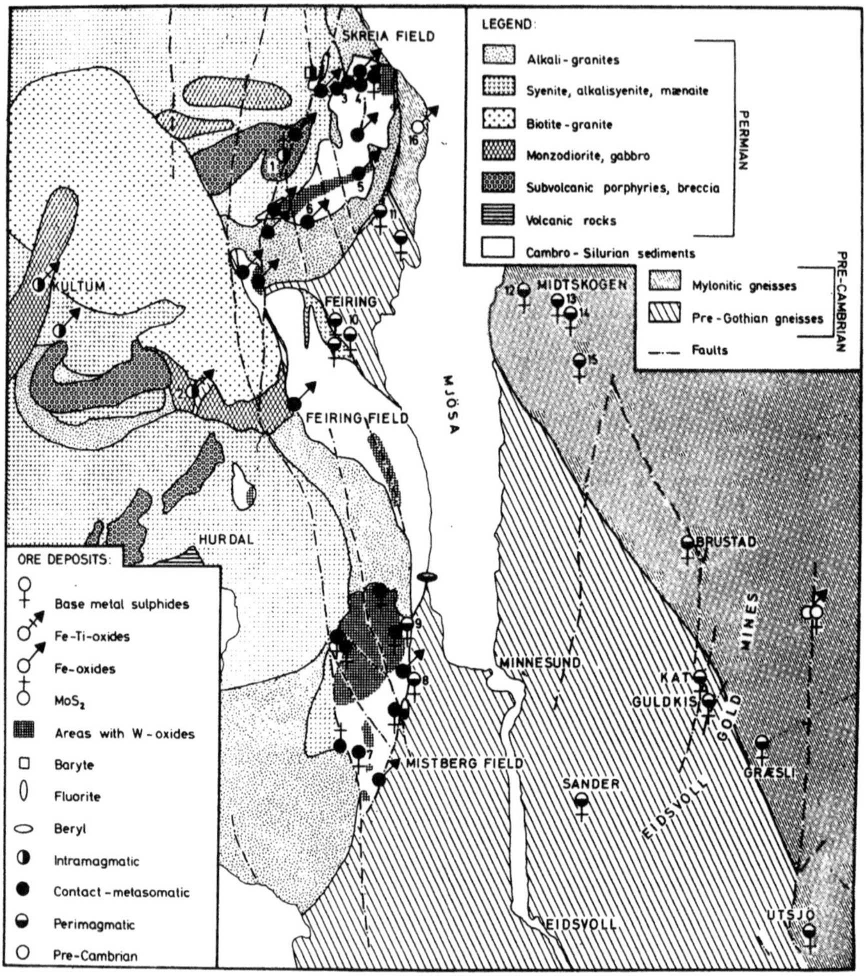

Figure 1. Geological map of the Eidsvoll-Skreia area, showing the distribution of different ore deposits. Numbers referred in the text (partly compiled from Nystuen [8]).

Mo-deposits and sulphide dominated skarn deposits, which are prominent elsewhere in the Oslo Region, are only sparsely represented in this area.

2. GEOLOGICAL OUTLINE

The Eidsvoll-Skreia area includes rocks ranging in age from Precambrian to Permian. The western part is occupied by the Permian ingenous complex comprising intrusions of gabbro, monzodiorite, syenite, alkalisyenite, granite and alkali-granite, which contain intrusive bodies of subvolcanic porphyries and subsided blocks of Permian volcanites, Lower Paleozoic sediments and Precambrian gneisses [8]. To the east the igneous complex borders on the crystalline Precambrian basement with its cover of folded Cambro-Silurian sediments which are commonly thermally metamorphosed into hornfelses, quartzites and marbles [2,10]. At the base of the Lower Paleozoic rocks, sills of mænaite have intruded Cambrian black shales (alumshales) a few meters above the sub-Cambrian peneplain.

During initial rifting and subsequent plutonic emplacement in Permian times a system of faults and breccia zones occurred, some representing reactivation of earlier Precambrian structures.

The Precambrian rocks to the east can be separated into two different mega-units, with a regional NW-SE strike. The southern mega-unit consists mostly of grey plagioclase-hornblende/biotite gneisses, usually referred to as the grey Romerike or Magnor gneisses [3]. They are possibly of svecofenian age [14]. North of this mega-unit occurs a nearly 20 km wide crush belt (Magnusson mylonite Zone), which developed during the Sveco-Norwegian orogeny (= Grenvillian). This unit includes Pre-Gothian gneisses and Gothian granites and supracrustal rocks (greenstones, rhyolites and sediments) and their mylonitized equivalents, such as augen-gneisses and blasto-mylonites [4,5]. The border zone between these mega-units is transected by a set of N-S trending fault zones.

3. METALLOGENY

In all, four main classes of ore mineralization can be distinguished and will be described in order of apparent genetic relationship with the Permian magmatism.

3.1 Ortho-magmatic deposits

Mafic association. At Kultum, Hurdalen, Fe-Ti-oxides occur together with minor sulphides, within a body of layered olivine-gabbro. The magmatic layering is caused by varying proportions of olivine, labradorite, clinopyroxene, amphibole, apatite and Fe-Ti-oxides (ilmenite, hemoilmenite, ilmenomagnetite and magnetite) usually intergrown with minor pyrrhotite, chalcopyrite and pyrite. Pentlandite is found as discrete grains and spindles intergrown with pyrrhotite. The following paragenesis have been observed in the ore-bearing layers:

I: Amphibole + olivine + labradorite + spinel + phlogopite + ore minerals.

II: Labradorite + clinopyroxene + ore minerals.

In the first of these, corroded crystals of olivine, labradorite, spinel and ore minerals occur embedded in a matrix of colorless non pleochroitic amphibole and phlogopite. The second paragenesis shows clinopyroxene and ore minerals interstitial to corroded labradorite crystals. These types apparently represent cumulus and inter-cumulus crystallization.

During subsequent hydrothermal activity amphibole (uralite), epidote and sphene were formed by alteration of clinopyroxene, labradorite and ilmenite, possibly involving solutions which had interacted with a nearby limestone xenolith.

Felsic association. Within the syenites and alkali-granites surrounding the Skreia area several small molybdenite mineralizations are present in which MoS_2 apparently occurs as true magmatic crystals. Among these, the Raumyr prospect west of Skreia, can be mentioned (No. 1, Fig. 1). At this locality disseminations of molybdenite, magnetite and ilmenomagnetite occur inside a small body of alkali-granite transecting subvolcanic porphyries. Accessory scheelite and fluorite, together with molybdenite and magnetite are locally observed in pegmatitic segregations.

At our present state of knowledge, one cannot rule out the possibility that this mineralization constitutes the K-silicate zone of a porphyry-molybdenum type mineralization situated in the central part of a previous cauldron which has now been destroyed by later intrusions.

3.2 Intramagmatic veins

Mafic association. 50 m outside the olivine-gabbro contact at Kultum ½ - 1 meter wide flatlying veins of partly serpentinized olivine occur in a matrix of Fe-Ti-oxides and spinel. These veins are rimmed by aggregates of amphibole and phlogopite towards the wallrock which consists of corroded labradorite with minor interstitial amphibole (anorthosite). Locally pegmatitic hornblende-phlogopite-plagioclase veins carry some Fe-Ti-oxides. This same paragenesis also constitutes the ore in the Ødemarken mine (No. 2), Hurdalen, located at the contact between monzodiorite and a xenolith of Precambrian gneisses

Felsic association. Except for local concentrations of magnetite-pyrite-plagioclase and barite-kaolinite veins and breccia fillings, the felsic intrusions are barren of this type of deposits.

3.3 Contact-metasomatic deposits

The contact-metasomatic deposits have a widespread distribution among the Cambro-Silurian rocks inside the thermal aureoles and are mainly associated with alkali-granites. These deposits occur either at the immediate contact with the intrusion or along favourable tectonic structures up to a few hundred meters away.

The ore minerals in these skarns are dominantly magnetite and sheelite. Based on the difference in host-rock lithology, the deposits can be subdivided into the following two groups:

1) Fe-deposits associated with skarn-altered limestones, and

2) W-deposits associocited with altered argillaceous and calcareous hornfelses.

The former group of deposits is especially concentrated in the Skreia area [9-11] associated with skarn-altered Upper-Ordovician and Silurian limestones. Among these deposits several paragenetic types of metasomatic alteration and ore deposition can be outlined.

The major type consists of magnetite bearing andradite-clino-pyroxene skarns and forms the basis for the mining activity at Skreikampen (No. 3). These mines are located along the outcrop of a Lower-Silurian (Llandoverian) limestone that has been totally altered to skarn for about 1½ km along strike. The magnetite with minor associated pyrite forms massive plates, lenses, and irregular pods reaching a length of 50 m and a width of 10 m.

The ore-bearing skarns have been formed by large-scale introduction mainly of Si and Fe. During the metasomatic alteration, magnetite partly crystallized simultaneously with the andradite, and partly as somewhat later fissure and replacement veins. During the alteration process, excess Fe and released Ca moved out following fractures from the ore zone into the neighbouring hornfelses and igneous dykes which also were transformed to garnet-clino-pyroxene skarns. This latter type of skarn rock is, however, usually barren of magnetite. The only exception is a magnetite mineralization in altered diabase dykes at Bekke mine (No. 4).

The igneous dykes always pre-date the ore deposition and seem to have no genetic connection with the latter process, except that both follow the same set of common large-scale tectonic structures.

The calcareous skarn in the Skreikampen ore zone usually shows signs of repeated brecciation and late deposition of plagioclase, epidote, scheelite, amphibole, hematite, quartz, calcite and chlorite along fractures and in cavities. Most of these minerals are the products of retrograde alteration of magnetite, garnet and clinopyroxene. Similar mineral assemblages (lacking only scheelite) are encountered locally where late introduction of fluorite or sulphides has occurred (chalcopyrite, sphalerite, pyrrhotite and galena).

In the southern part of the area, epidote is the dominating skarn mineral, which can be explained in terms of difference in temperature and CO_2 pressure. The metasomatic alteration in this area often shows zonation around the mineralization locus.

In the Torgundsrud mine (No. 5) magnetite, scheelite and flurite occur abundantly in a garnet-clinopyroxene-epidote skarn surrounded by successive zones of epidote-clinopyroxene and epidote-amphibole skarn. These outer zones carry excess quartz and calcite and subordinate amounts of scheelite.

Skarn zonation is also encountered in the Flesvik mine (No. 6), where the ore consists of hematite (specularite) disseminated in an amphibole skarn outside a magnetite-bearing andradite skarn at the contact with the alkali-granite. In addition similar mineral parageneses occur along scheelite bearing fissures in the surrounding hornfelses.

The second group of contact metasomatic deposits is very abundant in the Mistberg area. It is also encountered in the eastern part of the Skreia-area and in numerous hornfels xenoliths inside the igneous complex. Here scheelite invariably occurs as the major ore mineral in metasomatic veins and bodies in different types of hornfelses, but with a preference for banded argillaceous and calcareous hornfelses. This mineralization type, present in most thermal aureoles in the Oslo Region, is clearly different from most other contact deposits, in that the limestones present, including those bordering the mineralized hornfelses, show no signs of metasomatic alteration.

The strongest alteration and the tungsten deposition are connected with fault zones, especially strike-slip faults which give rise to stratiform scheelite ore bodies.

Detailed studies in the Mistberget area have revealed several metasomatic mineral parageneses based on crosscutting relationships between the veins. Early clinopyroxene-plagioclase veins are succeeded by the formation of grossular-clinopyroxene veins and bodies. The highest concentration of scheelite (usually high in Mo) is associated with the grossular stage, which to the east locally carries molybdenite instead of scheelite. Along these veins and bodies the argillaceous variety is altered to a light green clinopyroxene-alkalifeldspar hornfels.

These two parageneses occur mainly in the lower part of Mt.

Mistberget near the alkaligranite intrusion and within a Ca-rich lithology of Upper-Ordovician age. At a higher level or in a more remote position to the igneous rocks, veins of feldspar-clinopyroxene and feldspar-quartz-sulphides (Rødnabben prospect, No. 7) prevail. The sulphides are pyrrhotite, chalcopyrite and arsenopyrite. Most of the plagioclase in the lower part of the Mistberg area, both in calcareous hornfelses and in veins, is to varying degrees altered to chlor-scapolite (marialite). Accessory amounts of magnetite, pyrite, sphalerite, galena, chalcopyrite, millerite and violarite are irregularly distributed along the central part of the veins and within the skarn bodies. These minerals belong, however, to a late hydrothermal stage, together with quartz, calcite, fluorite, amphibole, epidote and chlorite, which mostly represent an alteration of earlier formed silicate-minerals. These parageneses also occur in some skarn-altered limestones along the eastern margin of the area, though scheelite is lacking in these Fe-Zn-deposits.

Although the above mentioned skarn parageneses are based on cross-cutting relationships, they must, both locally and regionally, be looked upon as part of a continuous metasomatic column. Due to repeated tectonic movements and closure of fractures by mineral deposition no wellformed metasomatic zonation appeared. Preliminary results from silicate analyses give indications that the alteration assemblages are caused by circulation of $CaCl_2$-rich solutions leading to mobilization and redistribution of the main components in the hornfelses. This process can be classified as desilicification by leaching of Si, Na and K and enrichment of Ca and Mn, while Al, Fe and Mg behaved partly as inert components. The leached Si, Na and K make up the major minerals in the veins encountered higher up in Mt. Mistberget. The scapolitization is due to increased activity of NaCl in the solutions at this stage.

4. Perimagmatic deposits

Pegmatites. At Byrud, in the southern end of the Feiring area, enrichment of beryl occurs inside a flatlying feldspar pegmatite transecting alumshales and mænaite sills which are strongly silicified along the contacts. Quartz, muscovite and fluorite are present in subordinate amounts. Goldschmidt [23] also reports the presence of topaz.

Quartz breccia veins. Quartz breccia veins carrying base metal sulphides are encountered in contact-metamorphic Cambrian sediments, Permian mænaite sills and in the Precambrian basement to the east. The major strike of these veins is N-S and E-W, and they all bear witness of repeated fracturing with subsequent formation of new minerals.

In the region of the Mistberget (No. 8 & 9) and Feiring (No. 10) mines [10] formation of minor garnet, epidote and plagioclase were succeeded by strong quartz infiltration, and mineralization of hematite and pyrite, followed by pyrrhotite sphalerite, galena and chalcopyrite. The ore deposition ceased during a late stage of mainly calcite-fluorite deposition on fractures and in cavities in the quartz veins. At one locality in the Mistberg area calcite-fluorite veins occur along a large regional breccia, but without any associated sulphides. In the Feiring mines these veins carry a conspicuous envelope of sericitic and chloritic-argillitic alteration, associated with the quartz and calcite stage, respectively.

The ore carries a great variety of accessory minerals such as hessite, tellurobismuthite, bismuthite, altaite, tetradymite, pentlandite, violarite, cubanite, aresenopyrite, gold and mackinawite. Gold and Ag-Bi-Te-S bearing minerals occur mostly as inclusions in galena and chalcopyrite, while most of the Ni-Fe-sulphides and mackinawite are located inside exsolution droplets of chalcopyrite in sphalerite. The Ni-Fe-sulphides are only observed in ore-bearing veins located above the sub-Cambrian peneplain.

Among the Stefferud (No. 11) and the Western Midtskogen prospects (No. 12 & 13) similar parageneses are encountered but here most of the carbonate is ankerite and siderite. In galena and chalcopyrite from Midtskogen prospects schabachite ($AgBiS_2$), hessite and unidentified Ag-Cu-sulphides are observed as minor constituents.

The Eidsvoll gold mines including Utsjøen [1], Sander, Brustad, Kat, Guldkis and Græsli mines located along the border of the Precambrian crush-belt to the east of L. Mjøsa, contain the most important quartz breccia veins in the area. The mines have been worked on a single, or a set of parallel quartz lodes having lengths of about 500 m and widths of 1-6 m.

Microscopic examination of ore and wallrocks has revealed early deposition of pyrite and minor chalcopyrite in a gangue of quartz, this was followed by a period of faulting and brecciation. Cataclastic pyrite formed during the tectonic period was later replaced by chalcopyrite and gangue minerals along numerous fractures. Most of the gold and accessory sulphides, such as sphalerite, galena and galenobismuthite are found as inclusions in, or as intergrowths with, the late formed chalcopyrite. Identical ore and gangue parageneses are found at the eastern Midtskogen prospects (No. 14 & 15).

Most of these mineralized lodes show silicification, sericitization and argillization of the wallrocks. The first alteration type is connected to the first mineralization event while the latter types developed contemporaneously with, or subsequently to, the brecciation. Thin carbonate veins with associated galena, sphalerite, chalcopyrite and chloritic wallrock alteration occur very sporadically.

To the east the gold-bearing lodes are succeeded by veins that also carry hematite and bornite, still further east in the Odalen - Kongsvinger area, these change to a bornite-chalcocite paragenesis with minor gold and silver associated. These veins together with the Fe-Ti-ore in the Langgård mine (No. 16), which are situated in a metagabbro, probably represent a Precambrian metallogenetic event.

4. CONCLUSION

The ore deposition in the Eidsvoll-Skreia area can be separated into a Precambrian and a Permian metallogenetic event. The latter event commenced with formation of the intramagmatic deposits which are related to different stages of magmatic differentiation and crystallization of the igneous complex. Orthomagmatic crystallization of Fe-Ti-oxides and minor associated Ni-Cu-sulphides in olivine-gabbro was succeeded by deposition of Fe-Ti-oxides and MoS_2 from residual liquides or solutions inside monzodiorites, syenites and alkali-granites. In addition minor beryl occurred during late magmatic crystallization of pegmatites inside the exocontact of an alkali-granite.

The Mistberg area W-mineralizations can be explained in terms of mobilization and redistribution of the primary contents of W in the Lower Paleozoic sediments by circulation of solutions along faults and associated fractures. These solutions were apparently released during, and subsequent to, the thermal metamorphism. One cannot, however, exclude the possibility that the W-bearing solutions have a magmatic (or mixed) origin, since minor scheelite occurs inside alkali-granite at Raumyr Mo-prospect and as a constituent of most of the Fe-ores in the Skreia area which are believed to have formed from post-magmatic solutions.

These post-magmatic solutions, mainly emanating from alkali-granites, led to skarn alteration of limestones. The different skarn and ore parageneses give evidence of an early post-magmatic stage dominated by Fe- and Si-rich solutions succeeded by solutions carrying mainly base metals, sulphur and fluorite.

The ore parageneses found among the peri-magmatic vein deposits west of L. Mjøsa probably reflect ore deposition from similar post-magmatic solutions though they show a predominance of the late sulphide stage. The great variety in types of wall-rock alteration among the skarn and peri-magmatic vein deposits is believed to be due to the diverse petrology of the host rocks.

The relationship between quartz lodes east of L. Mjøsa and the late Paleozoic rifting and magmatic activity is not well understood. The mineral assemblages in these distant quartz breccia veins show both resemblances with the near contact vein deposits and with those further to the east, which apparently are of Precambrian age.

The only undoubtedly Precambrian metallogenetic event is the occurrence of Fe-Ti-oxides in a meta-gabbro at Langgård mine, Skreia.

REFERENCES

1. S. Foslie, Norges geol. undersøkelse, 122, 73, 1924.
2. V.M. Goldschmidt, Vidensk.-selskaps skrift, Kristiania, I. Mat.-naturv. Kl., I, 1, 1911.
3. M. Gustavson, Norges geol. unders., 246, 5, 1967.
4. Ø. Gvein, Norges geol. unders., 246, 27, 1967.
5. A. Hjelle, Norges geol. unders., 211, 75, 1960.
6. P.M. Ihlen and F.M. Vokes, Metallogeny, in: Guidebook for the ASI meeting Paleorift Systems with Emphasis on the Permian Oslo Rift, ed. by J.A. Dons and B.T. Larsen, Norges geol. unders., 1978.
7. P.R. Ineson, J.G. Mitchell and F.M. Vokes, Econ. Geol., 70, 1426, 1975.
8. J.P. Nystuen, Norges geol. unders., 317, 1, 1975.
9. J.H.L. Vogt, Archiv f. mat. naturvitensk., 9, 231, 1884.
10. J.H.L. Vogt, Nyt mag. f. naturv., 215, 1884.
11. J.H.L. Vogt, Norges geol. unders., 6, 1892.
12. F.M. Vokes, Metallogeny possibly related to continental break-up in southwest Scandinavia, in: Implications of continental drift to the Earth Sciences, ed. by D.H. Tarling and S.K. Runcorn, Vol. I, London, Academic Press, 1973.
13. F.M. Vokes and G.H. Gale, Geol. Assoc. Canada Spec. Paper, 14, 413, 1976.
14. E. Welin and R. Gorbatschev, Geol. Fören. Stockh. Förh., 98, 378, 1976.

SUBJECT INDEX

9 789027 708663

GPSR Compliance

The European Union's (EU) General Product Safety Regulation (GPSR) is a set of rules that requires consumer products to be safe and our obligations to ensure this.

If you have any concerns about our products, you can contact us on ProductSafety@springernature.com

In case Publisher is established outside the EU, the EU authorized representative is:

Springer Nature Customer Service Center GmbH
Europaplatz 3
69115 Heidelberg, Germany

Batch number: 10816188

Printed by Printforce, the Netherlands